CODING AND CRYPTOLOGY

Series on Coding Theory and Cryptology

Editors: Harald Niederreiter *(National University of Singapore, Singapore)* and San Ling *(Nanyang Technological University, Singapore)*

Published

Vol. 1 Basics of Contemporary Cryptography for IT Practitioners
by B. Ryabko and A. Fionov

Vol. 2 Codes for Error Detection
by T. Kløve

Vol. 3 Advances in Coding Theory and Cryptography
eds. T. Shaska et al.

Vol. 4 Coding and Cryptography
eds. Yongqing Li et al.

Series on Coding Theory and Cryptology – Vol. 4

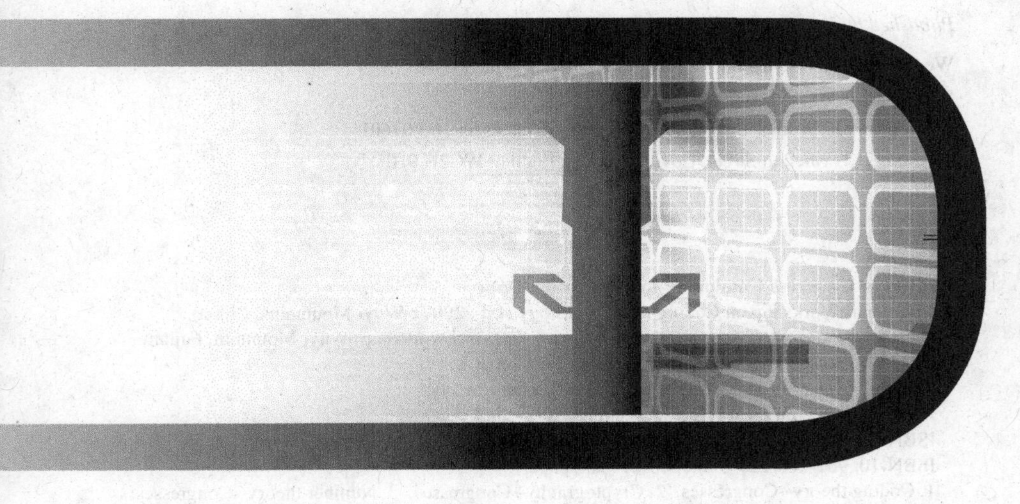

Proceedings of the First International Workshop on

CODING AND CRYPTOLOGY

Wuyi Mountain, Fujian, China 11 – 15 June 2007

Editors

Yongqing Li
Fujian Normal University, China

San Ling
Nanyang Technological University, Singapore

Harald Niederreiter
National University of Singapore, Singapore

Huaxiong Wang
Chaoping Xing
Nanyang Technological University, Singapore

Shengyuan Zhang
Fujian Normal University, China

NEW JERSEY • LONDON • SINGAPORE • BEIJING • SHANGHAI • HONG KONG • TAIPEI • CHENNAI

Published by

World Scientific Publishing Co. Pte. Ltd.

5 Toh Tuck Link, Singapore 596224

USA office: 27 Warren Street, Suite 401-402, Hackensack, NJ 07601

UK office: 57 Shelton Street, Covent Garden, London WC2H 9HE

Library of Congress Cataloging-in-Publication Data

International Workshop on Coding and Cryptology (1st : 2007 : Wuyi Mountains, China)

Coding and cryptology : proceedings of the international workshop, Wuyi Mountain, Fujian, China 11-15 June 2007 / edited by Yongqing Li ... [et al.].

p. cm. -- (Series on coding theory and cryptology ; v. 4)

Includes bibliographical references and index.

ISBN-13: 978-981-283-223-8 (hardcover : alk. paper)

ISBN-10: 981-283-223-8 (hardcover : alk. paper)

1. Coding theory--Congresses. 2. Cryptography--Congresses. 3. Number theory--Congresses. 4. Computer security--Congresses. I. Li, Yongqing. II. Title.

QA268.I59 2007

003'.54--dc22

2008022123

British Library Cataloguing-in-Publication Data

A catalogue record for this book is available from the British Library.

Printed in Singapore.

PREFACE

The first International Workshop on Coding and Cryptology was held in Wuyi Mountain, Fujian, China, June 11 - 15, 2007. The workshop was organised by Fujian Normal University, China and Nanyang Technological University, Singapore. We acknowledge with gratitude the financial support from the Key Laboratory of Network Security and Cryptology, Fujian Normal University.

The idea for this workshop grew out of the recognition of the recent development in various areas of coding theory and cryptology. Over the past years, we have seen the rapid growth of the Internet and World-Wide-Web, they have provided great opportunities for online commercial activities, business transactions and government services over open computer and communications networks. However, such developments are only possible if communications can be conducted in a secure and reliable way. The mathematical theory and practice of coding theory and cryptology underpin the provision of effective security and reliability for data communication, processing and storage. Theoretical and practical advances in the fields are therefore a key factor in facilitating the growth of data communications and data networks.

The aim of the workshop was to bring experts from coding theory, cryptology and their related areas for a fruitful exchange of ideas. We hoped (and achieved) the meeting would encourage and stimulate further research in telecommunications, information and computer security, the design and implementation of coding-related cryptosystems and other related areas. Another goal of the meeting was to stimulate collaboration and more active interaction between mathematicians, computer scientists, practical cryptographers and engineers.

This workshop post-proceedings consists of 20 papers submitted by the invited speakers of the workshop, each paper has been reviewed by at least two referees. They cover a wide range of topics in coding theory and cryptology such as theory, techniques, applications, practical experiences. They contain significant advances in the areas as well as very useful surveys.

We would like to express our thanks to World Scientific Publishing Co.

and in particular, E H Chionh for the advice and support. We thank Zhexian Wan for his encouragement and support, the staff and students from the Key Laboratory of Network Security and Cryptology, Fujian Normal University for their excellent organization of the event. Special thanks also go to Guo Jian for the LaTeX assistance.

ORGANIZING COMMITTEES

Chair of Organizing Committee

Zhexian Wan	– Academy of Math. and System Sciences, CAS, China

Organizing Committee

Yongqing Li	– Fujian Normal University, China
San Ling	– Nanyang Technological University, Singapore
Harald Niederreiter	– National University of Singapore, Singapore
Huaxiong Wang	– Nanyang Technological University, Singapore
Chaoping Xing	– Nanyang Technological University, Singapore
Shengyuan Zhang	– Fujian Normal University, China

CONTENTS

Fuzzy Identity-based Encryption: New and Efficient Schemes*

Joonsang Baek
Institute for Infocomm Research, Singapore
E-mail: jsbaek@i2r.a-star.edu.sg

Willy Susilo[§]
Centre for Computer and Information Security Research
University of Wollongong, Australia
E-mail: wsusilo@uow.edu.au

Jianying Zhou
Institute for Infocomm Research, Singapore
E-mail: jyzhou@i2r.a-star.edu.sg

In this paper we construct two new fuzzy identity-based encryption (IBE) schemes in the random oracle model. Not only do our schemes provide public parameters whose size is *independent* of the number of attributes in each identity (used as public key) but they also have useful structures which result in more efficient key extraction and/or encryption than the random oracle version of Sahai and Water's fuzzy IBE scheme, considered recently by Pirretti *et al.* We prove that the confidentiality of the proposed schemes is relative to the Decisional Bilinear Diffie-Hellman problem.

Keywords: Fuzzy IBE, random oracle model, BDBDH

1. Introduction

Motivation. The concept of fuzzy identity-based encryption (IBE) recently introduced by Sahai and Waters [12] is to provide an *error-tolerance property* for IBE. Namely, in fuzzy IBE, a user with the secret key for the identity ω can decrypt a ciphertext encrypted with the public key ω' if ω and ω' are within a certain distance of each other. We note that in contrast to the previous approaches [5,9], the biometric measurement in fuzzy IBE,

---

* A short version of this paper appeared at AsiaCCS 2007 [1]. This is an expanded version.
§ Corresponding Author

which is used as an identity, does not need to be kept secret [12]. However, it must be ensured that an attacker cannot convince the key issuing authority to believe that he owns a biometric identity that he does not possess. As noted in Sahai-Waters' work [12], fuzzy IBE can directly be applied to the situation where a user is traveling and another party wants to encrypt at an ad-hoc meeting between them. Another application of fuzzy IBE is "attribute-based encryption [7,10,12]" where a party can encrypt data to all users that have a certain set of attributes, e.g. {`company`, `division`, `department`}.

Related Work.
Since Sahai and Water's work, fuzzy IBE has been discussed in the context of the attribute-based encryption (ABE). Very recently, Goyal *et al.* [7] proposed an ABE scheme that provides fine-grained sharing of encrypted data. Piretti *et al.* [10] used Sahai and Waters' "large universe" construction of fuzzy IBE, which we simply call "Sahai-Waters construction", to realize their secure information management architecture. They also observed that if the random oracle [2] is employed, computational overhead of the Sahai-Waters construction can greatly be reduced. We remark that the random oracle not only reduces computational overhead but also provides a very short public parameters whose size is *independent* of the number of attributes associated with an identity or the number of attributes in the defined universe, which is crucial in the storage constrained applications.

Our Contribution.
In this paper, we go one step beyond Pirreti *et al.*'s results by presenting fuzzy IBE schemes in the random oracle model, which are structurally different from the Sahai-Waters construction. We show that the structural difference results in more efficient schemes than even the random oracle version of the Sahai-Waters construction considered by Pirretti *et al.* [10]. We prove that our schemes meet the security requirements as defined in [12] assuming that the Decisional Bilinear Diffie-Hellman (DBDH) problem is hard.

2. Preliminaries

Computational Primitives. We first review the definition of the admissible bilinear pairing [3,8], denoted by e. Let $\mathbb{G}_1$ and $\mathbb{G}_2$ be groups of the same order q which is prime. (By $\mathbb{G}_1^*$ and $\mathbb{Z}_q^*$, we denote $\mathbb{G}_1 \setminus \{1\}$ where 1 is the identity element of $\mathbb{G}_1$, and $\mathbb{Z}_q \setminus \{0\}$ respectively). Suppose that $\mathbb{G}_1$ is generated by g. Then, $e : \mathbb{G}_1 \times \mathbb{G}_1 \rightarrow \mathbb{G}_2$ has the following properties:

1) Bilinear: $e(g^a, g^b) = e(g, g)^{ab}$, for all $a, b \in \mathbb{Z}_q$ and 2) Non-degenerate: $e(g, g) \neq 1$.

A computational problem that will be used throughout this paper is the DBDH problem, a decisional version of the Bilinear Diffie-Hellman problem on which Boneh and Franklin's IBE scheme [3] is based. Informally, the DBDH problem refers to the problem where, given (g, g^a, g^b, g^c) for random $a, b, c \in \mathbb{Z}_q^*$, a polynomial-time attacker $\mathcal{A}$ is to distinguish $e(g, g)^{abc}$ from $e(g, g)^\gamma$ for random $\gamma \in \mathbb{Z}_q^*$.

Fuzzy IBE and Its Security.

The generic fuzzy IBE scheme [12] consists of the following algorithms.

- $\mathsf{Setup}()$: Providing some security parameter as input, the Private Key Generator (PKG) runs this algorithm to generate its master key mk and public parameters $params$ which contains an error tolerance parameter d. Note that $params$ is given to all interested parties while mk is kept secret.
- $\mathsf{Extract}(mk, \mathtt{ID})$: Providing the master key mk and an identity $\mathtt{ID}$ as input, the PKG runs this algorithm to generate a private key associated with $\mathtt{ID}$, denoted by $D_{\mathtt{ID}}$.
- $\mathsf{Encrypt}(params, \mathtt{ID}', M)$: Providing the public parameters $params$, an identity $\mathtt{ID}'$, and a plaintext M as input, a sender runs this algorithm to generate a ciphertext C'.
- $\mathsf{Decrypt}(params, D_{\mathtt{ID}}, C')$: Providing the public parameters $params$, a private key $D_{\mathtt{ID}}$ associated with the identity $\mathtt{ID}$ and a ciphertext C' encrypted with an identity $\mathtt{ID}'$ such that $|\mathtt{ID}' \cap \mathtt{ID}| \geq d$ as input, a receiver runs this algorithm to get a decryption, which is either a plaintext or a "*Reject*" message.

A first security requirement of fuzzy IBE is "indistinguishability of encryptions under fuzzy selective-ID, chosen plaintext attack (IND-FSID-CPA)" [12]. (Note that the "selective-ID attack" [4] refers to the attack in which an attacker commits ahead of time an identity that it intends to attack.) The formal definition based on the game between an attacker $\mathcal{A}$ and the "Challenger" is as follows.

In Phase 1, $\mathcal{A}$ outputs a challenge identity $\mathtt{ID}^*$. In Phase 2, the Challenger then runs the Setup algorithm to generate a master key mk and public parameters $params$. The Challenger gives $params$ to $\mathcal{A}$ while keeps mk secret from $\mathcal{A}$. In Phase 3, $\mathcal{A}$ issues private key extraction queries, each of which is denoted by $\mathtt{ID}$. A restriction here is that for all $\mathtt{ID}$, $|\mathtt{ID} \cap \mathtt{ID}^*| < d$. In Phase 4, $\mathcal{A}$ outputs equal-length messages M_0 and M_1. Upon receiving

(M_0, M_1), the Challenger picks $\beta \in \{0,1\}$ at random and creates a challenge ciphertext $C^* = \mathsf{Encrypt}(params, \mathtt{ID}^*, M_\beta)$. The Challenger returns C^* to $\mathcal{A}$. In Phase 5, $\mathcal{A}$ issues a number of private key extraction queries as in Phase 3. In Phase 6, $\mathcal{A}$ outputs its guess $\beta' \in \{0,1\}$.

We define $\mathcal{A}$'s guessing advantage by $|\Pr[\beta' = \beta] - \frac{1}{2}|$.

Notice that a stronger notion "indistinguishability of encryptions under fuzzy selective-ID, chosen ciphertext attack (IND-FSID-CCA)" can also be defined by giving $\mathcal{A}$ an access to a decryption oracle.

Another important security requirement for a fuzzy IBE scheme is the security against colluding attack, which implies that no group of users should be able to combine their keys in such a way that they can decrypt a ciphertext that none of them alone could [12].

3. Proposed Fuzzy IBE Schemes

In the rest of the paper, $\Delta_{a,S}$ denotes the Lagrange coefficient for $a \in \mathbb{Z}_q^*$ (q, a prime) and a set S of elements in $\mathbb{Z}_q^*$. Notice that

$$\Delta_{a,S}(x) = \prod_{b \in S, b \neq a} \frac{x-b}{a-b}.$$

Without loss of generality, we assume that an identity is a *set* of n *different* elements in $\mathbb{Z}_q^*$. For example, each of n strings of arbitrary length with an index $i \in \mathbb{Z}$ can be hashed using some collision-resistant hash function whose range is $\mathbb{Z}_q^*$.

Efficient Fuzzy IBE-I (EFIBE-I) Scheme.
As mentioned earlier the hash function H in our first fuzzy IBE scheme is assumed to be a random oracle, which gives rise to very short public parameters. However, we note that our scheme has a different structure compared to the random oracle version of the Sahai-Waters construction considered in [10]. The unique feature of EFIBE-I is that its private key extraction algorithm (Extract) is structurally simple and highly efficient.

- Setup(): Generate a group $\mathbb{G}_1$ of prime order q. Construct a bilinear map $e : \mathbb{G}_1 \times \mathbb{G}_1 \to \mathbb{G}_2$, where $\mathbb{G}_2$ is a group of the same order q. Pick a generator g of the group $\mathbb{G}_1$. Pick $g_1 \in \mathbb{G}_1$ at random. Pick $s \in \mathbb{Z}_q^*$ at random and compute $g_2 = g^s$. Choose a hash function $H : \mathbb{Z}_q^* \to \mathbb{G}_1$. Select a tolerance parameter d. Output a public parameter $params = (q, g, e, \mathbb{G}_1, \mathbb{G}_2, H, g_1, g_2, d)$ and a master key $mk = (q, g, e, \mathbb{G}_1, \mathbb{G}_2, H, g_1, g_2, s)$.

- $\mathsf{Extract}(mk, \mathtt{ID})$, where $\mathtt{ID} = (\mu_1, \ldots, \mu_n)$: Pick a random polynomial $p(\cdot)$ of degree $d-1$ over $\mathbb{Z}_q$ such that $p(0) = s$ and compute a private key $D_{\mu_i} = (\gamma_{\mu_i}, \delta_{\mu_i}) = (H(\mu_i)^{p(\mu_i)}, g^{p(\mu_i)})$ for $i = 1, \ldots, n$. Return $D_{\mathtt{ID}} = (D_{\mu_1}, \ldots, D_{\mu_n})$.
- $\mathsf{Encrypt}(params, \mathtt{ID}', M)$, where $\mathtt{ID}' = (\mu'_1, \ldots, \mu'_n)$ and $M \in \mathbb{G}_2$: Pick $r \in \mathbb{Z}_q^*$ at random and compute

$$\begin{aligned} C' &= (\mathtt{ID}', U, V_{\mu'_1}, \ldots, V_{\mu'_n}, W) \\ &= (\mathtt{ID}', g^r, (g_1 H(\mu'_1))^r, \ldots, (g_1 H(\mu'_n))^r, e(g_1, g_2)^r M) \end{aligned}$$

- $\mathsf{Decrypt}(params, D_{\mathtt{ID}}, C')$, where C' is encrypted with $\mathtt{ID}'$ such that $|\mathtt{ID}' \cap \mathtt{ID}| \geq d$ (Recall that $\mathtt{ID} = (\mu_1, \ldots, \mu_n)$).: Choose an arbitrary set $S \subseteq \mathtt{ID} \cap \mathtt{ID}'$ such that $|S| = d$ and compute

$$M = \frac{e(\prod_{\mu_j \in S} \gamma_{\mu_j}^{\Delta_{\mu_j, S}(0)}, U)}{\prod_{\mu_j \in S} e(V_{\mu_j}, \delta_{\mu_j}^{\Delta_{\mu_j, S}(0)})} \cdot W$$

(Here, notice that $\mu'_j = \mu_j$ if $\mu_j \in S$). Return M.

The above decryption algorithm is correct as

$$\begin{aligned} &\frac{e(\prod_{\mu_j \in S} \gamma_{\mu_j}^{\Delta_{\mu_j, S}(0)}, U)}{\prod_{\mu_j \in S} e(V_{\mu_j}, \delta_{\mu_j}^{\Delta_{\mu_j, S}(0)})} \cdot W \\ &= \frac{e(\prod_{\mu_j \in S} \gamma_{\mu_j}^{\Delta_{\mu_j, S}(0)}, g^r)}{\prod_{\mu_j \in S} e((g_1 H(\mu_j))^r, \delta_{\mu_j}^{\Delta_{\mu_j, S}(0)})} \cdot W \\ &= \frac{e(\prod_{\mu_j \in S} \gamma_{\mu_j}^{\Delta_{\mu_j, S}(0)}, g^r)}{\prod_{\mu_j \in S} e((g_1^{\Delta_{\mu_j, S}(0)} H(\mu_j)^{\Delta_{\mu_j, S}(0)})^r, g^{p(\mu_j)})} \cdot W \\ &= \frac{e(\prod_{\mu_j \in S} \gamma_{\mu_j}^{\Delta_{\mu_j, S}(0)}, g^r)}{\prod_{\mu_j \in S} e(g_1^{\Delta_{\mu_j, S}(0) p(\mu_j)} H(\mu_j)^{\Delta_{\mu_j, S}(0) p(\mu_j)}, g^r)} \cdot W \\ &= \frac{e(\prod_{\mu_j \in S} \gamma_{\mu_j}^{\Delta_{\mu_j, S}(0)}, g^r)}{e(\prod_{\mu_j \in S} g_1^{\Delta_{\mu_j, S}(0) p(\mu_j)}, g^r) e(\prod_{\mu_j \in S} \gamma_{\mu_j}^{\Delta_{\mu_j, S}(0)}, g^r)} \cdot W \\ &= \frac{1}{e(g_1^s, g^r)} \cdot e(g_1, g_2)^r M = \frac{1}{e(g_1, g_2)^r} \cdot e(g_1, g_2)^r M = M. \end{aligned}$$

We now prove the following theorem regarding the security of EFIBE-I in the IND-FSID-CPA sense.

Theorem 3.1. *The EFIBE-I scheme is IND-FSID-CPA secure in the random oracle model assuming that the DBDH problem is hard.*

Proof. Assume that an attacker $\mathcal{A}$ breaks IND-FSID-CPA of EFIBE-I with probability greater than ϵ within time t making q_H random oracle queries and q_{ex} private key extraction queries. We show that using $\mathcal{A}$, one can construct a DBDH attacker $\mathcal{B}$.

Suppose that $\mathcal{B}$ is given $(q,\ g,\ e,\ \mathbb{G}_1,\ \mathbb{G}_2,\ g^a,\ g^b,\ g^c,\ \tau)$, where τ is either $e(g,g)^{abc}$ or $e(g,g)^{\gamma}$ for random $\gamma \in \mathbb{Z}_q^*$, as an instance of the DBDH problem. By ϵ' and t', we denote $\mathcal{B}$'s winning probability and running time respectively. $\mathcal{B}$ can simulate the Challenger's execution of each phase of IND-FSID-CPA game for $\mathcal{A}$ as follows.

Simulation of Phase 1. Suppose that $\mathcal{A}$ outputs a challenge identity $\mathtt{ID}^* = (\mu_1^*, \ldots, \mu_n^*)$.

Simulation of Phase 2. $\mathcal{B}$ sets $g_1 = g^b$ and $g_2 = g^c$, and gives $\mathcal{A}$ $(q, g, e, \mathbb{G}_1, \mathbb{G}_2, H, g_1, g_2, d)$ as *params*, where $d \in \mathbb{Z}^+$ and H is a random oracle controlled by $\mathcal{B}$ as follows.

> Upon receiving a query μ to H:
>
> > If there exists $\langle \mu, (l, h) \rangle$ in HList, return h. Otherwise, do the following:
> >
> > > If $\mu = \mu_i^*$ for some $i \in [1, n]$, choose $l \in \mathbb{Z}_q^*$ at random and compute $h = g^l / g_1$.
> > > Else choose $l \in \mathbb{Z}_q^*$ at random and compute $h = g^l$.
> > > Add $\langle \mu, l, h \rangle$ to HList and return $H(\mu) = h$ as answer.

Simulation of Phase 3. $\mathcal{B}$ answers $\mathcal{A}$'s private key extraction queries as follows.

> Upon receiving a private key extraction query $\mathtt{ID} = (\mu_1, \ldots, \mu_n)$ such that $|\mathtt{ID} \cap \mathtt{ID}^*| < d$:
>
> > Let $\Gamma = \mathtt{ID} \cap \mathtt{ID}^*$; Let Γ' be any set such that $\Gamma \subseteq \Gamma' \subseteq \mathtt{ID}$ and $|\Gamma'| = d - 1$; Let $S = \Gamma' \cup \{0\}$.
> > For every $\mu_i \in \Gamma'$, run the above H-oracle simulator to get $\langle \mu_i, l_i, h_i \rangle$ in HList, pick $\lambda_i \in \mathbb{Z}_q^*$ at random and compute $D_i = (h_i^{\lambda_i}, g^{\lambda_i})$.
> > For every $\mu_i \in \mathtt{ID} \setminus \Gamma'$, run the above H-oracle simulator to

get $\langle \mu_i, l_i, h_i \rangle$ in HList and compute

$$D_i = \Big(\big(\prod_{\mu_j \in \Gamma'} h_i^{\Delta_{\mu_j, S(\mu_i)} \lambda_j} \big) g_2^{\Delta_{0, S(\mu_i)} l_i},$$
$$\big(\prod_{\mu_j \in \Gamma'} g^{\Delta_{\mu_j, S(\mu_i)} \lambda_j} \big) g_2^{\Delta_{0, S(\mu_i)}} \Big).$$

Return $(D_{\mu_1}, \ldots, D_{\mu_n})$.

Now define $\lambda_i = p(\mu_i)$ for a random polynomial $p(\cdot)$ of degree $d-1$ over $\mathbb{Z}_q^*$ such that $p(0) = c$. Notice that when $\mu_i \in \Gamma'$, the simulated D_i's and those of D_i's in the real attack are identically distributed. Notice also that even when $\mu_i \notin \Gamma'$, the above simulation is still correct. Since $\mu_i \notin \Gamma'$ means $\mu_i \notin \Gamma$, $h_i = H(\mu_i) = g^{l_i}$ by the simulation of H. Thus, noting that $g_2 = g^c$, we have

$$\begin{aligned}
D_i &= \\
&= \Big(\big(g^{l_i (\sum_{\mu_j \in \Gamma'} \Delta_{\mu_j, S(\mu_i)} p(\mu_j))} \big) g^{l_i \Delta_{0, S(\mu_i)} c}, \\
&\quad g^{\sum_{\mu_j \in \Gamma'} \Delta_{\mu_j, S(\mu_i)} p(\mu_j)} g^{\Delta_{0, S(\mu_i)} c} \Big) \\
&= \Big(g^{l_i (\sum_{\mu_j \in \Gamma'} \Delta_{\mu_j, S(\mu_i)} p(\mu_j) + \Delta_{0, S(\mu_i)} p(0))}, \\
&\quad g^{\sum_{\mu_j \in \Gamma'} \Delta_{\mu_j, S(\mu_i)} p(\mu_j) + \Delta_{0, S(\mu_i)} p(0)} \Big) \\
&= (g^{l_i p(\mu_i)}, g^{p(\mu_i)}) \\
&= (H(\mu_i)^{p(\mu_i)}, g^{p(\mu_i)}).
\end{aligned}$$

Consequently, the simulated key $(D_{\mu_1}, \ldots, D_{\mu_n})$ is distributed the same as the one in the real attack.

Simulation of Phase 4. $\mathcal{B}$ creates a challenge ciphertext C^* as follows.

Upon receiving (M_0, M_1):

Choose $\beta \in \{0, 1\}$ at random.
Search HList to get $l_1^*, \ldots, l_n^*$ that correspond to each of $\mathtt{ID}^* = (\mu_1^*, \ldots, \mu_n^*)$.
Compute $g^{a l_i^*}$ for $i = 1, \ldots, n$.
Return $C^* = (g^a, g^{a l_1^*}, \ldots, g^{a l_n^*}, \tau M_\beta)$ as a challenge ciphertext.

Simulation of Phase 5. $\mathcal{B}$ answers $\mathcal{A}$'s random oracle/private key extraction queries as in Phase 3.

Simulation of Phase 6. $\mathcal{A}$ outputs its guess β'. If $\beta' = \beta$, $\mathcal{B}$ outputs 1. Otherwise, it outputs 0.

Analysis. Notice in the above simulation thta if $\tau = e(g,g)^{abc}$ then $\tau M_\beta = e(g^b, g^c)^a M_\beta = e(g_1, g_2)^a M_\beta$. Notice also that $g^{al_i^*} = (g^{l_i^*})^a = (g_1 H(\mu_i^*))^a$ for $i = 1, \ldots, n$ from the construction of the random oracle H. Hence the challenge ciphertext C^* created above is distributed the same as the one in the real attack. On the other hand, if $\tau = e(g,g)^\gamma$ for $\gamma \in \mathbb{Z}_q^*$ chosen uniformly at random, τM_β is uniform in $\mathbb{G}_2$. As justified in the simulation of Phase 3, $\mathcal{B}$ perfectly simulates the random oracle H and the key private key extraction. Hence, we get $\Pr[\mathcal{B}(g, g^a, g^b, g^c, e(g,g)^{abc}) = 1] = \Pr[\beta' = \beta]$, where $|\Pr[\beta' = \beta] - \frac{1}{2}| > \epsilon$, and $\Pr[\mathcal{B}(g, g^a, g^b, g^c, e(g,g)^\gamma) = 1] = \Pr[\beta' = \beta] = \frac{1}{2}$, where γ is uniform in $\mathbb{G}_2$. Consequently, we get

$$|\Pr[\mathcal{B}(g, g^a, g^b, g^c, e(g,g)^{abc}) = 1] - \Pr[\mathcal{B}(g, g^a, g^b, g^c, e(g,g)^\gamma) = 1]| > \left|\left(\frac{1}{2} \pm \epsilon\right) - \frac{1}{2}\right| = \epsilon.$$

$\mathcal{B}$'s running time is computed as $t' < t + (q_H + q_{ex})O(T_e)$, where T_e denotes the computing time for an exponentiation in $\mathbb{G}_1$. □

By the same argument as [12], the EFIBE-I scheme prevents collusion attacks since each users' private key components are generated with different random polynomials.– Even if multiple users collude, they will not be able to combine their private key components to form a key which is useful to compromise the confidentiality of the scheme.

Efficient Fuzzy IBE-II (EFIBE-II) Scheme.
Our second fuzzy IBE scheme bears some similarities to the second scheme based on the DBDH problem [12]. However, its private key extraction has been simplified by using the outputs of the chosen random polynomial as random exponents for g_1, in contrast to the scheme in [12] which introduces extra random exponents and hence incurs extra exponentiations. More precisely, our scheme computes $((g_1 H(\mu_i))^{p(\mu_i)}, g^{p(\mu_i)})$ instead of $(g_1^{p(\mu_i)} H(\mu_i)^{r_i}, g^{r_i})$ [12] to generate a private key associated with an identity $\mathtt{ID}' = (\mu_1', \ldots, \mu_n')$.

A description of the scheme is as follows.

- $\mathsf{Setup}()$: Generate a group $\mathbb{G}_1$ of prime order q. Construct a bilinear map $e : \mathbb{G}_1 \times \mathbb{G}_1 \rightarrow \mathbb{G}_2$, where $\mathbb{G}_2$ is a group of the same order q. Pick a generator g of the group $\mathbb{G}_1$. Pick $g_1 \in \mathbb{G}_1$ at random. Pick $s \in \mathbb{Z}_q^*$ at random and compute $g_2 = g^s$. Choose a hash function

$H : \mathbb{Z}_q^* \to \mathbb{G}_1$. Select a tolerance parameter d. Output a public parameter $params = (q, g, e, \mathbb{G}_1, \mathbb{G}_2, H, g_1, g_2, d)$ and a master key $mk = (q, g, e, \mathbb{G}_1, \mathbb{G}_2, H, g_1, g_2, s)$.

- Extract(mk, ID), where ID $= (\mu_1, \ldots, \mu_n)$: Pick a random polynomial $p(\cdot)$ of degree $d-1$ over $\mathbb{Z}_q$ such that $p(0) = s$ and compute a private key $D_{\mu_i} = (\gamma_{\mu_i}, \delta_{\mu_i}) = ((g_1 H(\mu_i))^{p(\mu_i)}, g^{p(\mu_i)})$ for $i = 1, \ldots, n$. Return $D_{\mathtt{ID}} = (D_{\mu_1}, \ldots, D_{\mu_n})$.
- Encrypt($params$, ID$'$, M), where ID$' = (\mu'_1, \ldots, \mu'_n)$ and $M \in \mathbb{G}_2$: Pick $r \in \mathbb{Z}_q^*$ at random and compute

$$\begin{aligned} C' &= (\mathtt{ID}', U, V_{\mu'_1}, \ldots, V_{\mu'_n}, W) \\ &= (\mathtt{ID}', g^r, H(\mu'_1)^r, \ldots, H(\mu'_n)^r, e(g_1, g_2)^r M) \end{aligned}$$

- Decrypt($params$, $D_{\mathtt{ID}}$, C'), where C' is encrypted with ID$'$ such that $|\mathtt{ID}' \cap \mathtt{ID}| \geq d$ (Recall that ID $= (\mu_1, \ldots, \mu_n)$).: Choose an arbitrary set $S \subseteq \mathtt{ID} \cap \mathtt{ID}'$ such that $|S| = d$ and compute

$$M = \frac{\prod_{\mu_j \in S} e(V_{\mu_j}, \delta_{\mu_j}^{\Delta_{\mu_j, S}(0)})}{e(\prod_{\mu_j \in S} \gamma_{\mu_j}^{\Delta_{\mu_j, S}(0)}, U)} \cdot W$$

(Here, notice that $\mu'_j = \mu_j$ if $\mu_j \in S$). Return M.

The above decryption algorithm is correct as

$$\begin{aligned}
&\frac{\prod_{\mu_j \in S} e(V_{\mu_j}, \delta_{\mu_j}^{\Delta_{\mu_j, S}(0)})}{e(\prod_{\mu_j \in S} \gamma_{\mu_j}^{\Delta_{\mu_j, S}(0)}, U)} \cdot W \\
&= \frac{\prod_{\mu_j \in S} e(H(\mu_j)^r, g^{p(\mu_j)\Delta_{\mu_j, S}(0)})}{e(\prod_{\mu_j \in S} (g_1 H(\mu_j))^{p(\mu_j)\Delta_{\mu_j, S}(0)}, g^r)} \cdot W \\
&= \frac{\prod_{\mu_j \in S} e(H(\mu_j)^{p(\mu_j)\Delta_{\mu_j, S}(0)}, g^r)}{e(\prod_{\mu_j \in S} (g_1 H(\mu_j))^{p(\mu_j)\Delta_{\mu_j, S}(0)}, g^r)} \cdot W \\
&= \frac{\prod_{\mu_j \in S} e(H(\mu_j)^{p(\mu_j)\Delta_{\mu_j, S}(0)}, g^r)}{e(\prod_{\mu_j \in S} g_1^{p(\mu_j)\Delta_{\mu_j, S}(0)}, g^r)} \cdot \frac{W}{e(\prod_{\mu_j \in S} H(\mu_j)^{p(\mu_j)\Delta_{\mu_j, S}(0)}, g^r)} \\
&= \frac{1}{e(\prod_{\mu_j \in S} g_1^{p(\mu_j)\Delta_{\mu_j, S}(0)}, g^r)} \cdot e(g_1, g_2)^r M \\
&= \frac{1}{e(g_1^s, g^r)} \cdot e(g_1, g_2)^r M = \frac{1}{e(g_1, g_2)^r} \cdot e(g_1, g_2)^r M = M.
\end{aligned}$$

We then prove the following theorem regarding the security of EFIBE-II in the IND-FSID-CPA sense.

Theorem 3.2. *The EFIBE-II scheme is IND-FSID-CPA secure in the random oracle model assuming that the DBDH problem is hard.*

Proof. Assume that an attacker $\mathcal{A}$ breaks IND-FSID-CPA of EFIBE-II with probability greater than ϵ within time t making q_{ex} private key extraction queries. We show that using $\mathcal{A}$, one can construct a DBDH attacker $\mathcal{B}$.

Suppose that $\mathcal{B}$ is given $(q,\ e,\ \mathbb{G}_1,\ \mathbb{G}_2,\ g,\ g^a,\ g^b,\ g^c,\ \tau)$, where τ is either $e(g,g)^{abc}$ or $e(g,g)^\gamma$ for random $\gamma \in \mathbb{Z}_q^*$, as an instance of the DBDH problem. By ϵ' and t', we denote $\mathcal{B}$'s winning probability and running time respectively. $\mathcal{B}$ can simulate the Challenger's execution of each phase of IND-FSID-CPA game for $\mathcal{A}$ as follows.

Simulation of Phase 1. Suppose that $\mathcal{A}$ outputs a challenge identity $\mathtt{ID}^* = (\mu_1^*, \ldots, \mu_n^*)$.

Simulation of Phase 2. $\mathcal{B}$ sets $g_1 = g^b$ and $g_2 = g^c$, and gives $\mathcal{A}$ $(q, g, e, \mathbb{G}_1, \mathbb{G}_2, H, g_1, g_2, d)$ as *params*, where $d \in \mathbb{Z}^+$ and H is a random oracle controlled by $\mathcal{B}$ as follows.

> Upon receiving a query μ to H:
>
> > If there exists $\langle(\mu, l), h\rangle$ in HList, return h. Otherwise, do the following:
> >
> > > If $\mu = \mu_i^*$ for some $i \in [1, n]$, choose $l \in \mathbb{Z}_q^*$ at random and compute $h = g^l$.
> > > Else choose $l \in \mathbb{Z}_q^*$ at random and compute $h = g^l/g_1$.
> > > Add $\langle\mu, l, h\rangle$ to HList and return $h = H(\mu)$ as answer.

Simulation of Phase 3. $\mathcal{B}$ answers $\mathcal{A}$'s private key extraction queries as follows.

> Upon receiving a private key extraction query $\mathtt{ID} = (\mu_1, \ldots, \mu_n)$ such that $|\mathtt{ID} \cap \mathtt{ID}^*| < d$:
>
> > Let $\Gamma = \mathtt{ID} \cap \mathtt{ID}^*$; Let Γ' be any set such that $\Gamma \subseteq \Gamma' \subseteq \mathtt{ID}$ and $|\Gamma'| = d - 1$; Let $S = \Gamma' \cup \{0\}$.
> > For every $\mu_i \in \Gamma'$, run the above H-oracle simulator to get $\langle\mu_i, l_i, h_i\rangle$ in HList, pick $\lambda_i \in \mathbb{Z}_q^*$ at random and compute $D_i = ((g_1 h_i)^{\lambda_i}, g^{\lambda_i})$. Let $\lambda_i = p(\mu_i)$.
> > For every $\mu_i \in \mathtt{ID} \setminus \Gamma'$, run the above H-oracle simulator to

get $\langle \mu_i, l_i, h_i \rangle$ in HList and compute

$$D_i = \Big(\big(\prod_{\mu_j \in \Gamma'} (g_1 h_i)^{\Delta_{\mu_j, S(\mu_i)} \lambda_j} \big) g_2^{\Delta_{0,S(\mu_i)} l_i}, \big(\prod_{\mu_j \in \Gamma'} g^{\Delta_{\mu_j, S(\mu_i)} \lambda_j} \big) g_2^{\Delta_{0,S(\mu_i)}} \Big).$$

Return $(D_{\mu_1}, \ldots, D_{\mu_n})$.

Now define $\lambda_i = p(\mu_i)$ for a random polynomial $p(\cdot)$ of degree $d-1$ over $\mathbb{Z}_q^*$ such that $p(0) = c$. Notice that when $\mu_i \in \Gamma'$, the simulated D_i's and those of D_i's in the real attack are identically distributed. Notice also that even when $\mu_i \notin \Gamma'$, the above simulation is still correct. – Since $\mu_i \notin \Gamma'$ means $\mu_i \notin \Gamma$, $g_1 h_i = g^{l_i}$. Noting that $g_2 = g^c$, we have

$$\begin{aligned} D_i &= \\ &= \Big(\big(g^{l_i (\sum_{\mu_j \in \Gamma'} \Delta_{\mu_j, S(\mu_i)} p(\mu_j))} \big) g^{l_i \Delta_{0,S(\mu_i)} c}, \\ &\quad g^{\sum_{\mu_j \in \Gamma'} \Delta_{\mu_j, S(\mu_i)} p(\mu_j)} g^{\Delta_{0,S(\mu_i)} c} \Big) \\ &= \big(g^{l_i (\sum_{\mu_j \in \Gamma'} \Delta_{\mu_j, S(\mu_i)} p(\mu_j) + \Delta_{0,S(\mu_i)} p(0))}, \\ &\quad g^{\sum_{\mu_j \in \Gamma'} \Delta_{\mu_j, S(\mu_i)} p(\mu_j) + \Delta_{0,S(\mu_i)} p(0)} \big) \\ &= (g^{l_i p(\mu_i)}, g^{p(\mu_i)}) = ((g_1 h_i)^{p(\mu_i)}, g^{p(\mu_i)}) \\ &= ((g_1 H(\mu_i))^{p(\mu_i)}, g^{p(\mu_i)}). \end{aligned}$$

Consequently the simulated key $(D_{\mu_1}, \ldots, D_{\mu_n})$ is distributed the same as the one in the real attack.

Simulation of Phase 4. $\mathcal{B}$ creates a challenge ciphertext C^* as follows.

Upon receiving (M_0, M_1):

Choose $\beta \in \{0, 1\}$ at random.

Search HList to get $l_1^*, \ldots, l_n^*$ that correspond to each of $\mathtt{ID}^* = (\mu_1^*, \ldots, \mu_n^*)$.

Compute $g^{al_i^*}$ for $i = 1, \ldots, n$.

Return $C^* = (g^a, g^{al_1^*}, \ldots, g^{al_n^*}, \tau M_\beta)$ as a challenge ciphertext.

Simulation of Phase 5. $\mathcal{B}$ answers $\mathcal{A}$'s random oracle/private key extraction queries as in Phase 3.

Simulation of Phase 6. $\mathcal{A}$ outputs its guess β'. If $\beta' = \beta$, $\mathcal{B}$ outputs 1. Otherwise, it outputs 0.

Analysis. Note that if $\tau = e(g,g)^{abc}$, $\tau M_\beta = e(g^b, g^c)^a M_\beta = e(g_1, g_2)^a M_\beta$. Note also that $g^{al_i^*} = (g^{l_i^*})^a = H(\mu_i^*)^a$ for $i = 1, \ldots, n$ from the construction of the random oracle H. Hence the challenge ciphertext C^* created above is distributed the same as the one in the real attack. On the other hand, if τ is uniform and independent in $\mathbb{G}_2$, i.e. $\tau = e(g,g)^\gamma$ for some $\gamma \in \mathbb{Z}_q^*$ uniformly chosen at random, so is τM_β. As justified in the simulation of Phase 3, $\mathcal{B}$ perfectly simulates the random oracle H and the key private key extraction. Hence, we get $\Pr[\mathcal{B}(g, g^a, g^b, g^c, e(g,g)^{abc}) = 1] = \Pr[\beta' = \beta]$, where $|\Pr[\beta' = \beta] - \frac{1}{2}| > \epsilon$, and $\Pr[\mathcal{B}(g, g^a, g^b, g^c, e(g,g)^\gamma) = 1] = \Pr[\beta' = \beta] = \frac{1}{2}$, where γ is uniform in $\mathbb{G}_2$. Consequently, we get

$$|\Pr[\mathcal{B}(g, g^a, g^b, g^c, e(g,g)^{abc}) = 1] - \Pr[\mathcal{B}(g, g^a, g^b, g^c, e(g,g)^\gamma) = 1]| > \left|\left(\frac{1}{2} \pm \epsilon\right) - \frac{1}{2}\right| = \epsilon.$$

$\mathcal{B}$'s running time is calculated as $t' < t + q_H O(T_e)$, where T_e denotes the computing time for an exponentiation in $\mathbb{G}_1$. □

Finally we note that from the same reason as EFIBE-I, EFIBE-II is also secure against collusion attacks.

Finally we remark that EFIBE-I and EFIBE-II can be extended to achieve chosen ciphertext security, *i.e.* IND-FSID-CCA, using the Fujisaki-Okamoto transform [6] in the random oracle model or the simulation-sound NIZK proofs [11] without depending on the random oracle model, as discussed in [12].

4. Comparisons

Table 1 summarizes the size of various parameters and the cost of computing sub-algorithms of the proposed fuzzy IBE schemes and the random oracle version of the Sahai-Waters construction [10], which we denote by SW-RO.

Notice that both Extract and Encrypt algorithms of EFIBE-II are more efficient than those of SW-RO. The Extract algorithm of EFIBE-I is the most efficient among the three schemes but its Encrypt is slightly less efficient than those of EFIBE-II and SW-RO.

5. Concluding Remarks

We expect that our new fuzzy IBE schemes will serve as efficient building blocks for biometric authentication systems or attribute-based encryption systems.

	EFIBE-I	EFIBE-II	SW-RO
Size of $params\backslash$ $\{q, g, e, \mathbb{G}_1, \mathbb{G}_2, d\}$	$2\lvert\mathbb{G}_1\rvert$	$2\lvert\mathbb{G}_1\rvert$	$2\lvert\mathbb{G}_1\rvert$
Size of $D_{\mathtt{ID}}$	$2n\lvert\mathbb{G}_1\rvert$	$2n\lvert\mathbb{G}_1\rvert$	$2n\lvert\mathbb{G}_1\rvert$
Size of $C \setminus \mathtt{ID}$	$(n+1)\lvert\mathbb{G}_1\rvert$ $+\lvert\mathbb{G}_2\rvert$	$(n+1)\lvert\mathbb{G}_1\rvert$ $+\lvert\mathbb{G}_2\rvert$	$(n+1)\lvert\mathbb{G}_1\rvert$ $+\lvert\mathbb{G}_2\rvert$
Cost of Extract	$n(T_H + 2T_e)$	$n(T_H + T_m$ $+2T_e)$	$n(T_H + T_m$ $+3T_e)$
Cost of Encrypt	$n(T_m + T_e$ $+T_H) + 2T_e$ $+T_p + T'_m$	$n(T_e + T_H)$ $+2T_e + T_p$ $+T'_m$	$n(T_e + T_H)$ $+2T_e + T_p$ $+T'_m$
Cost of Decrypt	$d(T_e + T_m)$ $+d(T_e + T_p)$ $+T_p + T'_i$ $+T'_m$	$d(T_e + T_m)$ $+d(T_e + T_p)$ $+T_p + T'_i$ $+T'_m$	$d(T_e + T_m)$ $+ d(T_e + T_p)$ $+T_p + T'_i$ $+T'_m$
Security Rel. to	DBDH	DBDH	DBDH

Table 1. Comparisons of Various Fuzzy IBE Schemes. Abbreviations: $|S|$ – the bit-length of an element in set (or group) S; n – the number of elements in an identity; T_e – the computation time for a single exponentiation in $\mathbb{G}_1$; T_H – the computation time for a function H modeled as a random oracle; T_m – the computation time for a single multiplication in $\mathbb{G}_1$; T_i – the computation time for a single inverse operation in $\mathbb{G}_1$; T_p – the computation a single fora single pairing operation; T'_m – the computation time for a single multiplication in $\mathbb{G}_2$; T'_i – the computation time for a single inverse operation in $\mathbb{G}_2$; d – an error tolerance parameter

Construction of fuzzy IBE schemes that have the exactly the same structures as ours (that is, non-random oracle version of our schemes using Sahai-Waters' technique [12]) is an interesting open problem.

References

1. J. Baek, W. Susilo and J. Zhou, *New Constructions of Fuzzy Identity-Based Encryption*, In 2007 ACM Symposium on InformAtion, Computer and Communications Security (AsiaCCS 2007), pp. 368 - 370, 2007.
2. M. Bellare and P. Rogaway, *Random Oracles are Practical: A Paradigm for Designing Efficient Protocols*, In ACM CCS '93, pp. 62–73, ACM Press, 1993.
3. D. Boneh and M. Franklin, *Identity-Based Encryption from the Weil Pairing*, In Crypto '01, LNCS 2139, pp. 213–229, Springer-Verlag, 2001.
4. R. Canetti, S. Halevi, and J. Katz, *A Forward-Secure Public-Key Encryption Scheme*, Advances in Cryptology - In Eurocrypt 2003, LNCS 2656, pp. 255–271, Springer-Verlag, 2003.
5. Y. Dodis, L. Reyzin and A. Smith, *Fuzzy Extractors: How to Generate Strong*

Keys from Biometrics and Other Noisy Data, In Eurocrypt '04, LNCS 3027, pp. 523 – 540, Springer-Verlag, 2004.

6. E. Fujisaki and T. Okamoto, *Secure Integration of Asymmetric and Symmetric Encryption Schemes*, In Crypto '99, LNCS 1666, pp. 537 – 554, Springer-Verlag, 1999.
7. V. Goyal, O. Pandey, A. Sahai and B. Waters, *Attribute-Based Encryption for Fine-Grained Access Control of Encrypted Data*, In ACM CCS '06, 2006, to appear.
8. A. Joux: *The Weil and Tate Pairings as Building Blocks for Public Key Cryptosystems*, Algorithmic Number Theory Symposium (ANTS-V) '02, LNCS 2369, pp. 20–32, Springer-Verlag, 2002.
9. A. Juels and M. Wattenberg, *A Fuzzy Commitment Scheme*, In ACM CCS '99, pp. 28–36, ACM Press, 1999.
10. M. Pirretti, P. Traynor, P. McDaniel and B. Waters, *Secure Attribute-Based Systems*, In ACM CCS '06, 2006, to appear.
11. A. Sahai, *Non-Malleable Non-Interactive Zero Knowledge and Adaptive Chosen-Ciphertext Security*, In FOCS '99, pp. 543–553, IEEE Computer Society.
12. A. Sahai and B. Waters, *Fuzzy Identity-Based Encryption*, Advances in Cryptology - In Eurocrypt 2005, LNCS 3494, pp. 457–473, Springer-Verlag, 2005.

A Functional View of Upper Bounds on Codes

Alexander Barg[1,2] and Dmitry Nogin[2]

[1] *Department of ECE/ISR*
University of Maryland
College Park MD 20742, USA

[2] *Dobrushin Math. Laboratory*
Institute for Information Transmission Problems
Bol'shoj Karetnyj 19, Moscow 101447, Russia
E-mails: {*abarg@umd.edu, nogin@iitp.ru*}

Functional and linear-algebraic approaches to the Delsarte problem of upper bounds on codes are discussed. We show that Christoffel-Darboux kernels and Levenshtein polynomials related to them arise as stationary points of the moment functionals of some distributions. We also show that they can be derived as eigenfunctions of the Jacobi operator.

Keywords: Delsarte problem, Jacobi matrix, moment functional, stationary points

1. Introduction

In the problem of bounding the size of codes in compact homogeneous spaces, Delsarte's polynomial method gives rise to the most powerful universal bounds on codes. Many overviews of the method exist in the literature; see for instance Levenshtein [1]. In this note, which extends our previous work [2] we develop a functional perspective of this method and give some examples. We also discuss another version of the functional approach, a linear algebraic method for the construction of polynomials for Delsarte's problem. Our main results are new constructions of Levenshtein's polynomials.

Let $\mathfrak{X}$ be a compact metric space with distance function τ whose isometry group G acts transitively on it. The zonal polynomials associated with this action give rise to a family of orthogonal polynomials $\mathcal{P}(\mathfrak{X}) = \{P_\kappa\}$ where $\kappa = 0, 1, \ldots$ is the total degree. These polynomials are univariate if G acts on $\mathfrak{X}$ doubly transitively (the well-known examples include the

Hamming and Johnson graphs, their q-analogs and other Q-polynomial distance-regular graphs; the sphere $S^{d-1} \in \mathbb{R}^d$) and are multivariate otherwise.

First consider the univariate case. Then for any given value of the degree $\kappa = i$ the family $\mathcal{P}(\mathfrak{X})$ contains only one polynomial of degree i, denoted below by P_i. Suppose that the distance on $\mathfrak{X}$ is measured in such a way that $\tau(x, x) = 1$ and the diameter of $\mathfrak{X}$ equals -1 (to accomplish this, a change of variable is made in the natural distance function on $\mathfrak{X}$). We refer to the model case of $\mathfrak{X} = S^{d-1}$ although the arguments below apply to all spaces $\mathfrak{X}$ with the above properties. Let $\langle f, g\rangle = \int_{-1}^{1} fg d\mu$ be the inner product in $L_2([-1,1], d\mu)$ where $d\mu(x)$ is a distribution on $[-1,1]$ induced by a G-invariant measure on $\mathfrak{X}$. Let $\mathcal{F}(\cdot) \triangleq \langle \cdot, 1\rangle$ be the moment functional with respect to $d\mu$. We assume that this distribution is normalized, i.e., that $\mathcal{F}(1) = 1$.

Let C be a code, i.e., a finite collection of points in $\mathfrak{X}$. By Delsarte's theorem, the size of the code C whose distances take values in $[-1, s]$ is bounded above by

$$|C| \le \inf_{f \in \Phi} f(1)/\hat{f}_0, \tag{1}$$

where

$$\Phi = \Phi(s) \triangleq \{f : f(x) \le 0, x \in [-1, s]; \quad \hat{f}_0 > 0, \ \hat{f}_i \ge 0, i = 1, 2, \dots\} \tag{2}$$

is the cone of positive semidefinite functions that are nonpositive on $[-1, s]$ (here $\hat{f}_i = \langle f, P_i\rangle / \langle P_i, P_i\rangle$ are the Fourier coefficients of f).

2. Functional approach

The choice of polynomials for problem (1)-(2) was studied extensively in the works of Levenshtein [4–6]. In this section we give a new construction of his polynomials and their simplified versions.

2.1. *Notation.*

Let V be the space of real square-integrable functions on $[-1, 1]$ and let V_k be the space of polynomials of degree k or less. Let $p_i = P_i / \langle P_i, P_i\rangle, i = 0, 1, \dots$ be the normalized polynomials. The polynomials $\{p_i\}$ satisfy a three-term recurrence of the form

$$x p_i = a_i p_{i+1} + b_i p_i + a_{i-1} p_{i-1}, \tag{3}$$
$$i = 1, 2, \dots; p_{-1} = 0, p_0 = 1; \ a_{-1} = 0.$$

In other words, the matrix of the operator $x : V \to V$ (multiplication by the argument) in the orthonormal basis is a semi-infinite symmetric tridiagonal matrix, called the Jacobi matrix. Let $X_k = E_k \circ x$ where $E_k = \mathrm{proj}_{V \to V_k}$, and let J_k be the $(k+1) \times (k+1)$ submatrix of J,

$$J_k = \begin{bmatrix} b_0 & a_0 & 0 & 0 & \dots & 0 \\ a_0 & b_1 & a_1 & 0 & \dots & 0 \\ 0 & a_1 & b_2 & a_2 & \dots & 0 \\ \dots & \dots & \dots & \dots & \dots & a_{k-1} \\ 0 & 0 & \dots & \dots & a_{k-1} & b_k \end{bmatrix}.$$

Example 2.1. (*a*) For instance, let $\mathfrak{X}$ be the binary n-dimensional Hamming space. Then $p_i(x) = \tilde{k}_i(n/2(1-x))$, where $\tilde{k}_i(z)$ is the normalized Krawtchouk polynomial. The polynomials $p_i(x)$ are orthogonal on the finite set of points $\{x_j = 1-(2j/n), j = 0, 1, \dots, n\}$ with weight $w(x_j) = \binom{n}{j}2^{-n}$ and have unit norm. In this case,

$$a_i = (1/n)\sqrt{(n-i)(i+1)},\ b_i = 0, \quad 0 \le i \le n. \tag{4}$$

(*b*) Let $\mathfrak{X}$ be the unit sphere in d dimensions. Then $p_i(x)$ are the normalized Gegenbauer polynomials; in this case

$$a_i = \sqrt{\frac{(n-i+2)(i+1)}{(n+2i)(n+2i-2)}},\ b_i = 0, \quad i = 0, 1, \dots.$$

It is well known [7, p.243] that for $k \ge 1$ the spectrum of X_k coincides with the set $\mathfrak{X}_{k+1} = \{x_{k+1,1}, \dots, x_{k+1,k+1}\}$ of zeros of p_{k+1}. Below we denote the largest of these zeros by x_{k+1}. Let

$$K_k(x, s) \triangleq \sum_{i=0}^{k} p_i(s)p_i(x) \tag{5}$$

be the k-th reproducing kernel. By the Christoffel-Darboux formula,

$$(x-s)K_k(x,s) = a_k(p_{k+1}(x)p_k(s) - p_{k+1}(s)p_k(x)). \tag{6}$$

In particular, if $s \in \mathfrak{X}_{k+1}$ then $X_k K_k(x,s) = sK_k(x,s)$. Note that $K_k(x,y)$ acts on V_k as a delta-function at y:

$$\langle K_k(\cdot, y), f(\cdot)\rangle = f(y). \tag{7}$$

2.2. *Construction of polynomials.*

Without loss of generality let us assume that $f(1) = 1$. Then (1) is equivalent to the problem

$$\sup\{\mathcal{F}(f), f \in \Phi\}.$$

Let us restrict the class of functions to V_n. By the Markov-Lucacs theorem [8, Thm. 6.4], a polynomial $f(x)$ that is nonpositive on $[-1, s]$ can be written in the form

$$f_n(x) = (x - s)g^2 - (x + 1)\phi_1^2 \quad \text{or} \quad f_n(x) = (x + 1)(x - s)g^2 - \phi_2^2$$

according as its degree $n = 2k + 1$ or $2k + 2$ is odd or even. Here $g, \phi_1 \in V_k, \phi_2 \in V_{k+1}$ are some polynomials. Below the negative terms will be discarded. We use a generic notation c for multiplicative constants chosen to fulfill the condition $f(1) = 1$.

2.2.1. *The MRRW polynomial.*

Restricting our attention to odd degrees $n = 2k + 1$, let us seek $f(x)$ in the form $(x-s)g^2$. Let us write the Taylor expansion of $\mathcal{F}$ in the "neighborhood" of g. Let $h \in V_k$ be a function that satisfies $\|h\| \le \varepsilon$ for a small positive ε and the condition $h(1) = 0$. We obtain

$$\mathcal{F}((x - s)(g + h)^2) = \mathcal{F}((x - s)g^2) + \langle (x - s)(g + h), g + h\rangle - \langle (x - s)g, g\rangle$$

$$= \mathcal{F}(f) + \mathcal{F}'(h) + {}^1\!/\!{}_2\langle \mathcal{F}''h, h\rangle,$$

where $\mathcal{F}' = 2(x - s)g, \mathcal{F}'' = 2(x - s)$ are the Fréchet derivatives of $\mathcal{F}$. This relation shows that for f to be a stationary point of $\mathcal{F}$, the function g should satisfy $d\mathcal{F} = 2\langle g, (x-s)h\rangle = 0$ for any function h with the above properties. First assume that $s = x_{k+1}$. Then by (6), a stationary point of $\mathcal{F}$ is given by $g = K_k(x, s)$, and we obtain f in the form

$$f_n(x) = c(x - s)(K_k(x, s))^2.$$

Since $\hat{f}_0 = 0$, conditions (2) are not satisfied; however, it is easy to check that they are satisfied if $x_k < s < x_{k+1}$. For all such s, the polynomial f_n is a valid choice for problem (1), yielding

$$|C| \le -\frac{1 - s}{a_k p_{k+1}(s) p_k(s)} K_k^2(1, s). \tag{8}$$

The polynomial f_n was used by McEliece *et al.* [9] and Kabatiansky and Levenshtein [10] to derive their well known upper bounds on codes.

2.2.2. *Levenshtein polynomials, $n = 2k + 1$.*

So far in our optimization we did not use the condition $h(1) = 0$. To use it, let us write $h = (1-x)h_1, h_1 \in V_{k-1}$ and repeat the above calculation. We find that stationary points of $\mathcal{F}$ should satisfy

$$d\mathcal{F}^{(-)} = 2\langle (x-s)g, (1-x)h_1 \rangle = 0,$$

where $\mathcal{F}^{(-)}(\,.\,) = \int .\,(1-x)d\mu$ is the moment functional with respect to the distribution $d\mu^{(-)}(x) = (1-x)d\mu(x)$. A stationary point of $\mathcal{F}^{(-)}$ is given by the reproducing kernel $K_k^-(x, s)$ *with respect to this distribution*:

$$K_k^-(x, s) = \sum_{i=0}^{k} p_i^-(s) p_i^-(x), \tag{9}$$

where $\{p_i^-(x), i = 0, 1, \dots\}$ is the corresponding orthonormal system. To find the polynomials $p_i^-(x)$ observe that

$$\mathcal{F}^{(-)}(p_i^- p_j^-) = \mathcal{F}(p_i^-(x) p_j^-(x)(1-x)) = \delta_{i,j}$$

is satisfied for $p_i^-(x) = K_i(1, x)/(a_i p_{i+1}(1) p_i(1))^{1/2}$. Indeed, if $j < i$ then the function $(1-x)K_i(1, x)$ is in the subspace spanned by p_{i+1}, p_i and thus is orthogonal to $K_j(1, x)$. To conclude, the function sought can be taken in the form

$$f_n^-(x) = c(x-s)(K_k^-(x, s))^2.$$

2.2.3. *Levenshtein polynomials, $n = 2k + 2$.*

In this case we seek the polynomial in the form $f_n = (x-s)(x+1)g^2$. The necessary condition for the stationary point takes the form $\mathcal{F}^{\pm}((x-s)gh) \triangleq \langle (x-s)(1-x^2)g, h \rangle = 0$. From this, $g = K_{k-1}^{\pm}(x, s)$ where the kernel $K_{k-1}^{\pm}$ is taken with respect to the distribution $d\mu^{(\pm)}(x) = (1+x)(1-x)d\mu(x)$. The corresponding orthogonal polynomials $p_i^{\pm}(x)$ are also easily found: up to normalization they are equal

$$p_i^{\pm}(x) = K_i(x, -1)p_{i+1}(1) - K_i(x, 1)p_{i+1}(-1).$$

Then

$$f_n^{\pm}(x) = c(x-s)(x+1)(K_{k-1}^{\pm}(x, s))^2.$$

Let x_k^- ($x_k^{\pm}$) be the largest root of $p_k^-(x)$ (resp. of $p_k^{\pm}(x)$). Then $f_{2k+1}^-(x) \in \Phi$ if $x_k^{\pm} \le s \le x_{k+1}^-$ and $f_{2k+2}^{\pm}(x) \in \Phi$ if $x_{k+1}^- < s < x_{k+1}^{\pm}$.

Remarks.

1. The polynomials $f_n^-, f_n^{\pm}$ were constructed and applied to coding theory by Levenshtein [4–6]. Polynomials closely related to them were studied

in a more general context in the works of M. G. Krein *et al.*; see Krein and Nudelman [8]. The orthogonal systems $\{p_i^-\}, \{p_i^\pm\}$ are sometimes called *adjacent polynomials* of the original system $\{p_i\}$.

2. The stationary points found above are not true extremums because the second differential of the functionals $\mathcal{F}, \mathcal{F}^{(-)}, \mathcal{F}^{(\pm)}$ is undefined: for instance, $d^2\mathcal{F}(g) = 2\langle (x-s)h, h\rangle$. Nevertheless, the polynomials $f_n^-, f_n^\pm$ have been proved [11] to be optimal in the following sense: for any $n \geq 1$ and all $f \in \Phi, \deg f \leq n$

$$\mathcal{F}(f_n) \geq \mathcal{F}(f).$$

3. Asymptotic bounds derived from (1) relying upon the polynomials $f_n, f_n^-, f_n^\pm$ coincide. For the finite values of the parameters, better bounds are obtained from $f_n^-, f_n^\pm$.

3. Spectral method

This section is devoted to a different way of constructing polynomials for the Delsarte problem. The ideas discussed below originate in the work of C. Bachoc [12]. They were elaborated upon in an earlier work of the authors [2].

We develop the remark made after (6), namely that for any $i \geq 1$, $K_k(x, x_{k,i})$ is an eigenfunction of the Jacobi operator X_k. Since $K_k(x,s)$ is a good choice for the polynomial in Delsarte's problem, it is possible to construct polynomials as eigenvectors of X_k as opposed to the analytic arguments discussed above. In particular, $K_k(x,s)$ arises as an eigenfunction of the operator $T_k = T_k(s)$ defined by

$$\begin{aligned} T_k : & V_k \to V_k \\ & \phi \mapsto X_k\phi + \rho_k\hat{\phi}_k p_k \end{aligned}$$

where $\rho_k = a_k p_{k+1}(s)/p_k(s)$. Indeed, using (5) and (6) we obtain

$$(T_k - s)K_k(x,s) = (X_k - s)K_k(x,s) + a_k p_{k+1}(s)p_k(x) = 0.$$

On account of earlier arguments we should choose the polynomial for problem (1) in the form $F(x) = (x-s)f^2(x)$ where $f(x) = f(x,s)$ is an eigenfunction of T_k. The positive definiteness condition of f can be proved using the Perron-Frobenius theorem; for this we must take f to be the eigenfunction that corresponds to the *largest* eigenvalue of T_k. This condition defines the range of code distances s in which the method is applicable.

A variant of this calculation was performed in [2] to which we refer for details. The difference between [2] and the argument above is that there we took $\rho_k = a_k p_{k+1}(1)/p_k(1)$. This has the advantage of defining T_k independently of s but leads to a bound of the form

$$|C| \leq \frac{4a_k p_{k+1}(1) p_k(1)}{1-\lambda_k} \tag{10}$$

which is generally somewhat weaker than (8). Using the function F defined above we can improve this to recover the estimate (8).

We note that this argument does not depend on the choice of the functional space; in particular, the kernels $K_k^-, K_k^{\pm}$ arise if the operator X_k is written with respect to the basis of the corresponding adjacent polynomials ($\{p_i^-\}$ or $\{p_i^{\pm}\}$) and their generating distribution. To conclude, Levenshtein's polynomials and bounds on codes can be derived within the framework of the spectral method.

Example 3.1. Consider again Example 2.1(a). The adjacent polynomials up to a constant factor that does not depend on i are given by [5, p.81]

$$p_i^-(x) = \tilde{k}_i^{(n-1)}(z),\ p_i^{\pm}(x) = \tilde{k}_i^{(n-2)}(z) \quad \text{for } z = \frac{n}{2}(1-x) - 1,$$

where $\tilde{k}_i^{(n-1)}(z)$ for instance denotes the degree-i normalized Krawtchouk polynomial associated with the $(n-1)$-dimensional Hamming space. The Jacobi matrix J_k for the basis p_i^- can be computed from (4) as follows. Since

$$xp_i^-(x) = \Big(1 - \frac{2}{n}(z+1)\Big)\tilde{k}_i^{(n-1)}(z),$$

we find that the coefficients of three-term recurrence for the family $\{p_i^-\}$ are

$$a_i = (1/n)\sqrt{(n-k-1)(k+1)},\ b_i = -1/n,\ \ i = 0, 1, \ldots.$$

Constructing the operator T_k as described above, we obtain $K_k^-(x, s)$ as its eigenfunction. A similar construction can be pursued for the function $K_k^{\pm}$.

The approach outlined above has two advantages. First, it enables one to obtain simple estimates of the largest eigenvalue of X_k which is important in verifying the condition $f(x) \leq 0, x \in [-1, s]$. The second advantage is a more substantial one: this method can be extended to the case of *multivariate zonal polynomials* when the analytic alternative is not readily available. This case arises when the space $\mathfrak{X}$ is homogeneous but not 2-point homogeneous. Worked examples include the real Grassmann manifold $G_{k,n}$ ([12]; the P_i are given by the generalized k-variate Jacobi polynomials) and the so-called ordered Hamming space [3]. We provide a few more details on the latter case in order to illustrate the general method.

Let $\mathcal{Q}$ be a finite alphabet of size q. Consider the set $\mathcal{Q}^{r,n}$ of vectors of dimension rn over $\mathcal{Q}$. A vector $\boldsymbol{x}$ will be written as a concatenation of n blocks of length r each, $\boldsymbol{x} = \{x_{11}, \dots, x_{1r}; \dots; x_{n1}, \dots, x_{nr}\}$. For a given vector $\boldsymbol{x}$ let $e_i, i = 1, \dots, r$ be the number of r-blocks of $\boldsymbol{x}$ whose rightmost nonzero entry is in the ith position counting from the beginning of the block. The r-vector $e = (e_1, \dots, e_r)$ will be called the *shape* of $\boldsymbol{x}$. A shape vector $e = (e_1, \dots, e_r)$ defines a partition of a number $N \le n$ into a sum of r parts. Let $e_0 = n - \sum_i e_i$. Let $\Delta_{r,n} = \{e \in (\mathbb{Z}_+ \cup \{0\})^r : \sum_i e_i \le n\}$ be the set of all such partitions. The zonal polynomials associated to $\mathcal{Q}^{r,n}$ are r-variate polynomials $P_f(e), f, e \in \Delta_{r,n}$ of degree $\kappa = \sum_i f_i$. They are orthogonal on $\Delta_{r,n}$ according to the following inner product

$$\sum_{e \in \Delta_{r,n}} P_f(e) P_g(e) w(e) = 0 \quad (f \ne g).$$

The weight in this relation is given by the multinomial probability distribution

$$w(e_1, \dots, e_r) = n! \prod_{i=0}^{r} \frac{p_i^{e_i}}{e_i!} \qquad (p_i = q^{i-r-1}(q-1), i = 1, \dots, r; p_0 = q^{-r}),$$

so the polynomials $P_f(e)$ form a particular case of *r-variate Krawtchouk polynomials.*

Let $\boldsymbol{x} \in \mathcal{Q}^{r,n}$ be a vector of shape e. Define a norm on $\mathcal{Q}^{r,n}$ by setting $\mathrm{w}(\boldsymbol{x}) = \sum_i ie_i$ and let $d_r(\boldsymbol{x}, \boldsymbol{y}) = \mathrm{w}(\boldsymbol{x} - \boldsymbol{y})$ be the ordered Hamming metric (known also as the Niederreiter-Rosenbloom-Tsfasman metric). We note that in the multivariate case there is no direct link between the variables and the metric. For instance, for the space $\mathcal{Q}^{r,n}$ the polynomials (as well as relations in the corresponding association scheme) are naturally indexed by shape vectors e while the weight is some function e.

The Delsarte theorem in this case takes the following form: *The size of an (n, M, d) code $C \subset \mathcal{Q}^{r,n}$ is bounded above by* $M \le \inf_{f \in \Phi} f(0)/f_0$, where

$$\Phi = \{f(x) = f(x_1, \dots, x_r) = f_0 + \sum_{e \ne 0} f_e P_e(x) : f_0 > 0, f_e \ge 0 \; (e \ne 0);$$
$$f(e) \le 0 \;\; \forall e \text{ s.t. } \sum_{i=1}^{r} ie_i \le d\}$$

The argument for the univariate case given in this section can be repeated once we establish a three-term relation for the polynomials $P_f(e)$. Let $\mathbb{P}_\kappa$ be the column vector of the normalized polynomials P_f ordered lexicographically with respect to all f that satisfy $\sum_i f_i = \kappa$ and let $F(e)$ be a suitably chosen linear polynomial. Then

$$F(e)\mathbb{P}_\kappa(e) = A_\kappa \mathbb{P}_{\kappa+1}(e) + B_\kappa \mathbb{P}_\kappa(e) + A_{\kappa-1}^T \mathbb{P}_{\kappa-1}(e)$$

where A_κ, B_κ are matrices of order $\binom{\kappa+r-1}{r-1} \times \binom{\kappa+s+r-1}{r-1}$ and $s = 1, 0$, respectively. The elements of these matrices can be computed explicitly from combinatorial considerations. This gives an explicit form of the operator $S_\kappa = E_\kappa \circ F(e)$ in the orthonormal basis. Relying on this, it is possible to derive a bound on codes in the NRT space of the form (10) and perform explicit calculations, both in the case of finite parameters and for asymptotics. The full details of the calculations are given in [3].

Acknowledgments: The research of A. Barg is supported in part by NSF grants CCF0515124 and CCF0635271, and by NSA grant H98230-06-1-0044. The research of D. Nogin is supported in part by Russian Foundation for Basic Research through grants RFBR 06-01-72550-CNRS and RFBR 06-01-72004-MST. Parts of this research were presented at the International Workshop on Coding and Cryptology, The Wuyi Mountain, Fujian, China, June 11-15, 2007, and COE Conference on the Development of Dynamic Mathematics with High Functionality (DMHF2007), Fukuoka, Japan, October 1-4, 2007.

References

1. V. I. Levenshtein. Universal bounds for codes and designs. In eds. V. Pless and W. C. Huffman, *Handbook of Coding Theory*, vol. 1, pp. 499–648. Elsevier Science, Amsterdam, (1998).
2. A. Barg and D. Nogin, Spectral approach to linear programming bounds on codes, *Problems of Information Transmission.* **42**, 12–25, (2006).
3. A. Barg and P. Purkayastha, Bounds on ordered codes and orthogonal arrays. arxiv:CS/0702033.
4. V. I. Levenshtein. On choosing polynomials to obtain bounds in packing problems. In *Proc. 7th All-Union Conf. Coding Theory and Information Transmission,Part 2*, pp. 103–108, Moscow, Vilnius, (1978). (In Russian).
5. V. I. Levenshtein, Bounds for packings of metric spaces and some of their applications, *Problemy Kibernet.* **40**, 43–110 (In Russian), (1983).
6. V. I. Levenshtein, Designs as maximum codes in polynomial metric spaces, *Acta Appl. Math.* **29**(1-2), 1–82, (1992). ISSN 0167-8019.
7. G. Andrews, R. Askey, and R. Roy, *Special functions.* (Cambridge University Press, 1999).
8. M. G. Kreĭn and A. A. Nudel′man, *The Markov moment problem and extremal problems.* (American Mathematical Society, Providence, R.I., 1977).

9. R. J. McEliece, E. R. Rodemich, H. Rumsey, and L. R. Welch, New upper bound on the rate of a code via the Delsarte-MacWilliams inequalities, *IEEE Trans. Inform. Theory.* **23**(2), 157–166, (1977).
10. G. A. Kabatiansky and V. I. Levenshtein, Bounds for packings on the sphere and in the space, *Problems of Information Transmission.* **14**(1), 3–25, (1978).
11. V. M. Sidelnikov, Extremal polynomials used in bounds of code volume, *Problemy Peredachi Informatsii.* **16**(3), 17–30, (1980).
12. C. Bachoc, Linear programming bounds for codes in Grassmannian spaces, *IEEE Trans. Inform. Theory.* **52**, 2111–2126, (2006).

A Method of Construction of Balanced Functions with Optimum Algebraic Immunity

Claude Carlet

University of Paris 8, Department of Mathematics (MAATICAH), 2 rue de la liberté, 93526 Saint-Denis, Cedex France; Email: claude.carlet@inria.fr.

Because of the recent algebraic attacks, a high algebraic immunity is now an absolutely necessary (but not sufficient) property for Boolean functions used in stream ciphers. Very few examples of (balanced) functions with high algebraic immunity have been found so far. These examples seem to be isolated and no method for obtaining such functions is known. In this paper, we introduce a general method for proving that a given function, in any number of variables, has a prescribed algebraic immunity. We deduce an algorithm, valid for any even number of variables, for constructing functions with optimum (or, if this can be useful, with high but not optimal) algebraic immunity and which can be balanced if we wish. We also introduce a new example of an infinite class of such functions. We study their Walsh transforms. We finally give similar algorithm and infinite class in the case n is odd (which is different).

Keywords: Algebraic Attacks, Annihilators, Boolean Functions, Nonlinearity, Walsh Spectrum, Stream ciphers.

1. Introduction

The two main models of pseudo-random generators using Boolean functions in stream ciphers - the combiner model, in which the outputs to several LFSRs are combined by the nonlinear Boolean function to produce the keystream, and the filter model, in which the content of some of the flip-flops in a single (longer) LFSR constitute the input to the function - have been the targets of a lot of cryptanalyses. This has led to design criteria for these functions, mainly: balancedness, a high algebraic degree, a high nonlinearity and, in the case of the combiner model, a high resiliency order (the filter model is theoretically equivalent to the combiner model, but the attacks do not work similarly on each system). A recent attack uses the fact that it is possible to obtain a very over-defined system of multivariate nonlinear equations whose unknowns are the bits of the initialization of the LFSR(s).

This improvement of an idea due to C. Shannon [33] uses the existence of low degree multiples of the nonlinear function. It is called *algebraic attack* [4,16,19,30] and has deeply modified the situation with Boolean functions in stream ciphers. Given a Boolean function f on n variables, different kinds of scenarios related to low degree multiples of f have been studied in ref. 19,30. The core of the analysis is to find out minimum (or low) degree nonzero annihilators of f or of $1+f$, that is, functions g such that $f * g = 0$ or $(1+f) * g = 0$, where "$*$" is the multiplication of functions inheritated from the multiplication in F_2.

Since the introduction of algebraic attacks on stream ciphers (see ref. 19), the research of Boolean functions that can resist them has not given fully satisfactory results. It has produced only a small number of examples (excluding those results of ref. 8 which may be false; see how these results have been modified in ref. 5) and no method for obtaining such functions. The main results are:

1. In ref. 20, an iterative construction of a $2k$-variable Boolean function with algebraic immunity provably equal to k (that is, optimal). The produced function has been further studied in ref. 13. It has very high algebraic degree and there exists an algorithm giving a very fast way (whose complexity is linear in the number of variables) of computing the output to the function, given its input. But the function is not balanced and its nonlinearity is weak.
2. In ref. 21 and 8, examples of symmetric functions (that is, of functions whose outputs depend only on the Hamming weight of their input) or nearly symmetric functions, achieving optimum algebraic immunities. They present a risk if attacks using symmetry can be found in the future. Moreover, they do not have high nonlinearities either (see below).

Last but not least drawback of all these functions: they do not behave well with respect to fast algebraic attacks [16] : see ref. 1,22.

In the present paper, we give a general way of proving that a given function has algebraic immunity at least k, where k is any integer upper bounded by $\lceil \frac{n}{2} \rceil$, leading to a way of designing Boolean functions whose algebraic immunity is at least k and which can provide balanced functions if the parameters are chosen accordingly. After observing that this proves a result of ref. 21, but that the functions concerned by this result have not sufficient nonlinearities, we deduce an algorithm for designing numerous other functions in even number of variables n and with optimal algebraic immunity $\frac{n}{2}$, among which exist balanced functions having better nonlinearities. We also exhibit an infinite class of functions with optimal algebraic immunity. We study the Walsh transforms of these functions in Section 5. We give tables of the best nonlinearities obtained with the algorithm. We give similar results in Section 6 for the n odd case and also give the best obtained nonlinearities.

2. Preliminaries

A Boolean function on n variables is a mapping from F_2^n into F_2, the finite field with two elements. We denote by B_n the set of all n-variable Boolean functions. The basic representation of a Boolean function $f(x_1, \ldots, x_n)$ is by the output column of its *truth table*, i.e., a binary string of length 2^n, $f = [f(0,0,\ldots,0), f(1,0,\ldots,0), f(0,1,\ldots,0), f(1,1,\ldots,0), \ldots, f(1,1,\ldots,1)]$.

The *Hamming weight* $wt(f)$ of f is the weight of this string, that is, the size of the support $supp(f) = \{x \in F_2^n;\ f(x) = 1\}$ of the function. The *Hamming distance* $d(f,g)$ between two Boolean functions f and g is the Hamming weight of their difference $f+g$ (we use $+$ to denote also the addition in F_2, i.e., the XOR). We say that a Boolean function f is balanced if its truth table contains an equal number of 1's and 0's, that is, if its Hamming weight equals 2^{n-1}.

The truth table does not give an idea of the algebraic complexity of the function. Any Boolean function has a unique representation as a multivariate polynomial over F_2 (called the algebraic normal form, ANF), of the special form:

$$f(x_1, \ldots, x_n) = \sum_{I \subseteq \{1,\ldots,n\}} a_I \prod_{i \in I} x_i; \quad a_I \in F_2.$$

The algebraic degree, $\deg(f)$, is the degree of this polynomial, that is, the number of variables in the highest order term with non zero coefficient. A Boolean function is affine if it has degree at most 1 and the set of all affine functions is denoted by A_n.

Boolean functions used in cryptographic systems must have high nonlinearity to withstand linear and correlation attacks, see ref. 23. The *nonlinearity* of an n-variable function f is its distance from the set of all n-variable affine functions, i.e.,

$$nl(f) = \min_{g \in A_n} (d(f,g)).$$

This parameter can be expressed by means of the Walsh transform. Let $x = (x_1, \ldots, x_n)$ and $a = (a_1, \ldots, a_n)$ both belonging to F_2^n and $x \cdot a = x_1 a_1 + \ldots + x_n a_n$. Then the *Walsh transform* of $f(x)$ is the integer valued function over F_2^n defined as

$$W_f(a) = \sum_{x \in F_2^n} (-1)^{f(x)+x \cdot a}.$$

A Boolean function f is balanced if and only if $W_f(0) = 0$. The nonlinearity of f is given by $nl(f) = 2^{n-1} - \frac{1}{2} \max_{a \in F_2^n} |W_f(a)|$. It is upper bounded by $2^{n-1} - 2^{n/2-1}$, for every n and every function (this bound is the so-called covering radius bound).

Any Boolean function should have also high algebraic degree to be cryptographically secure, see ref. 23,32. In fact, it must keep high degree even if a few output bits are modified. In other words, it must have high nonlinearity profile, see ref. 12. Another notion plays a role, at least for the combiner model. A function is m-resilient if its Walsh transform satisfies $W_f(a) = 0$, for $0 \leq wt(a) \leq m$, where $wt(a)$ denotes the Hamming weight of the vector a. Any combining function should be highly resilient to withstand correlation attacks, see ref. 34.

Recently, it has been identified that any combining or filtering function should not have a low degree multiple. More precisely, it is shown in ref. 19 that, given any n-variable Boolean function f, it is always possible to find a nonzero Boolean function g with degree at most $\lceil \frac{n}{2} \rceil$ such that $f * g$ has degree at most $\lceil \frac{n}{2} \rceil$. While choosing a function f, the cryptosystem designer should avoid that the degree of $f * g$ falls much below $\lceil \frac{n}{2} \rceil$ with a nonzero function g whose degree is also much below $\lceil \frac{n}{2} \rceil$. Indeed, otherwise, resulting low degree multivariate relations involving key/state bits and output bits of the combining or filtering function f allow a very efficient attack, see ref. 19, if enough data from the pseudo-random sequence is known. In fact, as observed in ref. 30, it is necessary and sufficient to check that f and $f + 1$ do not admit nonzero annihilators of such low degrees.

Definition 2.1. Given $f \in B_n$, define $AN(f) = \{g \in B_n | \ f * g = 0\}$. A function $g \in AN(f)$ is called an annihilator of f.

Indeed, if f or $f+1$ has an annihilator g of low degree d, then $f*g$ either is null or equals g and therefore has degree at most d; conversely, if we have $f*g = h$ where $g \neq 0$ and where g and h have degrees at most d, then either $g = h$, and then g is an annihilator of $f + 1$, or $g \neq h$, and by multiplying both terms of the equality $f * g = h$ by f proves that $f * (g + h) = 0$ and shows that $g + h$ is a nonzero annihilator of f of degree at most d.

Definition 2.2. Given $f \in B_n$, the algebraic immunity of f is the minimum degree of all nonzero annihilators of f or $f + 1$. We denote it by

$AI(f)$.

Note that $AI(f) \leq \deg(f)$, since $f*(1+f) = 0$. Note also that the algebraic immunity, as well as the nonlinearity and the degree, is affine invariant (i.e. is invariant under composition by an affine automorphism). Because of the observation made in ref. 19 and recalled above, we have $AI(f) \leq \lceil \frac{n}{2} \rceil$.

If a function has optimal algebraic immunity $\lceil \frac{n}{2} \rceil$ with n odd, then it is balanced (see e.g. ref. 13). Whatever is n, a high value of $AI(f)$ automatically implies that the nonlinearity is not very low: M. Lobanov has obtained in ref. 28 (see a simple proof in ref. 12 and further results on the higher order nonlinearity) the following tight lower bound:

$$nl(f) \geq 2 \sum_{i=0}^{AI(f)-2} \binom{n-1}{i}.$$

However, this bound does not assure that the nonlinearity is high enough.
• For n even and $AI(f) = \frac{n}{2}$, it gives $nl(f) \geq 2^{n-1} - 2\binom{n-1}{n/2-1} = 2^{n-1} - \binom{n}{n/2}$ which is much smaller than the best possible nonlinearity $2^{n-1} - 2^{n/2-1}$ and, more problematically, much smaller than the asymptotic alsmost sure nonlinearity of Boolean functions, which is, when n tends to ∞, located in the neighbourhood of $2^{n-1} - 2^{n/2-1}\sqrt{2n \ln 2}$ (see ref. 31). The best nonlinearity reached by the known functions with optimum AI is that of the majority function recalled below (see ref. 21) and of the iterative construction recalled in ref. 13 : $2^{n-1} - \binom{n-1}{n/2} = 2^{n-1} - \frac{1}{2}\binom{n}{n/2}$. It is a little better than what gives Lobanov's bound but it is insufficient.
• For n odd and $AI(f) = \frac{n+1}{2}$, Lobanov's bound gives $nl(f) \geq 2^{n-1} - \binom{n-1}{(n-1)/2} \simeq 2^{n-1} - \frac{1}{2}\binom{n}{(n-1)/2}$ which is a little better than in the n even case, but still far from the average nonlinearity of Boolean functions. The best known nonlinearity matches this bound and is achieved by the majority function.

Hence, the algebraic immunity property takes care of three fundamental properties of a Boolean function, balancedness, algebraic degree and nonlinearity (and more generally nonlinearity profile, see ref. 12), but it does this incompletely in the case of nonlinearity (and also in the case of balancedness when n is even).

As shown in ref. 1,16, a high algebraic immunity is a necessary but not sufficient condition for robustness against all kinds of algebraic attacks. Indeed, if one can find g of low degree and $h \neq 0$ of reasonable degree such that $f * g = h$, then a fast algebraic attack is feasible, see ref. 4,16,25. Since $f * g = h$ implies $f * h = f * f * g = f * g = h$, we see that h is then an annihilator of $f+1$ and its degree is then at least equal to the algebraic immunity of f. This means that having high algebraic immunity is not a property that allows resisting all kinds of algebraic attacks, but that it is a necessary condition for a resistance to fast algebraic attacks as well. Even a high resistance to fast algebraic attacks is not sufficient, since algebraic attacks on the augmented function (see ref. 24) can be efficient when fast algebraic attacks are not. Finally, a new version of algebraic attack has been found recently by S. Rønjom and T. Helleseth [32] and is very efficient. Its time complexity is roughly $O(\mathcal{D})$, where $\mathcal{D} = \sum_{i=0}^{d^\circ f} \binom{N}{i}$. But it needs much more data than standard algebraic attacks: $O(\mathcal{D})$ also! When f has degree close to n and algebraic immunity close to $\frac{n}{2}$, this is the square of what is needed by standard algebraic attacks. Hence this new attack does not eliminate the interest of algebraic immunity.

3. The general method

The idea of our method for constructing functions with high algebraic immunity is simple but efficient. We use the fact that, if a function has degree strictly less than k and if it is null on an affine subspace (a flat) of dimension at least k, except maybe at one vector of this affine subspace, then it must be null on the whole affine subspace. Indeed, any Boolean function of degree less than k on a k-dimensional affine subspace has even Hamming weight (see e.g. ref. 11). We can exploit this idea for the annihilators of f and $f+1$: to show for instance that f has no nonzero annihilator of degree strictly less than k, we can try to exhibit a sequence of affine subspaces of dimensions at least k, such that each of them contains at most one vector lying outside the support of f and outside all those affine subspaces which come previously in the sequence, and such that with such vectors, we cover all the complement of the support of f:

Proposition 3.1. *Let k be any positive integer such that $k \leq \lceil \frac{n}{2} \rceil$. A sufficient condition for a function f to have no non-zero annihilator of degree strictly less than k is that there exists a sequence of flats (i.e. of affine subspaces of F_2^n) $(A_i)_{1\leq i\leq r}$ of dimensions at least k, such that:*

$$\forall i \leq r,\ card(A_i \setminus [\bigcup_{j<i} A_j \cup supp(f)]) \leq 1 \tag{1}$$

$$F_2^n \setminus supp(f) \subseteq \bigcup_{i\leq r} A_i. \tag{2}$$

Proof. Relation (1) allows proving by induction on i that any annihilator g of degree at most $k-1$ of f is null on A_i for every i, since we know that, for every affine subspace A of dimension at least k, the weight of the restriction of g to A is even. Then (2) shows that g must be null on F_2^n. □

We obtain by applying Proposition 3.1 to f and to $f+1$ (exhibiting a sequence of affine subspaces $(A_i)_{1\leq i\leq r}$ for f and a sequence of affine subspaces $(A'_i)_{1\leq i\leq r'}$ for $f+1$) a sub-class of the class of functions with algebraic immunity at least k. We do not know if this sub-class is in fact the whole class and we leave it as an *open problem.* Note that both class and sub-class are affine invariant.

Example 3.1. The simplest known example of a function with optimal algebraic immunity (whatever is n) is the *majority function*, which takes value 1 at all vectors of weights at least $\frac{n}{2}$ and 0 at all the other vectors[a]. We can take for the flats A'_i the vector spaces $\{x \in F_2^n \,/\, supp(x) \subseteq supp(a)\}$ where $supp(x) = \{j = 1, \cdots, n \,/\, x_j = 1\}$ and where a ranges over the set of vectors of weights at least $k = \lceil \frac{n}{2} \rceil$, the order being by increasing weights (with any order for vectors of the same weight), and for the flats A_i the affine subspaces $\{x \in F_2^n \,/\, supp(a) \subseteq supp(x)\}$ where a ranges over the set of vectors of weights at most $n-k$, the order being by decreasing weights. Then, for every i, the set $A_i \setminus supp(f)$ is at most a singleton when A_i has dimension k, and when A_i has dimension greater than k, the set $A_i \setminus \bigcup_{j<i} A_j$ equals the singleton containing the vector of minimum weight in A_i. Similarly, for every i, the set $A'_i \setminus supp(f+1)$ is a singleton when A'_i has dimension k, and when A'_i has dimension greater than k, the set $A'_i \setminus \bigcup_{j<i} A'_j$ equals the singleton containing the vector of maximum weight in A'_i. □

4. Constructing functions with optimum algebraic immunity

Note that, if the support S of a function satisfying the hypotheses of Proposition 3.1 contains a k-dimensional affine subspace A, then if we take off one vector of A from S, we obtain a function which still satisfies the hypotheses of Proposition 3.1: we can insert this affine subspace as the first item of

[a]Another possible choice of a majority function, which has been considered in ref. 21, takes value 1 at all vectors of weights strictly greater than $\frac{n}{2}$. It is different when n is even, but affinely equivalent up to addition of a constant.

the sequence of the A_i's. Similarly, if S is disjoint of a k-dimensional affine subspace A, then if we add one of its vectors to S, we do not change the fact that $f+1$ satisfies the hypothesis of Proposition 3.1. This can be applied iteratively. In the case that n is even, this directly implies the following result, already obtained in ref. 21:

Corollary 4.1. *Let f be any function in an even number of variables n, such that $f(x)=0$ if $wt(x)<\frac{n}{2}$; $f(x)=1$ if $wt(x)>\frac{n}{2}$ (or conversely) and where $f(x)$ can take any value when $wt(x)=\frac{n}{2}$. Then f has optimum algebraic immunity $\frac{n}{2}$.*

A similar result has been also obtained in ref. 3. The functions obtained there are those of Corollary 4.1 above, such that the set of vectors of weight $\frac{n}{2}$ in their support is stable under translation by $(1,\ldots,1)$. They satisfy then a somewhat stronger condition. Note that they all have the property that $f(x+(1,\ldots,1))=f(x)+1$ if $wt(x)\neq\frac{n}{2}$ and $f(x+(1,\ldots,1))=f(x)$ if $wt(x)=\frac{n}{2}$. This looks like a linear structure (a function f has the linear structure a if $f(x+a)$ equals $f(x)$ plus a constant, see ref. 11) though it is different.

Some of the functions of Corollary 4.1 are balanced. We show in Section 5 that these functions cannot have good nonlinearities. Hence, further constructions are necessary. Proposition 3.1 allows a more general construction that we describe now:

Corollary 4.2. *Let n be even and let $a^1,\ldots,a^{\binom{n}{n/2}}$ be an ordering of the set of all vectors of weight $\frac{n}{2}$ in F_2^n. For every $i\in\left\{1,\ldots,\binom{n}{n/2}\right\}$, let us denote by A_i the affine subspace $\{x\in F_2^n/\,supp(a^i)\subseteq supp(x)\}$ and by A'_i the vector space $\{x\in F_2^n/\,supp(x)\subseteq supp(a^i)\}$. Let I, J and K be three disjoint subsets of $\left\{1,\ldots,\binom{n}{n/2}\right\}$. Assume that, for every $i\in I$, there exists a vector $b^i\neq a^i$ such that $b^i\in A_i\setminus\bigcup_{j\in I;j<i}A_j$. Assume that, for every $i\in J$, there exists a vector $c^i\neq a^i$ such that $c^i\in A'_i\setminus\bigcup_{j\in J;j<i}A'_j$. Then the function whose support equals: $\{x\in F_2^n/\,wt(x)>\frac{n}{2}\}\cup\{c^i,\,i\in J\}\cup\{\,a^i,\,i\in I\cup K\}\setminus\{b^i,\,i\in I\}$ has algebraic immunity $\frac{n}{2}$.*

Proof. Let the sequence of the affine subspaces A_i of Proposition 3.1 begin with the affine subspaces A_i described above for $i\in I$ and be completed by all the other affine subspaces $\{x\in F_2^n/\,supp(a)\subseteq supp(x)\}$, ordered by decreasing weights of the vectors a of weights at most $\frac{n}{2}$. Let the sequence of the affine subspaces A'_i begin with the vector spaces A'_i described above for $i\in J$ and be completed by all the other vector spaces

$\{x \in F_2^n \,/\, supp(x) \subseteq supp(a)\}$ ordered by increasing weights of the vectors a of weights at least $\frac{n}{2}$. Then, as for Example 3.1, the hypotheses of Proposition 3.1 are satisfied. The only differences with Example 3.1 are that, for any $i \in I$, the set $A_i \setminus supp(f)$ equals $\{b^i\}$ instead of $\emptyset$ and for any $i \in J$, the set $A'_i \setminus supp(f+1)$ equals $\{c^i\}$ instead of $\{a^i\}$. □

We deduce now an algorithm for constructing functions with algebraic immunity $\frac{n}{2}$ (n even), allowing the functions to be balanced:

Algorithm

- Choose two positive integers $k \le l \le \binom{n}{n/2}$;
- For i ranging from 1 to k, choose a vector a^i of weight $\frac{n}{2}$, different from $a^1, \ldots, a^{i-1}$, and a vector b^i such that $supp(a^i) \subseteq supp(b^i)$ and $\forall j < i$, $supp(a^j) \not\subseteq supp(b^i)$;
- For i ranging from $k+1$ to l, choose a vector a^i of weight $\frac{n}{2}$, different from $a^1, \ldots, a^{i-1}$, and a vector c^i such that $supp(c^i) \subseteq supp(a^i)$ and for all j such that $k+1 \le j < i$, we have $supp(c^i) \not\subseteq supp(a^j)$;
- Output the function whose support equals $\{x \in F_2^n \,/\, wt(x) > \frac{n}{2}\} \setminus \{b^i, i = 1, \ldots, k\} \cup \{a^i, i = 1, \ldots, k\} \cup \{c^i, i = k+1, \ldots, l\}$.

Note that, since b^i and c^i are allowed to equal a^i, the algorithm never falls into a dead end where no more fitting vector would exist. The function output by the algorithm is that of Corollary 4.2 with $I = \{i = 1, \cdots, k \,/\, b^i \neq a^i\}$, $J = \{i = k+1, \cdots, l \,/\, c^i \neq a^i\}$, $K = \{1, \cdots, l\} \setminus (I \cup J)$. Its weight equals $2^{n-1} - \frac{1}{2}\binom{n}{n/2}$, plus the number of b^i of weight $\frac{n}{2}$, plus $l - k$. Hence, we can easily produce balanced functions.
The number of functions with optimal algebraic immunity that we can obtain this way seems large, contrary to the constructions of ref. 3,20,21. It is upper bounded by $\left(2^{1+n/2}\right)^{\binom{n}{n/2}}$, but this upper bound is approximately in $\Omega\left(2^{\frac{\sqrt{n}}{\sqrt{2\pi}} 2^{n/2}}\right)$, and is therefore asymptotically huge.

Since the assumption made by Corollary 4.2 is not automatically verified, we still need to exhibit a related infinite class of functions. In ref. 8 is asserted that the function whose support equals the union of the set of vectors of weight $\frac{n}{2} - 4$ and of the set of vectors of weights at least $\frac{n}{2}$ except those of weight $\frac{n}{2} + 4$ has optimum algebraic immunity. This result is probably false (see ref. 5) – it is true up to some (even) value of n. We exhibit an infinite class of functions, among which some differ slightly from the function just mentioned, and for which we can prove that the algebraic immunity equals $\frac{n}{2}$, thanks to Corollary 4.2. In this example of application of the corollary, the ordering of the set of vectors of weight $\frac{n}{2}$ plays no role. This construction gives functions which can be implemented very efficiently and which differ from symmetric functions more significantly than those of Corollary 4.1.

Corollary 4.3. *Let n be even and let u be any nonzero vector of weight at most $\frac{n}{2}$ in F_2^n. Let f be any function whose support contains:*
1. all vectors of weights strictly greater than $\frac{n}{2}$, except those of weight $wt(u) + \frac{n}{2}$ and whose supports contain the support of u,
2. all vectors of weight $\frac{n}{2} - wt(u)$ and whose supports are disjoint of the support of u,
3. all vectors of weight $\frac{n}{2}$ and whose supports are disjoint of the support of u,
4. any additional vectors of weight $\frac{n}{2}$ and whose supports neither are disjoint of the support of u nor contain it.
Then f has algebraic immunity $\frac{n}{2}$.

Proof: For every vector a of weight $\frac{n}{2}$ and whose support is disjoint of the support of u, let $b_a = a \vee u$ be the vector whose support equals the union of those of a and u. Obviously, b_a has weight $wt(u) + n/2$ and its support contains the support of u. For every vector a of weight $n/2$ and whose support contains the support of u, let $c_a = a \setminus u$ be the vector whose support equals the difference between those of a and u. Obviously, c_a has weight $n/2 - wt(u)$ and its support is disjoint of the support of u. For two distinct vectors a and a' of weight $n/2$ and whose supports are disjoint of the support of u, we have $supp(a') \not\subseteq supp(b_a)$, and for two distinct vectors a and a' of weight $n/2$ and whose supports contain the support of u, we have $supp(c_a) \not\subseteq supp(a')$. Corollary 4.2, applied with $A_i = \{x \in F_2^n / \, supp(a^i) \subseteq supp(x)\}$, where a^i is the i-th vector of weight $n/2$ in some predetermined order (any order will work), $A'_i = \{x \in F_2^n / \, supp(x) \subseteq supp(a^i)\}$, $b^i = a^i \vee u$, $c^i = a^i \setminus u$, and taking for I the indices of those vectors a^i whose supports are disjoint of the support of u, for J the indices of those vectors a^i whose supports contain the support of u, and for K the indices of some extra vectors a^i, proves then that f has algebraic immunity $n/2$. □

Notation: We shall denote by f_L the function described in Corollary 4.1, where L is the set of those vectors of weight $n/2$ at which f_L takes value 1, and by $f_{u,L}$ the function described in Corollary 4.3, where L is the set of those vectors of weight $n/2$, whose supports neither are disjoint of the support of u nor contain it, and at which $f_{u,L}$ takes value 1.

Lemma 4.1. *For every nonzero vector u of weight less than $n/2$ and every subset L of the set of those vectors of weight $n/2$, whose supports neither are disjoint of the support of u nor contain it, the weight of function $f_{u,L}$ equals $2^{n-1} - \frac{1}{2}\binom{n}{n/2} + \binom{n-wt(u)}{n/2} + |L|$. Hence, $f_{u,L}$ is balanced if and only if $|L| = \frac{1}{2}\binom{n}{n/2} - \binom{n-wt(u)}{n/2}$. Given $u \neq 0$, there always exists such L.*

We skip the easy proof of this lemma.

Generalization: If in the construction of Corollary 4.3, we take $b^i = a^i \vee w^i$ (resp. $c^i = a^i \setminus w^i$) where the support of w^i is included in the support of u, we obtain a more general construction. The support of the resulting function contains then:

1. all vectors of weights strictly greater than $n/2$, except some of the form $a \vee w$ where a has weight $\frac{n}{2}$, the support of a is disjoint of the support of u and the support of w is included in the support of u,
2. some vectors of the form $a \setminus w$ where a has weight $\frac{n}{2}$, the support of a contains the support of u and the support of w is included in the support of u,
3. the vectors a of weight $n/2$, whose supports are disjoint of the support of u, and such that the vector a is among the vectors considered in 1.
4. any additional vectors of weight $n/2$ and whose supports neither are disjoint of the support of u nor contain it.

5. Study of the Walsh transforms of the constructed functions and their nonlinearity

We study now the Walsh spectra of the functions f_L and $f_{u,L}$.

Lemma 5.1. *Let L be any set of vectors of weight $n/2$. Let f_L be the function in an even number of variables n whose support equals the union of the set $\{x \in F_2^n \,/\, wt(x) > n/2\}$ and of L (see Corollary 4.1). Let a be any vector and let $i = wt(a)$. Then*

$$W_{f_L}(a) = (-1)^{i+1} W_f(a) - 2 \sum_{x \in L} (-1)^{a \cdot x}$$

where f is the majority function (whose support equals $\{x \in F_2^n \,/\, wt(x) \geq n/2\}$).

Indeed, if we denote by f' the function whose support equals $\{x \in F_2^n \,/\, wt(x) > n/2\}$, we have $W_{f'}(a) = -W_{f'+1}(a) = -\sum_{x\in F_2^n}(-1)^{f(\overline{x})+a\cdot x} = -\sum_{x\in F_2^n}(-1)^{f(x)+a\cdot\overline{x}} = (-1)^{i+1}W_f(a)$. And $W_{f_L}(a) = W_{f'}(a) - 2\sum_{x\in L}(-1)^{a\cdot x}$.
We can see now that all functions f_L (including the majority function) have bad nonlinearities. Indeed, according to ref. 21, we know that the maximum of $|W_f(a)|$ is achieved when a has weight 1, that is, $nl(f) = 2^{n-1} - \frac{1}{2}|W_f(a)|$ with $wt(a) = 1$. We have, for every vector a of weight 1: $W_f(a) = -(n/2+1)\left(\binom{n-1}{n/2+1} - \binom{n-1}{n/2}\right) = 2\binom{n-1}{n/2}$, according to ref. 21.
We also have $\sum_{wt(a)=1}\sum_{x\in L}(-1)^{a\cdot x} = \sum_{x\in L}(n - 2wt(x)) = 0$. Hence, there exists at least one vector a of weight 1 such that $\sum_{x\in L}(-1)^{a\cdot x} \leq 0$ and one such that $\sum_{x\in L}(-1)^{a\cdot x} \geq 0$. The nonlinearity of f_L is therefore not better than the nonlinearity of f.

Lemma 5.2. *Let u be any nonzero vector of weight less than $n/2$ and L any set of vectors of weight $n/2$ whose supports neither are disjoint of the support of u nor contain it. Let $f_{u,L}$ be the function described in Corollary 4.3. Let a be any vector and let $i = wt(a)$. Then:*
- if i is even, then $W_{f_{u,L}}(a)$ equals

$$(-1)^{i+1}W_f(a) - 2\sum_{\substack{x\in F_2^n/wt(x)=n/2\\ supp(u)\cap supp(x)=\emptyset}}(-1)^{a\cdot x} - 2\sum_{x\in L}(-1)^{a\cdot x},$$

- if $i = wt(a)$ is odd and $a\cdot u = 0$, then it equals

$$(-1)^{i+1}W_f(a) + 2\sum_{\substack{x\in F_2^n/wt(x)=n/2\\ supp(u)\cap supp(x)=\emptyset}}(-1)^{a\cdot x} - 2\sum_{x\in L}(-1)^{a\cdot x},$$

- if $i = wt(a)$ is odd and $a\cdot u = 1$, then it equals

$$(-1)^{i+1}W_f(a) - 6\sum_{\substack{x\in F_2^n/wt(x)=n/2\\ supp(u)\cap supp(x)=\emptyset}}(-1)^{a\cdot x} - 2\sum_{x\in L}(-1)^{a\cdot x}.$$

Proof. The value at $a \in F_2^n$ of the Walsh transform of the indicator f' of the set of vectors of weights strictly greater than $n/2$ being equal to

$(-1)^{i+1}W_f(a)$, where $i = wt(a)$, the Walsh transform of $f_{u,L}$ equals

$$(-1)^{i+1}W_f(a) + 2 \sum_{\substack{x \in F_2^n / wt(x)=n/2+wt(u) \\ supp(u) \subseteq supp(x)}} (-1)^{a \cdot x}$$

$$-2 \sum_{\substack{x \in F_2^n / wt(x)=n/2-wt(u) \\ supp(u) \cap supp(x)=\emptyset}} (-1)^{a \cdot x} - 2 \sum_{\substack{x \in F_2^n / wt(x)=n/2 \\ supp(u) \cap supp(x)=\emptyset}} (-1)^{a \cdot x} - 2 \sum_{x \in L} (-1)^{a \cdot x}$$

$$= (-1)^{i+1}W_f(a) + 2 \sum_{\substack{x \in F_2^n / wt(x)=n/2 \\ supp(u) \cap supp(x)=\emptyset}} \left[(-1)^{a \cdot (x+u)} - (-1)^{a \cdot (\overline{x}+u)} - (-1)^{a \cdot x} \right]$$

$$-2 \sum_{x \in L} (-1)^{a \cdot x},$$

where $\overline{x} = x + (1, \ldots, 1)$. Indeed, for every vector $x \in F_2^n$, the condition ($wt(x) = n/2 + wt(u)$ and $supp(u) \subseteq supp(x)$) is equivalent to ($wt(x+u) = n/2$ and $supp(u) \cap supp(x+u) = \emptyset$) and the condition ($wt(x) = n/2 - wt(u)$ and $supp(u) \cap supp(x) = \emptyset$) is equivalent to ($wt(\overline{x}) = n/2 + wt(u)$ and $supp(u) \subseteq supp(\overline{x})$).
If $wt(a)$ is even, then we have $(-1)^{a \cdot (x+u)} - (-1)^{a \cdot (\overline{x}+u)} - (-1)^{a \cdot x} = -(-1)^{a \cdot x}$. If $wt(a)$ is odd, then we have $(-1)^{a \cdot (x+u)} - (-1)^{a \cdot (\overline{x}+u)} - (-1)^{a \cdot x} = 2(-1)^{a \cdot (x+u)} - (-1)^{a \cdot x}$. And if $a \cdot u = 0$, then we have $2(-1)^{a \cdot (x+u)} - (-1)^{a \cdot x} = (-1)^{a \cdot x}$; if $a \cdot u = 1$, then we have $2(-1)^{a \cdot (x+u)} - (-1)^{a \cdot x} = -3(-1)^{a \cdot x}$. This completes the proof. □

Note that the argument used for proving that the functions f_L have bad nonlinearities does not show that the functions $f_{u,L}$ cannot have good nonlinearities either: for every reals λ and μ, we have

$$\sum_{i=0}^{n} \left[\lambda \sum_{\substack{x \in F_2^n / wt(x)=n/2 \\ supp(u) \cap supp(x)=\emptyset}} (-1)^{a \cdot x} + \mu \sum_{x \in L} (-1)^{a \cdot x} \right] =$$

$$\lambda \sum_{\substack{x \in F_2^n / wt(x)=n/2 \\ supp(u) \cap supp(x)=\emptyset}} (n - 2wt(x)) + \mu \sum_{x \in L} (n - 2wt(x)) = 0.$$

Hence, there exists at least one vector a of weight 1 such that the number $\lambda \sum_{\substack{x \in F_2^n / wt(x)=n/2 \\ supp(u) \cap supp(x)=\emptyset}} (-1)^{a \cdot x} + \mu \sum_{x \in L} (-1)^{a \cdot x}$ is negative. But in the formulae of Lemma 5.2, the values of λ and μ differ according to whether $a \cdot u$ is null or not.

Further work is in progress to determine whether some infinite subclasses of balanced functions $f_{u,L}$ achieve high nonlinearities and are robust against fast algebraic attacks.

We give in Table 1, for n even ranging from 8 to 24, the values given by Lobanov's bound, the nonlinearity of the majority function, the best nonlinearity of balanced functions that we could obtain with our algorithm, the best nonlinearity of (balanced or non-balanced) functions given by the algorithm, and the covering radius bound. We can observe a gain of our functions over the majority function, but there is still a big gap with the covering radius bound.

n	Lobanov	maj func	our best bal nl	our best nl	$2^{n-1} - 2^{n/2-1}$
8	58	93	94	98	120
10	260	386	390	398	496
12	1124	1586	1600	1615	2016
14	4760	6476	6524	6566	8128
16	19898	26333	26498	26630	32640
18	82452	106762	107334	107763	130816
20	339532	431910	433912	435342	523776
22	1391720	1744436	1751508	1756370	2096128
24	5684452	7036530	7061724	7077558	8386560

6. Constructing functions with optimum algebraic immunity, in odd numbers of variables

In ref. 9, A. Canteaut has observed that, if a balanced function f in an odd number n of variables admits no non-zero annihilator of degree at most $\frac{n-1}{2}$, then it has optimum algebraic immunity $\frac{n+1}{2}$ (this means that we do not need to check also that $f+1$ has no non-zero annihilator of degree at most $\frac{n-1}{2}$ for showing that f has optimum algebraic immunity). For self-completeness, let us recall the reasons why this is true. Consider the Reed-Muller code of length 2^n and of order $\frac{n-1}{2}$. This code is self-dual (i.e. is its own orthogonal as a vector subspace of F_2^n, endowed with the usual inner product "$\cdot$") see ref. 29. Let G be a generator matrix of this code. Each column of G is labeled by a vector of F_2^n. Saying that f has no non-zero annihilator of degree at most $\frac{n-1}{2}$ is equivalent to saying that the matrix obtained by selecting those columns of G corresponding to the elements of the support of f has full rank $\sum_{i=0}^{\frac{n-1}{2}} \binom{n}{i} = 2^{n-1}$. Since f has weight 2^{n-1}, this is also equivalent to saying that the support of the function is an information set, that is (assuming for simplicity that the columns corresponding to the support of f are the 2^{n-1} first ones), that we can take $G = (Id\,|\,M)$. Then the complement of the support of f is also an information set (otherwise there would exist a vector $(z\,|\,0)$, $z \neq 0$, in the code and this is clearly impossible since G is also a parity-check matrix of the code). This completes Canteaut's proof.

We deduce the following corollary of Proposition 3.1:

Corollary 6.1. *Let n be odd. Let A_i, $i = 1, \cdots, 2^{n-1}$ be a sequence of affine subspaces of F_2^n, of dimensions at least $\frac{n+1}{2}$, and such that, for every $i = 1, \cdots, 2^{n-1}$, the set $A_i \setminus \bigcup_{j<i} A_j$ is non-empty. Then, for any choice of an element c^i in each set $A_i \setminus \bigcup_{j<i} A_j$, the Boolean function of support $C = \{c^i;\, i = 1, \cdots, 2^{n-1}\}$ and the function of support $F_2^n \setminus C$ are balanced functions of optimum algebraic immunity $\frac{n+1}{2}$.*

Proof. The function of support C is a balanced function of optimum algebraic immunity if and only if its complement, the function of support $F_2^n \setminus C$, has the same properties. So let us prove that this last function - let us call it f - has these properties. By construction, the vectors c^i are distinct since, for $j < i$, we have $c^j \in A_j$ and $c^i \in A_i \setminus A_j$. Hence, C has size 2^{n-1} and f is balanced. According to Canteaut's result, it is sufficient for completing the proof, to show that f has no nonzero annihilator of degree at most $\frac{n-1}{2}$. This is a direct consequence of Proposition 3.1 with $k = \frac{n+1}{2}$, since we have $A_i \setminus \left[\bigcup_{j<i} A_j \cup (F_2^n \setminus C)\right] = C \cap \left(A_i \setminus \bigcup_{j<i} A_j\right) = \{c^i\}$ for every $i = 1, \cdots, 2^{n-1}$ and $C \subseteq \bigcup_{i \le 2^{n-1}} A_i$. □

This simplification in the research of functions with optimum algebraic immunity leads to the following algorithm:

Algorithm
Let n be odd and let $a^1, \cdots, a^{2^{n-1}}$ be the list of all vectors in F_2^n of weights greater than or equal to $\frac{n+1}{2}$, sorted by increasing Hamming weights.

(1) For $i := 1$ to 2^{n-1}, choose a vector c^i such that $supp(c^i) \subseteq supp(a^i)$ and $\forall j < i$, $supp(c^i) \not\subseteq supp(a^j)$.
(2) Output the function whose support is the set $\{c^i\,;\, i = 1, \cdots, 2^{n-1}\}$.

Note that it is always possible to choose c^i satisfying the condition of 1, since $c^i = a^i$ satisfies it. In fact, it is possible to choose c^i different from

a^i only when a^i has weight $\frac{n+1}{2}$, since for greater weights, a^i is the only element in the vector space $\{b \in F_2^n \,/\, supp(b) \subseteq supp(a^i)\}$ whose support is included in the support of no vector a^j, $j < i$.
The support of the produced function contains all vectors of weights at least $\frac{n+3}{2}$, and $\binom{n}{\frac{n+1}{2}}$ vectors of weights at most $\frac{n+1}{2}$.

We give in Table 2, for n odd ranging from 7 to 19, the values given by Lobanov's bound (which equals the nonlinearity of the majority function), the best nonlinearity of (balanced) functions that we could obtain with this algorithm and the covering radius bound. Here again, we can observe a gain of our functions over the majority function, but there is still a big gap with the covering radius bound.

n	Lobanov	our best nl	$2^{n-1} - 2^{n/2-1}$
7	44	48	59
9	186	196	245
11	772	796	1002
13	3172	3248	4051
15	12952	13178	16294
17	52666	53350	65355
19	213524	215676	261782

We give now an infinite class of functions deduced from Corollary 6.1, and similar to Corollary 4.3.

Corollary 6.2. *Let n be odd and let u be any nonzero vector of weight at most $\frac{n+1}{2}$ in F_2^n. Let f be any function whose support contains:*
1. all vectors of weight $\frac{n+1}{2}$, except some of those whose supports include the support of u,
2. all vectors of weight strictly greater than $\frac{n+1}{2}$,
3. for each vector a^i excluded at step 1., a vector c^i whose support contains the support of $a^i \setminus u$ and is included in the support of a^i.
Then f has algebraic immunity $\frac{n+1}{2}$.

Proof. For every vector a of weight $\frac{n+1}{2}$ and whose support contains the support of u, let $c_a = a \setminus w$ where w is a vector covered by u. Corollary 6.1, applied with $A_i = \{x \in F_2^n / \, supp(x) \subseteq supp(a^i)\}$, where a^i is the i-th vector of weight at least $\frac{n+1}{2}$ in some predetermined order, beginning with the vectors of weight $\frac{n+1}{2}$ and whose supports contain the support of u

(any such order will work), and $c^i = c_{a^i}$ if a^i has weight $\frac{n+1}{2}$ and if its support contains the support of u, $c^i = a^i$ otherwise, proves then that f has algebraic immunity $\frac{n+1}{2}$. □

Conclusion We have introduced, for the first time, a method for designing wide classes of functions with optimum algebraic immunity, in any numbers of variables. The balancedness of the functions is easily achieved as well. The nonlinearities of the examples of classes which could be deduced in the present paper are strictly better than for the sporadic functions with optimum algebraic immunity which had been found before. But it is not yet sufficient. Further work has to be done, using the method and specifying it, to design classes of highly nonlinear balanced functions with optimum algebraic immunity. A first encouraging step in this direction is done in the preprint given in ref. 14.

Acknowledgement We thank Rafael Fourquet for the computation of Tables 1 and 2 and Xiangyong Zeng and Chunlei Li for useful information.

References

1. F. Armknecht, C. Carlet, P. Gaborit, S. Kuenzli, W. Meier and O. Ruatta. Efficient computation of algebraic immunity for algebraic and fast algebraic attacks. Proceedings of EUROCRYPT 2006. Lecture Notes in Computer Science 4004, Springer, pp. 147-164, 2006.
2. F. Armknecht and M. Krause. Algebraic Attacks on combiners with memory. In *Advances in Cryptology - CRYPTO 2003*, number 2729 in Lecture Notes in Computer Science, pp. 162–175. Springer Verlag, 2003.
3. F. Armknecht and M. Krause. Constructing single- and multi-output boolean functions with maximal immunity. Proceedings of *ICALP 2006*, Lecture Notes in Computer Science 4052, Springer, pp. 180-191, 2006.
4. F. Armknecht. Improving Fast Algebraic Attacks. In *FSE 2004*, number 3017 in Lecture Notes in Computer Science, pp. 65–82. Springer Verlag, 2004.
5. A. Braeken. Cryptographic properties of Boolean functions and S-boxes. PhD thesis available at URL http://homes.esat.kuleuven.be/ abraeken/thesisAn.pdf.
6. A. Braeken, J. Lano, N. Mentens, B. Preneel and I. Verbauwhede. SFINKS: A Synchronous stream cipher for restricted hardware environments. SKEW - Symmetric Key Encryption Workshop, 2005.
7. A. Braeken, J. Lano and B. Preneel. Evaluating the Resistance of Filters and Combiners Against Fast Algebraic Attacks. Eprint on ECRYPT, 2005.
8. A. Braeken and B. Preneel. On the Algebraic Immunity of Symmetric Boolean Functions. In *Indocrypt 2005*, number 3797 in LNCS, pp. 35–48. Springer Verlag, 2005. Also available at Cryptology ePrint Archive, http://eprint.iacr.org/,

No. 2005/245, 26 July, 2005.

9. A. Canteaut. Open problems related to algebraic attacks on stream ciphers. Workshop on Coding and Cryptography 2005. Lecture Notes in Computer Science 3969, 2006 Springer, pp. 120-134. Paper available on the web http://www-rocq.inria.fr/codes/Anne.Canteaut/Publications/canteaut06a.pdf
10. A. Canteaut and M. Trabbia. Improved fast correlation attacks using parity-check equations of weight 4 and 5. In *EUROCRYPT 2000*, number 1807 in Lecture Notes in Computer Science, pp. 573–588. Springer Verlag, 2000.
11. C. Carlet. Boolean Functions for Cryptography and Error Correcting Codes. Chapter of the monography *Boolean Methods and Models*, Y. Crama and P. Hammer eds, Cambridge University Press, to appear in 2006. Prelimnary version available at http://www-rocq.inria.fr/codes/Claude.Carlet/pubs.html
12. C. Carlet. On the higher order nonlinearities of algebraic immune functions. Proceedings of CRYPTO 2006. Lecture Notes in Computer Science 4117, pp. 584-601, 2006.
13. C. Carlet, D. Dalai, K. Gupta and S. Maitra. Algebraic Immunity for Cryptographically Significant Boolean Functions: Analysis and Construction. IEEE Transactions on Information Theory, vol. 52, no. 7, pp. 3105-3121, July 2006.
14. C. Carlet, X. Zeng, C. Li and L. HU. Further properties of several classes of Boolean functions with optimum algebraic immunity. IACR e-print archive 2007/370.
15. J. Y. Cho and J. Pieprzyk. Algebraic Attacks on SOBER-t32 and SOBER-128. In *FSE 2004*, number 3017 in Lecture Notes in Computer Science, pp. 49–64. Springer Verlag, 2004.
16. N. Courtois. Fast algebraic attacks on stream ciphers with linear feedback. In *Advances in Cryptology - CRYPTO 2003*, number 2729 in Lecture Notes in Computer Science, pp. 176–194. Springer Verlag, 2003.
17. N. Courtois. Cryptanalysis of SFINKS. In *ICISC 2005*. Also available at Cryptology ePrint Archive, http://eprint.iacr.org/, Report 2005/243, 2005.
18. N. Courtois and J. Pieprzyk. Cryptanalysis of block ciphers with overdefined systems of equations. In *Advances in Cryptology - ASIACRYPT 2002*, number 2501 in Lecture Notes in Computer Science, pp. 267–287. Springer Verlag, 2002.
19. N. Courtois and W. Meier. Algebraic attacks on stream ciphers with linear feedback. In *Advances in Cryptology - EUROCRYPT 2003*, number 2656 in Lecture Notes in Computer Science, pp. 345–359. Springer Verlag, 2003.
20. D. K. Dalai, K. C. Gupta and S. Maitra. Cryptographically Significant Boolean functions: Construction and Analysis in terms of Algebraic Immunity. In *Workshop on Fast Software Encryption, FSE 2005*, pp. 98–111, number 3557, Lecture Notes in Computer Science, Springer-Verlag.
21. D. K. Dalai, S. Maitra and S. Sarkar. Basic Theory in Construction of Boolean Functions with Maximum Possible Annihilator Immunity. Cryptology ePrint Archive, http://eprint.iacr.org/, No. 2005/229, 15 July, 2005. To be published in Designs, Codes and Cryptography.
22. D. K. Dalai, K. C. Gupta and S. Maitra. Notion of Algebraic Immunity and Its evaluation Related to Fast Algebraic Attacks. In *2nd International*

Workshop on Boolean Functions: Cryptography and Applications, BFCA 2006, University of Rouen, France, March 13-15, 2006. Cryptology ePrint Archive, eprint.iacr.org, Report 2006/018, January 2006.
23. C. Ding, G. Xiao, and W. Shan. *The Stability Theory of Stream Ciphers.* Number 561 in Lecture Notes in Computer Science. Springer-Verlag, 1991.
24. S. Fischer and W. Meier. Algebraic Immunity of S-boxes and Augmented Functions. Proceedings of Fast Software Encryption 2007. To appear.
25. P. Hawkes and G. G. Rose. Rewriting Variables: The Complexity of Fast Algebraic Attacks on Stream Ciphers. In M. K. Franklin, editor, *Advances in Cryptology - CRYPTO 2004, LNCS 3152*, pp. 390–406. Springer Verlag, 2004.
26. D. H. Lee, J. Kim, J. Hong, J. W. Han and D. Moon. Algebraic Attacks on Summation Generators. In *FSE 2004*, number 3017 in Lecture Notes in Computer Science, pp. 34–48. Springer Verlag, 2004.
27. F. Liu and K. Feng. On the 2^m-variable symmetric Boolean functions with maximum algebraic immunity 2^{m-1}. *Proceedings of the Workshop on Coding and Cryptography 2007*, pp. 225-232, 2007.
28. M. Lobanov. Tight bound between nonlinearity and algebraic immunity. Paper 2005/441 in http://eprint.iacr.org/
29. F. J. MacWilliams and N. J. A. Sloane. *The theory of error-correcting codes*, Elsevier, North-Holland, 1977.
30. W. Meier, E. Pasalic and C. Carlet. Algebraic attacks and decomposition of Boolean functions. In *Advances in Cryptology - EUROCRYPT 2004*, number 3027 in Lecture Notes in Computer Science, pp. 474–491. Springer Verlag, 2004.
31. F. Rodier. Asymptotic nonlinearity of Boolean functions. *Designs, Codes and Cryptography*, no 40:1 2006, pp 59-70.
32. S. Rønjom and T. Helleseth. A new attack on the filter generator. *IEEE Transactions on Information theory* 53 (5), pp. 1752-1758, 2007.
33. C. E. Shannon. Communication theory of secrecy systems. *Bell system technical journal*, 28, pp. 656-715, 1949.
34. T. Siegenthaler. Correlation-immunity of nonlinear combining functions for cryptographic applications. *IEEE Transactions on Information Theory*, IT-30(5):776–780, September 1984.
35. L. Qu, K. Feng and C. Li. On the Boolean functions With Maximum Possible Algebraic Immunity : Construction and A Lower Bound of the Count. IACR e-print archive 2005/449.
36. L. Qu, C. Li and K. Feng. A note on symmetric Boolean functions with maximum algebraic immunity in odd number of variables. *IEEE Trans. on Inf. Theory* 53, pp. 2908-2910, 2007.

Enumeration of a Class of Sequences Generated by Inversions

A. Çeşemlioğlu, W. Meidl and A. Topuzoğlu*

MDBF, Sabancı University,
Orhanlı, Tuzla, 34956 İstanbul , Turkey
**E-mail: alev@sabanciuniv.edu*

Dedicated to Prof. Jürgen Lehn on the occasion of his 65th birthday

Any permutation of a finite field $\mathbb{F}_q$ can be represented by a polynomial $\mathcal{P}_n(x) = (\ldots((a_0x + a_1)^{q-2} + a_2)^{q-2} \ldots + a_n)^{q-2} + a_{n+1}$, for some $n \geq 0$. In this note we present the number of distinct permutations of the types $\mathcal{P}_2(x)$ and $\mathcal{P}_3(x)$ with full cycle. These results extend earlier work on the inversive pseudorandom number generator and on $\mathcal{P}_1$.

Keywords: Pseudorandom number generators; Inversive generator; Sequences over finite fields; Permutation Polynomials; Enumeration of Permutation Polynomials.

1. Introduction

Let p be an odd prime, r a positive integer and $\mathbb{F}_q$ be the finite field with $q = p^r$ elements. The *inversive pseudorandom number generator* (u_n) over $\mathbb{F}_q$ is defined as

$$u_n = \theta(u_{n-1}) \tag{1}$$

with an initial value $u_0 \in \mathbb{F}_q$, where θ is the permutation of $\mathbb{F}_q$ given by

$$\theta(x) = c_1 x^{q-2} + c_2 \tag{2}$$

for some $c_1 \neq 0$, $c_2 \in \mathbb{F}_q$. Starting with [5], the inversive generator has been studied extensively, and has been shown to have favourable behaviour with respect to most measures of randomness. The reader may like to look up [12], and the references therein for many interesting features of inversive generators. In particular, results on the period length of the inversive generator can be found in [3,4,6].

By a classical result of Carlitz [1], S_q, the symmetric group on q letters, which is isomorphic to the group of permutation polynomials of $\mathbb{F}_q$ of degree less than $q-1$ under the operation of composition and reduction modulo

$x^q - x$, is generated by the linear polynomials $ax+b$, for $a, b \in \mathbb{F}_q$, $a \neq 0$, and x^{q-2} (see [10,11] for a detailed exposition of permutation polynomials of finite fields). Consequently, as pointed out in [2], with $\mathcal{P}_0(x) = a_0x + a_1$, any permutation of a finite field $\mathbb{F}_q$ can be represented by a polynomial

$$\mathcal{P}_n(x) = (\ldots((a_0x + a_1)^{q-2} + a_2)^{q-2} \ldots + a_n)^{q-2} + a_{n+1},\ n \geq 0, \qquad (3)$$

where $a_1 \in \mathbb{F}_q$, $a_i \in \mathbb{F}_q^*$ for $i = 0, 2, \ldots, n$. With this notation, the permutation θ in (2) above is $\mathcal{P}_1$ with $a_1 = 0$. The cycle structure of $\mathcal{P}_1$ therefore, yields possible period lengths of the sequence defined by (1). See [4] for the cycle structure of the permutation θ. More generally, the cycle structure of $\mathcal{P}_n$, for $n \geq 1$ is studied in [2].

Note that iterations of permutations with full cycle give rise to sequences over $\mathbb{F}_q$ with full period q, hence are of primary interest for applications. The number of distinct permutations θ with full cycle can be found in [3] and [2] gives the number of distinct permutations $\mathcal{P}_1$ with full cycle. Here we present the number of distinct permutations of the types $\mathcal{P}_2$ and $\mathcal{P}_3$ with full cycle.

Throughout this work, we shall follow the terminology and notation used in [2], and accordingly we put $\mathcal{P}_n(x) = \bar{P}_n(x)$ if $a_{n+1} \neq 0$ and $\mathcal{P}_n(x) = P_n(x)$ if $a_{n+1} = 0$. We state the main results now and give the proofs in the third section. In what follows ϕ denotes the Euler ϕ-function.

Theorem 1.1. *Let* $\mathcal{P}_2(x) = ((a_0x+a_1)^{q-2}+a_2)^{q-2}+a_3 \in \mathbb{F}_q[x]$ *and* $q > 5$. *The number of distinct permutations of the form* $\mathcal{P}_2(x)$ *with full cycle is* $\frac{1}{4}\phi(\frac{q+1}{2})(q+1)q(q-1)$ *when* $q = p^r$ *for a prime* p *with* $r > 1$, *and* $\frac{1}{4}\phi(\frac{p+1}{2})(p+1)p(p-1) + p(p-1)$ *when* $q = p$ *is prime.*

Theorem 1.2. *Let* $P_3(x) = (((a_0x + a_1)^{q-2} + a_2)^{q-2} + a_3)^{q-2} \in \mathbb{F}_q[x]$, *and* $q > 5$. *The number of distinct permutations of the form* $P_3(x)$ *with full cycle is*
$\frac{1}{4}\phi(q+1)(q-1)^2(q-2) + 3\phi(q-1)(q-1)$ *if* $3 \nmid (q+1)$, *and*
$\frac{1}{4}\phi(q+1)(q-1)^2(q-2) + 3\phi(q-1)(q-1) + \frac{1}{9}\phi\left(\frac{q+1}{3}\right)(q-1)(q+1)^2$ *if* $3 \mid (q+1)$.

2. Preliminaries

As pointed out in [2], the rational transformation $\mathcal{R}_n(x)$ can be associated with the permutation $\mathcal{P}_n(x)$ as follows:

$$\mathcal{R}_n(x) = \frac{\alpha_{n+1}x + \beta_{n+1}}{\alpha_n x + \beta_n}, \tag{4}$$

where

$$\alpha_k = a_k\alpha_{k-1} + \alpha_{k-2} \quad \text{and} \quad \beta_k = a_k\beta_{k-1} + \beta_{k-2},$$

for $k \geq 2$ and $\alpha_0 = 0, \alpha_1 = a_0, \beta_0 = 1, \beta_1 = a_1$.
Then we have $\mathcal{P}_n(x) = \mathcal{R}_n(x)$ for $x \in \mathbb{F}_q \setminus \mathbf{O}_n$ where $\mathbf{O}_n$ is the set of *poles* which are not necessarily distinct:

$$\mathbf{O}_n = \{x_i : x_i = \frac{-\beta_i}{\alpha_i}, i = 1, \ldots, n\} \subset \mathbb{P}^1(\mathbb{F}_q) = \mathbb{F}_q \cup \{\infty\}.$$

Now a permutation $\mathcal{F}_n(x)$ can be defined by $\mathcal{F}_n(x) = \mathcal{R}_n(x)$ for $x \neq x_n$ and $\mathcal{F}_n(x) = \alpha_{n+1}/\alpha_n$ when $x_n \in \mathbb{F}_q$. It is easy to see that the cycle structure of $\mathcal{P}_n$ depends on the cycle structure of $\mathcal{F}_n$ and the positioning of the poles in the cycles of $\mathcal{F}_n$. For instance, in case the poles are distinct elements of $\mathbb{F}_q$, $\mathcal{P}_n$ can be expressed as a product of the n-cycle $(\mathcal{F}_n(x_{n-1}) \;\cdots\; \mathcal{F}_n(x_1)\; \mathcal{F}_n(x_n))$ with the permutation $\mathcal{F}_n$, i.e.

$$\mathcal{P}_n(x) = (\mathcal{F}_n(x_{n-1}) \;\cdots\; \mathcal{F}_n(x_1)\; \mathcal{F}_n(x_n))\mathcal{F}_n(x), \tag{5}$$

(multiplying in right-to-left order) (see Lemma 1 in [2]).

In order to understand the cycle structure of permutations $\mathcal{P}_n$ and to enumerate them, it is therefore crucial to understand the cycle structure of permutations of the form

$$F(x) = \begin{cases} R(x) & \text{if } x \neq \frac{-d}{c} \\ \frac{a}{c} & \text{if } x = \frac{-d}{c} \end{cases} \tag{6}$$

for a nonconstant rational transformation $R(x) = (ax + b)/(cx + d) \in \mathbb{F}_q(x), c \neq 0$. We naturally associate the matrix

$$A = \begin{pmatrix} a\, b \\ c\, d \end{pmatrix}$$

in $GL(2, q)$, and its characteristic polynomial $f(x) = x^2 - \text{tr}(A)x + \det(A)$ with $R(x)$ (or $F(x)$). In what follows, $\text{ord}(z)$ denotes the order of an element z in the multiplicative group of $\mathbb{F}_{q^2}$.

Proposition 2.1. *Let F be the permutation of $\mathbb{F}_q$ defined by (6), $q = p^r$. Suppose that $f(x)$ is the characteristic polynomial of the matrix A associated* *with F and $\alpha, \beta \in \mathbb{F}_{q^2}$ are the roots of $f(x)$. The decomposition of F into disjoint cycles is as follows:*

(1) Suppose $f(x)$ is irreducible. If $k = \operatorname{ord}(\frac{\alpha}{\beta}) = \frac{q+1}{t}$, $1 \le t < \frac{q+1}{2}$, then F is a composition of $t-1$ cycles of length k and one cycle of length $k-1$. In particular F is a full cycle if $t = 1$.
(2) Suppose $\alpha, \beta \in \mathbb{F}_q$ and $\alpha \neq \beta$. If $k = \operatorname{ord}(\frac{\alpha}{\beta}) = \frac{q-1}{t}$, $t \geq 1$, then F is a composition of $t-1$ cycles of length k, one cycle of length $k-1$, and two cycles of length 1.
(3) Suppose $f(x) = (x-\alpha)^2$, $\alpha \in \mathbb{F}_q^* = \mathbb{F}_q \setminus \{0\}$, then F is a composition of one cycle of length $p-1$, $p^{r-1}-1$ cycles of length p and one cycle of length 1.

Remark 2.1. One can see from the proof (in [2]) of Proposition 2.1 that in cases *(1)*,*(2)* above, $x = -d/c$ is in the cycle of length $k-1$.

Now we focus on the permutations $\mathcal{P}_2$ and P_3. We first recall that

$$\mathcal{P}_2(x) = ((a_0x + a_1)^{q-2} + a_2)^{q-2} + a_3.$$

By (4), we have

$$\mathcal{R}_2(x) = \frac{a_0(a_2a_3+1)x + a_1(a_2a_3+1) + a_3}{a_0a_2x + a_1a_2 + 1}$$

and the poles $x_1 = -a_1/a_0$, $x_2 = -(a_1a_2+1)/a_0a_2$. Viewing $\mathcal{R}_2(x)$ as the rational transformation $R(x)$, which defines $F(x)$ in (6), the associated characteristic polynomial becomes

$$f(x) = x^2 - (a_0(a_2a_3+1) + a_1a_2 + 1)x + a_0. \tag{7}$$

By (5) we have

$$\mathcal{P}_2(x) = (\mathcal{F}_2(x_1)\ \mathcal{F}_2(x_2))\mathcal{F}_2(x) \tag{8}$$

with $\mathcal{F}_2(x) = \mathcal{R}_2(x)$ if $x \neq x_2$ and $\mathcal{F}_2(x_2) = (a_2a_3+1)/a_2$.
The second class of permutations, which we consider here consists of permutations of the form

$$P_3(x) = (((a_0x + a_1)^{q-2} + a_2)^{q-2} + a_3)^{q-2}. \tag{9}$$

Note that $a_4 = 0$ and since we deal with this special case, we put $\mathcal{F}_3(x) = F_3(x)$ and $\mathcal{R}_3(x) = R_3(x)$. We get by (4)

$$R_3(x) = \frac{a_0a_2x + a_1a_2 + 1}{a_0(a_2a_3+1)x + a_1(a_2a_3+1) + a_3} \tag{10}$$

and the characteristic polynomial, which is associated with $R_3(x)$ is

$$f(x) = x^2 - (a_0a_2 + a_1(a_2a_3 + 1) + a_3)x - a_0. \tag{11}$$

We have to distinguish two cases. In the first case we suppose that $a_2a_3 + 1 \neq 0$. Then the poles are distinct elements of $\mathbb{F}_q$ given by $x_1 = -a_1/a_0$, $x_2 = -(a_1a_2 + 1)/a_0a_2$, $x_3 = -(a_1(a_2a_3 + 1) + a_3)/(a_0(a_2a_3 + 1))$. In this case we have by (5) that

$$P_3(x) = (F_3(x_2)\ F_3(x_1)\ F_3(x_3))F_3(x) \tag{12}$$

with $F_3(x) = R_3(x)$ if $x \neq x_3$ and $F_3(x_3) = a_2/(a_2a_3 + 1)$.
If $a_2a_3 + 1 = 0$ then $x_1 = -\frac{a_1}{a_0}, x_2 = -\frac{a_1a_2+1}{a_0a_2}$, $x_3 = \infty$ and

$$R_3(x) = -a_2(a_0a_2x + a_1a_2 + 1) \tag{13}$$

is linear, thus $F_3(x)$ and $R_3(x)$ coincide, giving, (see [2], Section 4)

$$P_3(x) = (F_3(x_1)\ F_3(x_2))F_3(x) = (-a_2\ 0)F_3(x). \tag{14}$$

The associated characteristic polynomial becomes reducible in this case:

$$f(x) = x^2 - (a_0a_2 - a_2^{-1})x - a_0 = (x - a_0a_2)(x + a_2^{-1}).$$

Since we aim at counting the permutations $\mathcal{P}_2$ and P_3 with full cycle, we state two relevant results, which we take from [2].

Proposition 2.2. *The permutation $\mathcal{P}_2$ is a full cycle if and only if*

(1) (i) the polynomial $f(x)$ in (7) *is irreducible,*
(ii) the roots $\alpha, \beta \in \mathbb{F}_{q^2}$ of $f(x)$ satisfy ord$(\alpha/\beta) = (q+1)/2$,
(iii) $\gamma_0 = (\beta - 1)/(\alpha - 1)$ satisfies $\gamma_0^{(q+1)/2} \neq 1$, or
(2) $\mathbb{F}_q$ is a prime field and $f(x) = (x-1)^2$ (then $a_0 = 1$, $a_3 = -a_1$).

Remark 2.2. It is easy to see why $\mathcal{P}_2$ is a full cycle when *(1)* or *(2)* holds. We refer to [2] for details. For instance, one knows by *(1-ii)* and Proposition 2.1 that $\mathcal{F}_2$ is composed of two cycles $\mathcal{C}_1, \mathcal{C}_2$ of lengths $(q+1)/2$ and $(q-1)/2$, respectively. The parameter γ_0 is introduced in connection with the distribution of the poles, more precisely, $x_1, x_2 \in \mathcal{C}_2$ if and only if $\gamma_0^{(q+1)/2} = 1$. Hence $x_1 \in \mathcal{C}_1$, $x_2 \in \mathcal{C}_2$ by the condition *(1-iii)*, and (8) implies that $\mathcal{C}_1, \mathcal{C}_2$ join together to yield a full cycle.

In case of *(2)*, Proposition 2.1 *(3)* implies that $\mathcal{F}_2$ is composed of the fixed point x_1 (a cycle of length 1) and a cycle of length $p-1$ which contains x_2. By (8), they join up to give a cycle of length p.

Now that the reader is familiar with the reasoning used for the proof, we include the explanation of the technical conditions in the statement of the next proposition.
When P_3 is considered, the relative positions of the poles x_1, x_2, x_3 in the cycles of F_3 become significant. The condition *(1)* below addresses such a case, where F_3 itself is a full cycle, i.e. $F_3 = (s_0 \ldots s_{q-1})$, for $s_n = F_3^n(s_0)$, $s_0 \in \mathbb{F}_q$, and the positioning of the poles in $(s_0 \ldots s_{q-1})$ determines the cycle structure of P_3. Here, as usual, τ^n denotes the nth iterate of $\tau \in S_q$, with $\tau^0(a) = a$ for $a \in \mathbb{F}_q$.

Proposition 2.3. *The permutation P_3 in (9) is a full cycle if and only if one of the following conditions (1)-(4) is satisfied.*

(1) *(i) The polynomial $f(x)$ in* (11) *is irreducible,*
(ii) the roots $\alpha, \beta \in \mathbb{F}_{q^2}$ of $f(x)$ satisfy ord$(\alpha/\beta) = q+1$ so that F_3 is a full cycle, and
(iii) the pole x_1 lies between the poles x_2, x_3 in the cycle F_3.

(2) *(i) The polynomial $f(x)$ in* (11) *is irreducible,*
(ii) 3 *divides $q+1$, and the roots $\alpha, \beta \in \mathbb{F}_{q^2}$ of $f(x)$ satisfy ord$(\alpha/\beta) = (q+1)/3$, i.e. F_3 is composed of* 2 *cycles of length $(q+1)/3$ and* 1 *cycle of length $(q-2)/3$,*
(iii) the elements $\gamma_1 = (\beta - a_3)/(\alpha - a_3)$, $\gamma_2 = (a_2\beta + 1)/(a_2\alpha + 1)$, $\gamma_3 = (\beta - a_1)/(\alpha - a_1) \in \mathbb{F}_{q^2}$ satisfy $\gamma_1^{(q+1)/3}, \gamma_2^{(q+1)/3}, \gamma_3^{(q+1)/3} \neq 1$, i.e. the poles x_1, x_2, x_3 are in distinct cycles of F_3.

(3) *(i) The polynomial $f(x)$ in* (11) *has two distinct roots $\alpha, \beta \in \mathbb{F}_q$,*
(ii) ord$(\alpha/\beta) = q-1$, i.e. F_3 is composed of one cycle of length $q-2$ and two cycles of length 1*,*
(iii) $a_2a_3 + 1 \neq 0$, i.e. the pole x_3 is in $\mathbb{F}_q$,
(iv) $a_3 = -a_0/a_1$ and $a_2 = -1/a_1$, i.e. x_1, x_2 are the fixed points of F_3.

(4) *(i) $a_2a_3 + 1 = 0$, i.e. $F_3(x)$ is linear,*
(ii) ord$(-a_0a_2^2) = q-1$, and
(iii) either $a_1 = a_0a_2$ or $a_2 = -1/a_1$, i.e. either x_1 or x_2 is the fixed point of F_3.

Remark 2.3. As seen in part *(2-iii)*, the parameters $\gamma_1, \gamma_2, \gamma_3$ are concerned with the distribution of the poles in the cycles of F_3. Indeed, x_i, x_3 are in the same cycle if and only if $\gamma_i^k = 1$ for $i = 1, 2$, and x_1, x_2 are in the same cycle if and only if $\gamma_3^k = 1$.

3. Proofs of Theorems 1.1 and 1.2

Before we prove Theorem 1.1 and Theorem 1.2 we present two lemmas. In the first lemma we recall a result of [3], the second lemma is on some simple properties of the parameters $\gamma_0, \gamma_1, \gamma_2, \gamma_3$ introduced in Section 2.

Lemma 3.1. *Let $k > 1$ be a divisor of $q+1$ or $q-1$. The number of monic quadratic polynomials $f(x) = x^2 - Tx + D \in \mathbb{F}_q[x]$ with two distinct roots $\alpha, \beta \in \mathbb{F}_{q^2}$ and $ord(\alpha/\beta) = k$ is given by $\frac{\phi(k)}{2}(q-1)$.*

Proof. For the convenience of the reader we sketch the proof of the case $k|(q+1), k \neq 2$, i.e. $f(x)$ is irreducible.
The number of irreducible polynomials $g(x) = x^2 + Cx + 1 \in \mathbb{F}_q[x]$ of order k is known to be $\phi(k)/2$, see [10, Theorem 3.5]. Suppose $g(\delta) = 0$. The polynomials $f(x) = x^2 - Tx + D = (x-\alpha)(x-\beta) \in \mathbb{F}_q[x]$ with $(T^2/D) - 2 = C$ are exactly the polynomials that satisfy $\alpha/\beta = \delta$ (see [3, Theorem 3]). Since $C \neq -2$, the parameter T can be chosen in $q-1$ different ways and then D is uniquely determined. □

Lemma 3.2. *Suppose $f(x) = x^2 - Tx + D \in \mathbb{F}_q[x]$ is irreducible with roots $\alpha, \beta \in \mathbb{F}_{q^2}$, and let γ_i, $i = 0, \ldots, 3$ be defined as in Section 2. Then*

(i) $ord(\gamma_i)$ divides $q+1$ for $i = 0, \ldots 3$,
(ii) $\gamma_i \neq 1$ and $\gamma_i \neq \beta/\alpha$ for $i = 0, 1, 2$ and $\gamma_3 \neq 1$,
(iii) $\gamma_1 = \gamma_2$ if and only if $a_2 a_3 + 1 = 0$, and $\gamma_2 = \gamma_3$ if and only if $a_1 a_2 + 1 = 0$.

Proof. (i) With $\beta = \alpha^q$ and the observation that

$$\gamma_1^{q+1} = \left(\frac{\beta - a_3}{\alpha - a_3}\right)^{q+1} = \left(\frac{\beta^q - a_3^q}{\alpha^q - a_3^q}\right)\left(\frac{\beta - a_3}{\alpha - a_3}\right) = 1,$$

the assertion follows for γ_1 and also similarly for $\gamma_0, \gamma_2, \gamma_3$.
(ii) Easily follows from the assumptions $a_2 \neq 0$, $a_3 \neq 0$ (for P_3) and $\alpha \neq \beta$.
(iii) Trivial. □

The connection between the permutations $\mathcal{P}_n$ and $\mathcal{F}_n$ brings up the question of whether or not a one-to-one correspondence can be formed between the set of permutations $\mathcal{P}_n$ and specific subsets of the set $\mathcal{S}_R$ of all formal expressions $\mathcal{S}_R = \{R(x) = (ax+b)/(cx+d) : (a,b,c,d) \in \mathbb{F}_q^4\}$.

As expected, the answer is affirmative only if n is small, namely for $\mathcal{P}_1$, $\mathcal{P}_2$, and P_3. It is clear for instance that, in case of $P_1(x) = (a_0 x + a_1)^{q-2} =$

$F_1(x)$ one needs to consider the subset $\mathcal{S}_R^{(1)} = \{1/(a_0x + a_1) : a_0 \neq 0, a_1 \in \mathbb{F}_q\}$. It is also easy to see that the identification $a = 1/a_2, b = a_1/(a_0a_2), d = (a_1a_2 + 1)/(a_0a_2)$ describes a one-to-one correspondence between the set of permutations of the form P_2 and the set of rational transformations $(ax + b)/(x + d)$ with $a \neq 0$. The permutation P_2 is then given by $P_2(x) = ((a_0x + a_1)^{q-2} + a_2)^{q-2} = (\frac{1}{a_2}\ 0)\frac{ax+b}{x+d}$, where a, b, d are defined as above.
We show a similar result for P_3 which will be used in the sequel. A corresponding result for $\bar{P}_2$ is also stated.

Proposition 3.1. *Let $q > 5$.*

(i) There is a one-to-one correspondence between the set of permutations of the form $\bar{P}_2$ and the set of the formal expressions $(ax + b)/(cx + d)$ with $c \neq 0$, $ad - bc \neq 0$ and $a - ad + bc \neq 0$.

(ii) There is a one-to-one correspondence between the set of permutations of the form P_3 and the set of the formal expressions $(ax + b)/(cx + d)$ with $a \neq 0$, $ad - bc \neq 0$ and $c + ad - bc \neq 0$.

Proof. Considering (10), we put

$$a = a_0a_2, b = a_1a_2 + 1, c = a_0(a_2a_3 + 1), d = a_1(a_2a_3 + 1) + a_3.$$

Then it immediately follows that $a = a_0a_2 \neq 0$, $c + ad - bc = a_0a_2a_3 \neq 0$, and $ad - bc = -a_0 \neq 0$. For given a, b, c, d satisfying the above conditions, one gets the unique solution of the above system of equations as

$$a_0 = -D, a_1 = \frac{(1 - b)D}{a}, a_2 = -\frac{a}{D} \text{ and } a_3 = \frac{c + D}{a},$$

with $D = ad - bc$. Note that the cardinality of the set $\Delta = \{(a, b, c, d) \in \mathbb{F}_q^4 \mid a \neq 0, ad - bc \neq 0, c + ad - bc \neq 0\}$ (and that of the set of possible expressions for P_3) is $q(q - 1)^3$.
Now that a one-to-one correspondence is established between the set $\mathcal{S}_R^{(3)} = \{R(x) = (ax + b)/(cx + d) : (a, b, c, d) \in \Delta\}$, and the set $\mathcal{S}_{P_3} = \{(((a_0x + a_1)^{q-2} + a_2)^{q-2} + a_3)^{q-2} : a_0a_2a_3 \neq 0\}$, it remains to show that two elements $P_3(x) = (((a_0x + a_1)^{q-2} + a_2)^{q-2} + a_3)^{q-2}$ and $P_3'(x) = (((a_0'x + a_1')^{q-2} + a_2')^{q-2} + a_3')^{q-2}$ of $\mathcal{S}_{P_3}$ induce the same permutation if and only if $a_i = a_i'$ for $i = 0, 1, 2, 3$. Clearly, if P_3 and P_3' correspond to rational transformations $R_3 \neq R_3'$ and $q > 5$, then the permutations are different. Now suppose that P_3, P_3' are mapped to the same rational function (but distinct elements of $\mathcal{S}_R^{(3)}$), i.e. $R_3 = (ax + b)/(cx + d)$ and $R_3' = (\varepsilon ax + \varepsilon b)/(\varepsilon cx + \varepsilon d)$, $\varepsilon \neq 1$, by the injection above. Then the corresponding poles are given by

$x_1 = -a_1/a_0 = \frac{1-b}{a}$, $x_1' = -a_1'/a_0' = \frac{1-\varepsilon b}{\varepsilon a}$, and clearly $x_2 = x_2' = -b/a$ and $x_3 = x_3' = -d/c$ (if $c = 0$, then $x_3 = x_3'$ is the pole at infinity). Hence by (12) and (14), P_3 and P_3' induce different permutations. □

Remark 3.1. The set of permutations P_3 with $x_3 \neq \infty$, i.e. $a_2a_3 + 1 \neq 0$, corresponds to the set $\mathcal{S}_R^{(3')} = \{R(x) = (ax+b)/(cx+d) : (a,b,c,d) \in \Delta, c \neq 0\}$, which is of cardinality $q(q-1)^2(q-2)$.
We emphasize that if $(a_0, a_1, a_2, a_3) \neq (a_0', a_1', a_2', a_3')$, then the permutations induced by $P_3(\bar{P}_2)$ and $P_3'(\bar{P}_2')$ are actually distinct. This is not true anymore for $\bar{P}_3$.

For $u, v \in \mathbb{F}_q^*$, the transposition $(u\ v)$, for instance, can be expressed as $\bar{P}_3$ for both choices of $a_0 = -1/(u-v)^2$, $a_1 = -ua_0$, $a_2 = v - u$, $a_3 = 1/(u-v)$, $a_4 = v$, and $a_0' = a_0$, $a_1' = -va_0$, $a_2' = -a_2$, $a_3' = -a_3$, $a_4' = u$.

Proof of Theorem 1.1. We first count those $\mathcal{P}_2$ which satisfy condition *(1)* of Proposition 2.2.
We fix a polynomial $g(x) = (x-\delta)(x-\delta^{-1})$ with $\mathrm{ord}(\delta) = (q+1)/2$. Among the $q-1$ polynomials $f_i(x) = x^2 - T_ix + D_i = (x-\alpha_i)(x-\beta_i)$, $i = 1, \ldots, q-1$ with $\alpha_i/\beta_i = \delta$ (see the proof of Lemma 3.1), we need to count the ones, which satisfy $ord(\frac{\beta_i - 1}{\alpha_i - 1}) \nmid (q+1)/2$.

We put $\gamma_{0(i)} = \frac{\beta_i-1}{\alpha_i-1}$ and show that $\gamma_{0(i)} \neq \gamma_{0(j)}$ for $i \neq j$. If $\gamma_{0(i)} = \gamma_{0(j)}$, then $\alpha_j\beta_i - \alpha_j - \beta_i = \alpha_i\beta_j - \alpha_i - \beta_j$. Multiplying both sides by $\delta = \alpha_i/\beta_i = \alpha_j/\beta_j$ yields $\alpha_i\delta + \alpha_j = \alpha_j\delta + \alpha_i$ which is equivalent to $\alpha_i(\delta - 1) = \alpha_j(\delta-1)$. Hence we have $\alpha_i = \alpha_j$ and therefore $\beta_i = \beta_j$, i.e. $i = j$. Consequently, the sets $\tilde{\Gamma} = \{\gamma_{0(i)}, i = 1, \ldots, q-1\}$ and $\Gamma = \{\eta \in \mathbb{F}_{q^2} : \eta^{q+1} = 1, \eta \neq 1, \eta \neq \delta^{-1}\}$ are the same by Lemma 3.2. The cardinality of the set $\Gamma_0 = \{\eta \in \Gamma : ord(\eta) | (q+1)/2\}$ is easily seen to be $(q+1)/2 - 2$ and hence $|\Gamma \setminus \Gamma_0| = (q+1)/2$. Therefore, exactly $(q+1)/2$ polynomials in $\{f_i(x), i = 1, \ldots, q-1\}$ satisfy $\mathrm{ord}(\alpha_i/\beta_i) = \mathrm{ord}(\delta) = (q+1)/2$ and $\gamma_{0(i)}^{(q+1)/2} \neq 1$. Given such a polynomial $f_i(x) = x^2 - T_ix + D_i$, the coefficient a_0 in (7) is uniquely determined by $a_0 = D_i$, we have q choices for $a_1 \in \mathbb{F}_q$ and $q-1$ choices for $a_2 \in \mathbb{F}_q^*$. The coefficient a_3 is then uniquely determined by $T = a_0(a_2a_3+1) + a_1a_2 + 1$. Since we have $\frac{\phi((q+1)/2)}{2}$ distinct choices for the polynomial $g(x)$, we obtain $\frac{\phi(\frac{q+1}{2})}{2}\frac{q+1}{2}q(q-1)$ for the total number of permutations $\mathcal{P}_2(x)$ with full cycle in case $q = p^r, r > 1$. (Note that here we use the one-to-one correspondence between the parameters (a_0, a_1, a_2, a_3) describing $\mathcal{P}_2$ and permutations induced by them.)

For the case that $\mathbb{F}_q = \mathbb{F}_p$ is a prime field we additionally obtain permutations $\mathcal{P}_2$ with a full cycle if $f(x) = (x-1)^2$. As can be seen easily we then have

$$\mathcal{P}_2(x) = ((x+a_1)^{p-2} + a_2)^{p-2} - a_1$$

for some arbitrary $a_1 \in \mathbb{F}_q$ and $a_2 \in \mathbb{F}_q^*$. $\square$

We now prove Theorem 1.2 on the number of permutations P_3 with full cycle. The proof consists of four parts, corresponding to each condition in Proposition 2.3.

Proof of Theorem 1.2. We start with case *(1)* of Proposition 2.3 and fix a polynomial $f(x) = x^2 - Tx + D \in \mathbb{F}_q[x]$ with roots $\alpha, \beta \in \mathbb{F}_{q^2}$ satisfying $ord(\frac{\alpha}{\beta}) = q+1$. Then any associated rational function of the form $R(x) = (ax+b)/(cx+d)$ satisfies $a+d = T$ and $ad - bc = D$. We recall that the corresponding permutation is always a full cycle. This cycle can be expressed as $(s_0\ s_1\ \dots\ s_{q-1})$ with $s_{q-1} = -d/c = x_3$. Equations (17), (28) in [2] show that

$$s_n = \frac{a}{c} - D\frac{\alpha^n - \beta^n}{c(\alpha^{n+1} - \beta^{n+1})}$$

for $0 \le n \le q-2$. In order that the pole x_1 lies between x_2 and x_3, we fix a pair of integers (j_1, j_2) with $0 \le j_2 < j_1 \le q-2$, and put $s_{j_1} = x_1, s_{j_2} = x_2$. Since $x_1 = -a_1/a_0 = (1-b)/a$ and $x_2 = -b/a$, we have $s_{j_2+1} = R(-b/a) = 0$ and $s_{j_1+1} = R((1-b)/a) = a/(D+c)$. Here we note that $D + c \neq 0$. Consequently we have

$$s_{j_2+1} = 0 = \frac{a}{c} - D\frac{\alpha^{j_2+1} - \beta^{j_2+1}}{c(\alpha^{j_2+2} - \beta^{j_2+2})},$$

which uniquely yields $a = D(\alpha^{j_2+1} - \beta^{j_2+1})/(\alpha^{j_2+2} - \beta^{j_2+2})$. We note that $a \neq 0$, otherwise $\alpha^{j_2+1} = \beta^{j_2+1}$ which contradicts with $\text{ord}(\alpha/\beta) = q+1$. With

$$s_{j_1+1} = \frac{a}{D+c} = \frac{a}{c} - D\frac{\alpha^{j_1+1} - \beta^{j_1+1}}{c(\alpha^{j_1+2} - \beta^{j_1+2})}$$

we obtain $c = a((\alpha^{j_1+2} - \beta^{j_1+2})/(\alpha^{j_1+1} - \beta^{j_1+1})) - D$. Finally we get $d = T - a$ and $b = \frac{ad-D}{c}$. Summarizing, with the choice of the characteristic polynomial $f(x)$ and the positions j_1, j_2 for the poles x_1, x_2 in the cycle of F_3, we obtain a, b, c, d uniquely, where $a \neq 0, ad - bc + c \neq 0$, and of course $ad - bc \neq 0$. It is easy to see that different choices of the triples $f(x), j_1, j_2$ give different elements of the set Δ, defined in the proof of Proposition 3.1.

By the same proposition we know that in order to enumerate the set of permutations P_3 satisfying condition (*1*) of Proposition 2.3, it is sufficient to count the possible choices for f, j_1, j_2. But there are $\frac{\phi(q+1)}{2}(q-1)$ choices for f and $(q-1)(q-2)/2$ choices for the pairs (j_1, j_2).

We now turn our attention to the third case of Proposition 2.3. In this case P_3 is of the form

$$P_3(x) = (((a_0 x + a_1)^{q-2} - \frac{1}{a_1})^{q-2} - \frac{a_0}{a_1})^{q-2}, \tag{15}$$

and the associated characteristic polynomial is given by

$$f(x) = x^2 - (a_1 - \frac{a_0}{a_1})x - a_0. \tag{16}$$

It is sufficient to determine the number of choices for the pair (a_0, a_1), $a_0 a_1 \neq 0$, for which the roots α, β of the polynomial (16) satisfy $\text{ord}(\alpha/\beta) = q-1$. We recall that there are $\frac{\phi(q-1)(q-1)}{2}$ polynomials $f(x) = x^2 - Tx + D$ with distinct roots α, β satisfying $\text{ord}(\alpha/\beta) = q-1$. For a fixed polynomial $f(x)$ with these properties, a_0 in (16) is determined to be $a_0 = -D \neq 0$. With $T = a_1 - \frac{a_0}{a_1}$ we get exactly two nonzero solutions for a_1, namely

$$a_1 = \frac{T \pm \sqrt{T^2 - 4D}}{2}.$$

Clearly $T^2 - 4D$ is a nonzero square in $\mathbb{F}_q$ since $f(x)$ has two distinct roots in $\mathbb{F}_q$. Thus we have $\phi(q-1)(q-1)$ permutations P_3 of the form (15) with a full cycle.

We now consider the fourth case of Proposition 2.3. First suppose that $a_1 = a_0 a_2$, i.e. x_1 is the unique fixed point of the linear function $F_3(x) = R_3(x)$ given in (13). Then we require $ord(-a_0 a_2^2) = ord(-\frac{a_1^2}{a_0}) = q-1$. For each of the $\phi(q-1)$ choices for $-a_1^2/a_0$ we have $q-1$ choices for a_1. The coefficients a_0 and a_2 are then uniquely determined as nonzero elements of $\mathbb{F}_q$, and hence a_3 is uniquely given by $a_2 a_3 + 1 = 0$. When x_2 is the fixed point of F_3 we similarly get the same number, $\phi(q-1)(q-1)$.

Therefore in case $a_2 a_3 + 1 = 0$, the total number of $P_3(x)$ with full cycle is given by $2\phi(q-1)(q-1)$.

This completes the proof of the Theorem 1.2 if 3 does not divide $q+1$.

Finally we assume that 3 divides $q+1$ and consider the case *(2)* of Proposition 2.3. For each of the $\frac{\phi(\frac{q+1}{3})}{2}(q-1)$ distinct irreducible polynomials $f(x) = x^2 - Tx + D = (x-\alpha)(x-\beta)$ with $ord(\frac{\alpha}{\beta}) = \frac{q+1}{3}$ we can determine the number of permutations P_3 as follows. By (11) the parameters a_0, a_1, a_2, a_3 satisfy $a_0 = -D$ and $a_0 a_2 + a_1(a_2 a_3 + 1) + a_3 = T$. We also recall that $a_2 a_3 \neq 0$. Hence we have $q-1$ choices for a_2. The parameter a_1 is uniquely determined by

$$a_1 = \frac{T + Da_2 - a_3}{a_2 a_3 + 1}, \tag{17}$$

if and only if $a_3 \neq -1/a_2$. Consequently, for each f we obtain precisely $(q-1)(q-2)$ possible parameters (a_0, a_1, a_2, a_3), and hence distinct permutations P_3. We therefore have the cardinality of the set S_F of permutations P_3, satisfying the conditions *(2-i, ii)* of Proposition 2.3:

$$|S_F| = \frac{\phi(\frac{q+1}{3})}{2}(q-1)^2(q-2).$$

We recall that for $P_3 \in S_F$, the permutation F_3 is composed of exactly 3 cycles.
Our aim, of course, is to obtain the cardinality of the set $S = \{P_3 \in S_F : \gamma_i^{(q+1)/3} \neq 1, i = 1, 2, 3\}$, i.e. we wish to enumerate $P_3 \in S_F$, for which the poles x_1, x_2, x_3 lie in distinct cycles of F_3. For this purpose, we shall evaluate $|S_F \setminus S|$ by considering the partition:

$$S_F \setminus S = S_{1,3} \cup S_{2,3} \cup S_{1,2} \cup S_{1,2,3}, \tag{18}$$

where $S_{i,j}$ is the set referring to the case of the two poles x_i, x_j being in the same cycle of F_3, which does not contain the third pole, $1 \leq i < j \leq 3$. The set $S_{1,2,3}$, obviously refers to the remaining P_3 with all three poles lying in the same cycle of F_3.

In parts *(i)- (iv)* below, we calculate the cardinalities of the four sets in (18), partitioning $S_F \setminus S$.

(i): We recall that this case is equivalent with $\gamma_1^{(q+1)/3} = 1$ and $\gamma_2^{(q+1)/3} \neq 1$, which implies $\gamma_3^{(q+1)/3} \neq 1$. Since we have $\gamma_1 \neq 1, \frac{\beta}{\alpha}$ by Lemma 3.2, out of the $\frac{q+1}{3}$ elements of $\mathbb{F}_{q^2}$ whose order divides $(q+1)/3$, γ_1 can have only $\frac{q+1}{3} - 2$ values. For each choice of $\gamma_1 = (\beta - a_3)/(\alpha - a_3)$ we uniquely obtain $a_3 = (\alpha\gamma_1 - \beta)/(\gamma_1 - 1)$. Note that a_3 is in fact an element of $\mathbb{F}_q$. Now γ_2 is among the $2(q+1)/3$ elements of $\mathbb{F}_{q^2}$ whose order divides $q+1$ but not $(q+1)/3$. The coefficient a_2 is then uniquely given by $a_2 = (\gamma_2 - 1)/(\beta - \alpha\gamma_2)$, again an element in $\mathbb{F}_q$. Since $a_0 = -D$ is determined by $f(x)$, we finally obtain a_1 by equation (17) which is well-defined by Lemma 3.2. Therefore $|S_{1,3}| = \tau_1 = \phi(\frac{q+1}{3})(q-1)\frac{(q+1)(q-5)}{9}$.

(ii): This case is essentially same as (i), with γ_1, γ_2 are interchanged. Hence $|S_{2,3}| = \tau_2 = \tau_1 = \phi(\frac{q+1}{3})(q-1)\frac{(q+1)(q-5)}{9}$.

(iii): This case applies for $\gamma_2^{(q+1)/3} \neq 1$ and $\gamma_3^{(q+1)/3} = 1$. Then $\gamma_1^{(q+1)/3} \neq 1$ follows. Since we only have to exclude $\gamma_3^{(q+1)/3} = 1$, we have $\frac{q-2}{3}$ choices for γ_3, each choice uniquely defines $a_1 = (\alpha\gamma_3 - \beta)/(\gamma_3 - 1) \in \mathbb{F}_q$. For γ_2 we have $2(q+1)/3$ choices, again each choice uniquely gives a_2. From $T = a_0a_2 + a_1(a_2a_3 + 1) + a_3$ we obtain $a_3 = (T - a_1 - a_0a_2)/(a_1a_2 + 1)$ which by Lemma 3.2 is well-defined since $\gamma_2 \neq \gamma_3$. Consequently, $|S_{1,2}| = \tau_3 = \phi(\frac{q+1}{3})(q-1)\frac{(q+1)(q-2)}{9}$.

(iv): This is equivalent to $\gamma_1^{(q+1)/3} = \gamma_2^{(q+1)/3} = 1$ and consequently also $\gamma_3^{(q+1)/3} = 1$. Again by choosing γ_1 and γ_2 appropriately we obtain a_3 and a_2, respectively, and then by equation (17) we get a_1. Here we need to exclude the possibility $\gamma_2 = \gamma_1$ in order to avoid $a_2a_3 + 1 = 0$. Consequently for each of the $(q-5)/3$ possible choices for γ_1 we have exactly $(q-8)/3$ choices for γ_2. This yields $|S_{1,2,3}| = \tau_4 = \phi(\frac{q+1}{3})\frac{(q-1)(q-5)(q-8)}{18}$.

Finally we can calculate

$$|S| = \frac{\phi(\frac{q+1}{3})}{2}(q-1)^2(q-2) - \sum_{i=1}^{i=4} \tau_i = \phi(\frac{q+1}{3})(q-1)\frac{(q+1)^2}{9},$$

and the proof is complete. □

4. Remarks

Since this note is a continuation of previous work on pseudorandom sequences, the assessment of randomness of the sequences,

$$v_n = \mathcal{P}_2(v_{n-1}) \text{ and } w_n = P_3(w_{n-1})$$

would be adequate. We remark that v_n and w_n perform well under two frequently used quality measures for pseudorandom number generators: the linear complexity profile and the discrepancy. With a straightforward generalization of the proof of the linear complexity profile for the inversive generator in [7] one obtains similar lower bounds. It is easy to check that the methods, that are used to find bounds for exponential sums involving inversive generators, can be applied to the case of sequences (v_n) and (w_n) to yield results in agreement with those on inversive generators. One has to point out of course that (v_n) and (w_n) have the disadvantage over (u_n) of being slower to generate due to more inversions involved. However results presented here may also be of theoretical interest on account of open problems concerning permutation polynomials of finite fields and their enumeration (see, for instance [8,9]).

References

1. L. Carlitz, Permutations in a finite field. *Proc. Amer. Math. Soc.* **4** (1953), 538.
2. A. Çeşmelioğlu, W. Meidl, A. Topuzoğlu, On the cycle structure of permutation polynomials, *Finite Fields Appl.*, to appear.
3. W.-S. Chou, On inversive maximal period polynomials over finite fields, *Appl. Algebra Eng. Comm. Comput.* **6** (1995), 245–250.
4. W-S. Chou, The period lengths of inversive pseudorandom vector generations, *Finite Fields Appl.* **1** (1995), 126–132.
5. J. Eichenauer, J. Lehn, A non-linear congruential pseudo random number generator, *Statist. Papers* **27** (1986), 315–326 .
6. M. Flahive, H. Niederreiter, On inversive congruential generators for pseudorandom numbers, *Finite Fields, Coding Theory and Advances in Communications and Computing, (Las Vegas,NV,1991)*, 75–80, Lecture Notes in Pure and Appl. Math. **141**, Marcel-Dekker, New York, 1993.
7. J. Gutierrez, I.E. Shparlinski and A. Winterhof, On the linear and nonlinear complexity profile of nonlinear pseudorandom number generators, *IEEE Transactions on Information Theory* **49** (2003), 60–64.
8. R. Lidl, G. L. Mullen, Unsolved Problems: When Does a Polynomial Over a Finite Field Permute the Elements of the Field?, *Amer. Math. Monthly* **95** (1988), No. 3, 243–246.
9. R. Lidl, G. L. Mullen, Unsolved Problems: When Does a Polynomial Over a Finite Field Permute the Elements of the Field?, II, *Amer. Math. Monthly* **100** (1993), No. 1, 71–74.
10. R. Lidl, H. Niederreiter, Finite fields, 2nd Ed. *Encyclopedia of Mathematics and its Applications* **20**, Cambridge University Press, Cambridge (1997).
11. I.E. Shparlinski, Finite fields: theory and computation. The meeting point of number theory, computer science, coding theory and cryptography. *Mathematics and its Applications* **477**, Kluwer Academic Publishers, Dordrecht (1999).
12. A. Topuzoğlu, A. Winterhof, Pseudorandom sequences, *in* "Topics in Geometry, Coding Theory and Cryptography", (A. Garcia and H. Stichtenoth, Eds.) Algebra and Applications **6**, pp. 135–166, Springer-Verlag, 2007.

A Critical Look at Cryptographic Hash Function Literature

Scott Contini, Ron Steinfeld, Josef Pieprzyk, and Krystian Matusiewicz
Centre for Advanced Computing, Algorithms and Cryptography,
Department of Computing, Macquarie University
Email: {scontini,rons,josef,kmatus}@ics.mq.edu.au

The cryptographic hash function literature has numerous hash function definitions and hash function requirements, and many of them disagree. This survey talks about the various definitions, and takes steps towards cleaning up the literature by explaining how the field has evolved and accurately depicting the research aims people have today.

1. Introduction

The literature on cryptographic hash functions abounds with numerous different definitions and requirements. There is no universal agreement on what a cryptographic hash function is, or what it is supposed to achieve. The purpose of this terminology survey is to call the research community together to agree upon common definitions and requirements so that we can move forward with a clear set of goals.

As an illustration of the problems, we note that most researchers agree that a cryptographic hash function should compress data, in particular, it should map very large domains to fixed size outputs. However, some researchers do not insist that compression is necessary [51] and there exists at least one well known example that does not compress [9]. Furthermore, excluding such non-compressing hash functions, we note that the research community attempts to bifurcate hash functions into ordinary hash functions (meaning a single, fixed function) and hash function families, though this distinction is often blurred. Hash function families were introduced by Damgård[19] in order to make the security requirements very precise in the complexity theory model. He specifically looked at collision resistant* hash function families, where one can aim for an algorithm such that no

---

*He actually used the term *collision free*, but this is no longer in use.

polynomially bounded (in time and size) circuit can find collisions. Nowadays, some people define collision resistant hash functions to be collision resistant [41, §9.2] while others define them to be both collision resistant and preimage resistant [46, Def. 2.2]. Despite the fact that Damgård's definition [19] is usually cited, the current definitions do not seem to make a distinction between ordinary hash functions and hash function families. Apparently these definitions are supposed to cover both.

Within the literature of ordinary hash functions, we typically find security requirements such as the following:

- Preimage resistance: Given y, it must be computationally infeasible to find x such that $h(x) = y$.
- Second preimage resistance: Given y and x_1 such that $h(x_1) = y$, it must be computationally infeasible to find $x_2 \neq x_1$ such that $h(x_2) = y$.
- Collision resistance: It must be computationally infeasible to find any x_1 and x_2 such that $h(x_1) = h(x_2)$.

These definitions are very informal, and attempts at formalizing them have had differing consequences. For instance, it is often stated that collision resistance implies second preimage resistance, but according to some interpretations of these requirements (such as in [41]), the claim is not true (details given later). Moreover, we note that these ordinary hash functions are used in cryptography in ways that require more complex properties than the three stated above. For example, the standardized HMAC [4] requires that the keyed compression function of the hash function is a pseudo random function family (PRF)[2], and the theory behind the standardized RSA-OAEP[6] and RSA-PSS[7] assumes the hash functions are random oracles. Since it is well known that random oracles do not exist [15], usually people settle for some type of emulation of random oracle goal. For instance, in NIST's recent draft call for a new hash standard, one of the security evaluation criterion is

> "The extent to which the algorithm output is indistinguishable from a random oracle."

What does this mean? We know that any ordinary hash function can be distinguished from a random oracle in a single hash function query since random oracles are secret objects whereas hash functions are entirely public. Since we have no definition of what random oracle emulation is, we have no objective way of telling how well candidate hash functions for NIST's new standard are behaving.

The main body of this paper elaborates on these and other problems with the cryptographic hash function literature. Although we do not have all the answers, we hope that it is a step forward in cleaning up the literature. However, we would like to emphasize that our criticisms of certain definitions and terminology should not be interpreted as a criticism of the people that said them. Cryptographic hash functions has been an evolving subject since its origin. Many people had good ideas as a step forward, but the ideas have now become obsolete as further research has been developed. One of the goals of this paper is to point out ideas which are obsolete so that new research can focus with less ambiguity on the current understanding of the way hash functions are intended to be used.

Before we begin, we remark that [48] deals with related issues of hash function terminology. In fact, we will reference this paper many times, since our goals are overlapping with theirs. However, our focus is more broad and less technical, and is more of a down-to-earth survey rather than original research.

2. A Brief History of Cryptographic Hashing

Here we outline a brief history on how we arrived to where we are today in hashing, according to the history we were able to puzzle together. We do not attempt to cover all hashing issues, only the ones that are most relevant to our discussions. Note that we are mainly interested in unkeyed hash functions, but we will talk about some applications where these functions are keyed one way or another.

2.1. *Definitions of Hash Functions*

The use of one-way functions (which may be non-compressing) for authentication was noted in Diffie and Hellman's New Directions in Cryptography [26], although they did not use the terminology "hash function". They claimed to need preimage resistance (i.e. one-wayness) and second preimage resistance. The term "hash function" to imply a compressing one-way function seems to come from Merkle [42] or Rabin [47]. Both were specifically concerned with message authentication, though not necessarily through public key cryptosystems. At this time, neither considered collision resistance, but the importance of the concept appeared soon afterward [55]. As far as we are aware, most (though not all) subsequent definitions of hash functions have required the compressing property, so for brevity we will not mention it any more.

Around the same time, the RSA public key cryptosystem was invented and soon after attempts at attacking the system were published. Most notably, Davida [21] demonstrated how to break the RSA signature scheme by a chosen message attack, and improved attacks followed [25]. As a consequence, hash functions were suggested by a few researchers as a tool for preventing such attacks as well as speeding up signatures. In [22], Denning states that the hash function should destroy homomorphic structure in the underlying public key cryptosystem, it should be one-way, and it should be second preimage resistant. Note that collision resistance is not mentioned. In [54], it was specifically stated that collision resistance is a requirement for hashing in digital signature schemes, including RSA. However, ironically, he defines *one-way hash functions* to be collision-resistant only – the preimage resistance requirement is not stated. Merkle used similar definitions in 1989 [43] where he defined *weak one-way hash functions* (second preimage resistant) and *strong one-way hash functions* (collision resistant) without explicitly stating the one-way requirement (thus one could include that they are misnomers). One might assume that the researchers had a reason to believe that collision resistance implies preimage resistance, but the implication holds only under certain conditions [19,49,51]. Also in 1989, Damgård formally defined *families of collision free hash functions* [19], which are collision resistant in the complexity theory model. However, he remarks that collision resistance does not necessarily imply preimage resistance. Note that both Denning and Damgård explicitly accepted the *heuristic* use of hash functions to securing public key systems since at that time, there was a lack of systems with a *security proof.* Later, in Preneel's PhD Thesis, he notes that there are many different definitions of hash functions [46], and attempts to clean up the literature. Citing Merkle's work as well as Rabin's, Preneel defines a *one-way hash function* to be preimage resistant and second preimage resistant. Additionally, citing Damgård's work and Merkle's work, he defines *collision resistant hash function* to be preimage resistant, second preimage resistant, and collision resistant. The Handbook of Applied cryptography [41] accepts Preneel's definition of one-way hash function (this may be the first two definitions that agree), yet they drop the preimage resistance requirement from the definition of collision resistant hash function. A cursory glance might lead one to think that this definition of collision resistant hash is the same as Damgård's families of collision free hash functions, but they are not. One of the major differences is that Damgård's definitions involve *families* of functions, which makes formalizing the definitions in the complexity theory model possible. The other definitions are informal, and attempts at formalizing them have given different results [49]. We will talk more about this later. A summary of this paragraph is in Table 1.

source	function name	pre. resist.	2 pre. resist.	coll. resist.	fam. funcs.
Winternitz [54]	o.w. hash			X	
Merkle [43]	weak o.w. hash		X		
Merkle [43]	strong o.w. hash			X	
Damgård [19]	coll. free hash			X	X
Preneel [46]	o.w. hash	X	X		
Preneel [46]	coll. resist. hash	X	X	X	
HAC [41]	o.w. hash	X	X		
HAC [41]	coll. resist. hash		X	X	

Nowadays, when a new design is created, people typically refer to it as simply a "hash function" with the intended three properties of collision resistance, second preimage resistance, and preimage resistance. In some cases, people say informally that the outputs are also intended to look random. There are also some designs which are called "collision-resistant hash functions" which, in addition to collision resistance, may or may not intend to have preimage resistance and may or may not be a family of functions (depending upon the definition the inventors use). In general, the more theoretical community does not require preimage resistance as part of the requirement for collision resistant hashing (for example, the Merkle-Damgård construct turns a collision resistant compression function into a collision resistant hash function, but nothing is proved about preimage resistance). On the other hand, the cryptanalysis literature often looks at these designs from the viewpoint of an "ideal hash" (more details will be given below), which should have all of the above properties (except possibly the family of functions property) and more. In summary, we have practical designers, theoreticians, and cryptanalysts all speaking different languages, and even within a single group they do not necessarily speak the same.

2.2. *Random Oracles*

As the importance of hash functions became clear, the theoreticians tried to formalize concepts such as destroying the homomorphic structure in the underlying public key cryptosystem. The way forward seemed to be viewing the hash function as a randomly chosen function from the space of all functions with the same domain and range. From here came the closely related concepts of random oracles [28] and pseudo random function families (PRFs) [33]. The random oracle in particular was aimed at modelling the *ideal hash function* – one that behaves like a truly random function. The concept was used by Bellare and Rogaway [5] as a first step between bridging the theory and practice of public key cryptography. They developed the first practical versions of RSA encryption and signatures that had some notion of a proof of security – a concept that was long sought after, since heuristic defenses against attacks on RSA were very ad hoc. Systems which have security proofs (i.e. security reductions) involving random oracles are said to be *proven secure in the random oracle model.* Some interpret these proofs as an argument that the design is secure against any adversary who treats the hash function as a black box. The ideas of Bellare and Rogaway came at the right time since a few years later Bleichenbacher developed a clever attack on the RSA PKCS #1 encryption standard [10]. Consequently, the standard was able to be quickly updated to use the ideas of Bellare and Rogaway, namely RSA-OAEP and RSA-PSS, which have more of a theoretical foundation than the previous version.

Around the same time as Bleichenbacher's attack, a new way of performing public key encryption was developed by Cramer and Shoup [18] that is provably secure without the use of random oracles. The only thing they require is a collision resistant hash function, where in this case the exact requirement is collision resistance only (i.e. not preimage resistance). The Cramer and Shoup designs are practical, though not as efficient as RSA-OAEP or RSA-PSS. The theoreticians considered this a great breakthrough – a practical and provably secure design that relied only on plausible number theory assumptions. Adding to its significance is research casting doubt on the random oracle model, starting with the work of Canetti, Goldreich, and Halevi [15] who showed that there are protocols that are secure in the random oracle model but immediately become insecure as soon as any concrete hash function is substituted for the random oracle. Some have criticized this work since the examples are quite artificial, but newer related results are closer to the real world [3].

Today there is a research trend of avoiding random oracles, though a few researchers are resisting [38]. Regardless of whether one agrees or disagrees with random oracle proofs, there are widely used standards whose security foundation is based upon them. We do know that random oracles can never be realized in practice since these functions are private whereas hash functions are entirely public objects. Despite this, right now much of the hash function community is aiming for a function that emulates a random oracle. Unfortunately, nobody has yet said what this means. Although the topic has been looked at by theoreticians (example: [14]), it remains unresolved today. We remark that talking about distinguishing a hash function from a random oracle seems to be the wrong concept – one can always distinguish in a single computation since hash functions are computable whereas random oracles are not. For instance, given an input/output pair (x, y), we can easily check whether $\text{SHA1}(x) = y$ just by running it through a computer program that implements SHA1. On the other hand, the probability that some random oracle $\mathcal{O}$ also has $\mathcal{O}(x) = y$ is 2^{-160}. So if the SHA1 computation matches, there is no reasonable doubt that it came from SHA1 instead of a random oracle.

2.3. *Other Requirements for Hash Functions*

Nowadays, hash functions are used in many different ways in practice. One of the most common ways is with Message Authentication Codes (MACs) which are a means that two users with a shared secret key can authenticate between each other. The widely used HMAC standard comes with a security proof [4]– it is secure provided that the underlying keyed compression function of the hash is a PRF. Hash functions are also widely used for pseudo random number generation. In one instance, security can be proved provided that the hash function behaves as a PRF where the secret involves adding a key to the message space [23]. See Table **??** later in our paper.

One of the reasons hash functions have taken on such a diverse role is because they have not been subject to export regulations, contrary to other cryptographic primitives. Additionally, they offered speed advantages and could be used without a license (unlike the IDEA block cipher). Today, export regulations are less strict and there are plenty of efficient alternative public domain cryptographic primitives available, so the mentioned benefits have disappeared. If we are going to continue to apply hashing to the various scenarios that we are using today, then we require a single function that satisfies all of these security requirements. An alternative is to cut back in the way we are using these functions.

2.4. *Summary*

Despite Preneel's efforts at cleaning up the hash literature, we still have multiple definitions and no clear vision of what exactly the requirements are. There seems to a few reasons for this. First, we need to carefully distinguish between ordinary hash functions and hash function families, and what they are aiming to achieve. Second, the names of the objects we are defining should accurately reflect the security properties they are intended to have.

Third, we need precise definitions on exactly what properties we require.

3. Ordinary Hash Functions Versus Hash Function Families

Hash function families were introduced by Damgård[19] in order to make a formal definition of collision resistance in the complexity theory model. Sometimes the word "ensembles" is used instead of "families." Damgård's functions had a single specific purpose rather than trying to solve all problems at once, contrary to the current approach used for ordinary hash functions.

Damgård required that there is no polynomially bounded circuit that could compute a collision with non-negligible probability, where the probability is evaluated over all members of the family. The problem with ordinary hash functions is that collision resistance cannot be defined in this way, since there is only a single member of the family. In other words, there is always a polynomially bounded circuit which will compute the collision for an ordinary hash function – the difficulty is finding that circuit, which depends upon human ignorance [48] rather than complexity theory. This point must be emphasized. When people design ordinary hash functions that are intended to be collision resistant, what they mean is that they have a design for which they believe it is tricky enough that nobody will be able to come up with a clever way of finding collisions – *not* that such algorithms *don't exist.*†

Thus, we see that there are different definitions used for ordinary cryptographic hash functions than there are for hash function families and that there are different research goals. With this is mind, it is a mistake to try to clump these together and use a single definition to represent both of them. We therefore shall treat the two topics separately. As we will justify below, ordinary hash functions meet engineering requirements but are lacking in a theoretical foundation, while hash function families have the theoretical foundation but are lacking in practical solutions. These gaps need to be closed.

3.1. *Ordinary Hash Functions*

By "ordinary hash functions," we mean the designs like MD5 and SHA-1 that are used today. We have not yet given a precise definition of what these

†Rogaway [48] also considers a hybrid variant, where unkeyed hash functions have a secuity parameter n representing the hash length. Most ordinary hash functions are currently not built in this way, but it may be advisable to do so in the future. Another option is custom designed families of functions having a security parameter.

are – we shall return to that later. But generally, we are speaking of single functions (not families) that compress and have additional goals such as collision resistance, second preimage resistance, and preimage resistance. We are essentially clumping together the so-called custom designed hash functions with hash functions based upon block ciphers.

Those who have done research in ordinary hash functions have been more concerned with engineering requirements than scientific definitions. They aim for a single solution that has numerous security properties, so that it can substituted anywhere without having to think about it. However, this view has come at a cost, since it is not clear that such goals are achieveable. Note that although the stated goals of such functions are usually only collision resistance, second preimage resistance, and preimage resistance, in fact the functions are being used in ways that require additional properties.

3.1.1. *Definitions and Implications of Security Properties*

Despite the fact that collision resistance is a central requirement of hash function research, nearly all analyses involving this concept are in the family of function setting. In practice we are not at all using that setting. Only recently has the model of human ignorance been proposed by Rogaway [48] in order to formalize collision resistance for ordinary hash functions. The next step is to develop results analagous to [49] in either the *code-constructive* or the *blackbox-constructive* form (see [48]). Note that the relation between aPre and aSec [49] carries directly over to this setting. Presumably, collision resistance carries over as well.

It is a convenient time to emphasize the importance of formalizing these definitions. The definition of second preimage resistance from the Handbook of Applied Cryptography [41] is:

> 2nd preimage resistance – it is computationally infeasible to find any second input which has the same output as any specified input, i.e., given x to find a 2nd preimage $x' \neq x$ such that $h(x) = h(x')$.

In [49], it is asked whether it is really meant that the specified x can be *any* domain point. In fact, this definition is better than most informal definitions in the sense that it does tell us how x is chosen. But, returning to Rogaway and Shrimpton's question, do we really mean *any* or is it sufficient to require *all except a negligible portion*? Or does it not matter at all?

In fact it does matter when one considers the relation between collision resistance and second preimage resistance. Informally, one would argue that the former trivially implies the latter, since if one can compute second preimages, then they have found a collision. But this argument is actually not valid for the Handbook of Applied Cryptography definition since it may be possible that there is a subset of negligible size where one can trivially compute second preimages, though in practice it may be impossible to find that subset [24]. For instance, consider the number theoretic hash of the form $h(x) = a^x \bmod N$, where N is a product of two large primes and a is a parameter having certain required properties – details in [45]. With proper padding (ignored here for simplicity) this function is provably collision resistant, with a security reduction from integer factorization. Actually, this is as a family of functions indexed by the modulus, N, but if we fix N then it becomes an ordinary hash function and we can use the code- or blackbox-constructive formalization. Despite having a security reduction from factoring N to collision finding, one can always trivially compute second preimages for any value of x that is a multiple of $\phi(N)$ since any other multiple of x is a second preimage. Of course, these pathological messages give away a break of that hash function, making it so that it is no longer collision resistant. Obviously, if factoring is really a hard problem, then it is safe to assume that we would never encounter such messages in practice, and therefore such counterexamples are of no interest. So we really want a definition that reflects that these negligible probability events should have no real security impact.

With proper definitions and formalizations, it appears that one can indeed show that collision resistance implies second preimage resistance for ordinary hash functions, which seemingly would simplify our list of goals for hash functions. Unfortunately, the argument only shows that finding second preimages is at least as hard as finding collisions, so it only guarantees that finding a second preimage is at least order $2^{n/2}$ effort for a hash with an $n-$bit output (assuming the best collision attack is a generic one). Those who do research in ordinary hash functions often want that second preimages should take 2^n effort, though it usually isn't explicitly stated. In that case, the security relation between collision resistance and second preimage resistance does not matter to them. A further setback is [37], which showed that nearly all ordinary hash function constructions that are being seriously considered today cannot achieve 2^n second preimage resistance. They suggest that perhaps we should not expect more than $2^{n/2}$ security for any aspect of an $n-$bit hash. On the other hand, new constructions are being considered which *might* resist the Kelsey and Schneier attack [39]. In any event, the hash function community has to come to a common agreement on their security goals, and to be very precise about those requirements.

Furthermore, for some reason the hash community has left out the notions of PRF and random oracle emulation when considering the implications of various security notions, despite the fact that hash functions are being used for this purpose. In regard to PRF, the current folklore indicates that we could build one by taking a hash function and putting a secret element into one of the inputs (IV or message). But there is no theoretical justification for this idea. In regard to random oracle emulation, someone should formally define what random oracle emulation means, and that definition presumably would imply the other security requirements we have so far listed. Then researchers would have the single objective goal of designing a hash function that "looks like" a random oracle, rather than a list of many smaller goals that do not accurately depict all the necessary security requirements.

In the absence of a satisfactory definition for random oracle emulation, we face a dilemma for evaluating proposals for new hash function standards. On the one hand, we would like a new function to support existing random oracle applications. On the other hand, we do not have a simple well-defined security requirement on the hash function to support all such applications. In that case, we could deal with this problem by identifying the main important applications of a new hash function standard, and then, for each such application, specify in a well-defined way: (1) the precise details of the application and how it makes use of the hash function (with the hash function treated as a black box), and (2) the security requirement on the application incorporating the hash function. For example, consider the OAEP-RSA public-key encryption application[50], which makes use of two "Mask Generation Functions" (treated as random oracles in the security analysis[6,32]). The precise details of the application and how it makes use of the hash functions (in particular how the "Mask Generation Functions" are constructed from a given hash function h) are specified in the PKCS standard[50] (see in particular the definition of MGF1 in Appendix B.2.1 in[50]), and the precise security requirement on the OAEP-RSA application is the well-known IND-CCA2 security requirement for public-key encryption schemes[32]. The important advantage of this approach is that it allows a hash standard to support the most important random oracle applications of hash functions with *well-defined* security requirements for the hash function, allowing objective evaluation of proposals. The disadvantages of this approach are that these security requirements are complex (involving details of the applications), and that the approach introduces a new security requirement for each random oracle application that is to be supported by the hash standard. However, at present we do not see any better way available to support all the uses of random oracles with well-defined security requirements.

An open research problem is find simpler security requirements on the hash functions, which are sufficient at least for the most popular current applications of random oracles. Some theoretical progress has been made in this direction over the last few years[3,11–16,27], but more research is needed for such results to be applicable in practice (e.g. to support the full security requirements of popular applications provable in the random-oracle model).

3.2. *Hash Functions Families*

In contrast to ordinary hash functions, the security properties and implications of hash function families are better understood [49]. Although not explicitly stated, hash function families seem to have traditionally been aimed at designing specific solutions for specific problems rather than a one solution for all problems approach. Where they are lacking is in practical solutions, both at the hash function design level and the protocol design level, which is entirely the reason they have so far not been used in practice today.

At the hash function design level, there has been some recent progress towards making this approach practical. Examples include VSH [17] and FFT Hash [40], which both achieve provable collision resistance and are approaching practicality. A preliminary implementation of VSH is reported to be within a factor of 25 of the speed of SHA-1. While both results are encouraging, more needs ot be done. Solutions need to be developed that have smaller output sizes (especially for FFT-hash, which requires very large parameters for the security reduction to hold) and faster speeds. Moreover, there is still no provable and practical hash solutions for other security needs, such as PRF.

At the protocol level, there have been many new designs that have eliminated random oracles. The down side is that they usually do not achieve the efficiency of random oracle based approaches.

4. What to Do about all the Cryptographic Hash Function Definitions?

We have so far eluded the question of how these hash functions should be defined. Given that so many people have cluttered the literature with different definitions, it is against our judgement to offer new definitions that are attempts at overriding others. However, we do recommend rejecting certain definitions.

Generally, we would like to avoid definitions that do not accurately reflect the security properties that they are intended to have. Examples include Merkle's weak and strong one-way hash functions, Winternitz's version of a one-way hash function, and Preneel's version of a collision-resistant hash function (since the name does not specify preimage resistance also). We would also like to avoid definitions that are not mainstream, such as definitions that do not involve compression.

Many hash function designers today simply call their design a "hash function" without adjectives such as "one-way" or "collision resistant." An interpretation of this vernacular is a compressing and easy to compute function that has additional security properties. We do not oppose such a definition (although it is informal), but we do strongly recommend that researchers say *exactly* what those additional properties are for their design. In particular, if researchers are proposing a single solution to be used everywhere like the way we are using SHA-1 today, then they should include PRF and random oracle emulation in their stated security goals, and at least include a reference to how to interpet such definitions formally. Moreover, when researchers develop a hash function family, do not omit the word "family."

5. Moving Forward

We recommend that standards bodies involved in selecting new standards for cryptographic hash functions, in collaboration with the cryptographic community, should aim to specify a set of *well-defined* security requirements for cryptographic hash functions. This set of security requirements would then allow an objective assessment of the security of candidate functions submitted for standardization. Such requirements are defined by specifying an interactive *computational game* between an adversary and a challenger, and defining a condition on the outcome of the game which defines *success* of the adversary in 'winning' the game, and a quantity called the *advantage* of the adversary (which is determined by its success probability). The security requirement can then be quantified by the maximal advantage of the adversary in the game given bounds on its computational resources (run-time/program, size, number of oracle queries, etc). In the standard complexity-theory model, the maximal adversary advantage is taken over *all* adversaries with the given resource bound. An alternative approach is the human ignorance model where details can be found in [48].

The security requirements needed from cryptographic hash functions are ultimately determined by their *applications.* Therefore, as a step towards the above stated goal, we present below a list of the main current practical applications of cryptographic hash functions. For each such application, we cite known well-defined security requirements on the underlying hash function which guarantee the security of the application (if such requirements are known). We also list other requirements from the hash function (whether a function *family* is needed, whether the family key is secret or public, the function input/output domain) and the relevant references to the literature where the security requirement was defined and shown sufficient for the application. To make this survey self-contained we also provide a definition of the relevant security requirements (via the games and adversary success conditions) in the Appendix.

We note that this list is not intended to be exhaustive but is presented to indicate the variety of requirements needed from hash functions today. Other applications are expected to add yet other new requirements. We also note that it is debatable whether all the listed applications (and corresponding security requirements) should in fact be supported by a new hash standard. For example, applications which need a PRF are usually implementable using a block cipher as the underlying cryptograpic primitive, rather than a hash function. Indeed, block ciphers are usually designed to achieve a PRF-like design goal (more precisely the closely related PRP family goal). On the other hand, it is possible that hash functions may be preferred for such applications for various (not necessarily technical) reasons.

We use the following conventions in Table ??. Set B denotes binary set $\{0,1\}$ and B^* denotes the set of all finite binary strings. Column 'App' lists the applications. For each application, column 'Fam?' indicates whether a function family is required (Y) or not (N). If a function family is needed, column 'Key Sec?' indicates whether the function key is secret (Y) or public (N). Columns 'In' and 'Out' indicate the required hash function input and output domains respectively (with typical values). Column 'Security' indicates the sufficient security requirement required from the hash for the application. Column 'Ref' gives a reference in the literature for relevant

App	Fam?	Key Sec?	In	Out	Security	Ref.
HMAC[31] (a)	N	–	$B^b \times B^c$	B^c	$h(\cdot, K)$ and $h(K, \cdot)$ are PRF families	[2], 1
Deterministic Message Hashing for Secure Sig (b)	Y	N	B^*	B^c	$H_K(\cdot)$ is CR Family	[19], 2
	Y	N	$B^b \times B^c$	B^c	$h_K(\cdot)$ is CR Family	[19,20], 2
Randomized Message Hashing for Secure Sig (b),(c)	Y	N	B^*	B^c	$H_K(\cdot)$ is TCR Family	[44], 3
	Y	N	B^*	B^c	$H_K(\cdot)$ is eTCR Family	[35], 4
	N	–	$B^b \times B^c$	B^c	$h(\cdot, \cdot)$ is e-SPR	[35], 5
RSA-PSS[50] (d)	N	N	B^*	B^c	"random oracle"	[7], ???
	Y	N	B^*	B^c	RSA-PSS (using $H_K(\cdot)$) is EF-CMA[34]	
RSA-OAEP[50] (e)	N	N	B^*	B^c	"random oracle"	[6,32], ???
	Y	N	B^*	B^c	RSAES-OAEP (using $H_K(\cdot)$) is IND-CCA2[32]	
FIPS 186 PRG[29,30]	N	–	$B^b \times B^c$	B^c	$h((\cdot + K) \bmod 2^b, IV)$ is PRF family (fixed IV)	[23], 1
DSA[29,30] (f)	Y	N	B^*	B^c	DSA (using $H_K(\cdot)$) is EF-CMA[34]	
ECDSA[1] (f)	Y	N	B^*	B^c	ECDSA (using $H_K(\cdot)$) is EF-CMA[34]	
Password Hashing (g)	N	–	B^b	B^c	preimage resistant	[53],6
Commitments	Y	N	$B^b \times B^c$	B^c	$h_K(\cdot, \cdot)$ is CR Family	[36], 2

security analysis, followed by a security requirement number, referring to our list of security requirement definitions in the appendix; if such a well-defined security requirement on the hash is not known, we write '???'. Note that for some applications there are several alternative hash function requirements, any one of which is sufficient for the application; in such cases, we list the alternatives on separate rows.

Remarks on table entries:

(a) The security requirement stated here are for the "2-key" variant of HMAC which uses two independent keys[2]. The security proof for the standard "1-key" variant of HMAC requires an additional 'related key' pseudorandomness assumption on h, see[2].

(b) In these applications the hash function is used to hash a long message prior to signing the hash value using a secure signature scheme which accepts short fixed-length input messages (of length at least equal to the output length c of the hash function). It is assumed that the given signature scheme is *fully* secure, i.e. is existentially unforgeable under adaptive chosen message attack (EF-CMA)[34]. In other words, we assume the hash function is only used to allow signing arbitrary length messages, and not for strengthening the security of the underlying signature scheme. This excludes many "classic" signature schemes such as plain RSA and ElGamal-type signatures (which are existentially forgeable and not secure against chosen message attacks). To strengthen such 'weak' signature schemes into secure ones, the hash function requires additional properties beyond collision-resistance. In many such cases, modelling the hash function as a "random oracle" is enough (see 'PSS-RSA' table entry for example). The hash security requirements listed here are also sufficient for other 'integrity checking' applications, where the message hash value is write-protected in some other way (rather than being signed), e.g. by storing it in a public read-only memory device, and is used to verify integrity of the message by hashing and comparing with the stored write-protected hash value.

(c) The alternative in the first row requires signing both the hash value and the hash randomizer, whereas in the alternatives in the last two rows, only the hash value needs to be signed.

(d) We refer here to the RSASSA-PSS signature scheme in the PKCS standard[50]. We denote by EF-CMA the standard security requirement for signature schemes, namely Existential Unforgeability under adaptive Chosen Message Attack[34].

(e) We refer here to the RSAES-OAEP public-key encryption scheme in the PKCS standard[50]. We denote by IND-CCA2 the standard security requirement for public-key encryption schemes, namely Indistinguishability under Adaptive Chosen Ciphertext Attack[32].

(f) Some security results for *variants* of DSA and ECDSA are known in the random oracle and generic group models, see[52] for a survey.

(g) For password hashing, applying the results of[53], if maximal adversary advantage against preimage resistance of $h(\cdot) : B^b \rightarrow B^c$ is $Adv(t)$ for run-time t then maximal adversary advantage against preimage resistance of $h(\cdot)$ when applied to a a uniformly random password from a subset $D \subset B^b$ of size $|D|$ is $Adv'(t) \leq \frac{Adv(t)2^b}{|D|}$, hence to guarantee $Adv'(t) \leq 1/2^s$ (password security level of 's-bits') we need a password set D of size at least $|D| \geq (Adv(t) \cdot 2^b) \cdot 2^s$. We note that we assume here (following [53]) that no 'salting' is used.

6. Conclusion

The field of cryptographic hash functions has been evolving since its origin approximately 30 years ago, and will continue to do so for quite some time. The informality of hash function terminology has resulted in cluttered literature, lacking a clean list of goals summarizing our security requirements. Consequently, there is no objective way of evaluating the security of new hash function proposals, except designs that are very obviously broken.

This survey has emphasized the importance of formal terminology and a clear set of objectives. The hope is that researchers will take our view into consideration as a first step of trying to clean up the cryptographic hashing literature.

References

1. ANSI X9.62. Public Key Cryptography for the Financial Services Industry: The Elliptic Curve Digital Signature Algorithm (ECDSA). American National Standards Institute. American Bankers Association, 1998.
2. M. Bellare. New proofs for NMAC and HMAC: Security without collision-resistance. In *Advances in Cryptology – CRYPTO'06*, volume 4117 of *LNCS*, pages 602–619. Springer, 2006.
3. M. Bellare, A. Boldyreva, and A. Palacio. An uninstantiable random-oracle-model scheme for a hybrid-encryption problem. In C. Cachin and J. Camenisch, editors, *Advances in Cryptology – EUROCRYPT '04*, volume 3027 of *LNCS*, pages 171–188. Springer, 2004.
4. M. Bellare, R. Canetti, and H. Krawczyk. Keying hash functions for message authentication. *Lecture Notes in Computer Science*, 1109, 1996.
5. M. Bellare and P. Rogaway. Random oracles are practical: A paradigm for designing efficient protocols. In *First ACM Conference on Computer and Communications Security*, pages 62–73, Fairfax, 1993. ACM.
6. M. Bellare and P. Rogaway. Optimal asymmetric encryption. In *Advances in Cryptology – EUROCRYPT'94*, volume 950 of *LNCS*, pages 92–111. Springer, 1995.
7. M. Bellare and P. Rogaway. The exact security of digital signatures – how to sign with RSA and Rabin. In *Advances in Cryptology – EUROCRYPT'96*, volume 1070, pages 399–416. Springer, 1996.
8. M. Bellare and P. Rogaway. Collision-Resistant hashing: Towards making UOWHFs Practical. In *CRYPTO '97*, volume 1294 of *LNCS*, pages 470–484, Berlin, 1997. Springer-Verlag.
9. D. J. Bernstein. The Salsa20 hash function. Web page, `http://cr.yp.to/salsa20.html`.
10. D. Bleichenbacher. Chosen ciphertext attacks against protocols based on rsa encryption standard pkcs #1. In *Advances in Cryptology – CRYPTO'98*, volume 1462 of *LNCS*, pages 1–12. Springer-Verlag, 1998.
11. A. Boldyreva and M. Fischlin. Analysis of random-oracle instantiation sce-

narios for OAEP and other practical schemes. In *Advances in Cryptology – CRYPTO '05*, volume 3621 of *LNCS*, pages 412–429. Springer-Verlag, 2005.

12. A. Boldyreva and M. Fischlin. On the security of OAEP. In *Advances in Cryptology – ASIACRYPT '06*, volume 4284 of *LNCS*, pages 210–225. Springer-Verlag, 2006.
13. D. Brown. Unprovable security of RSA-OAEP in the standard model. Cryptology ePrint Archive, Report 2006/223, June 2006. `http://eprint.iacr.org/`.
14. R. Canetti. Towards realizing random oracles: Hash functions that hide all partial information. In *CRYPTO '97*, volume 1294, pages 455–469, London, UK, 1997. Springer-Verlag.
15. R. Canetti, O. Goldreich, and S. Halevi. The random oracle methodology, revisited. *J. ACM*, 51(4):557–594, 2004.
16. R. Canetti, D. Micciancio, and O. Reingold. Perfectly one-way probabilistic hash functions. In *STOC '98*, pages 131–140. ACM Press, 1998.
17. S. Contini, A. Lenstra, and R. Steinfeld. VSH, an efficient and provable collision-resistant hash function. In S. Vaudenay, editor, *EUROCRYPT '06*, volume 4004 of *Lecture Notes in Computer Science*. Springer, 2006.
18. R. Cramer and V. Shoup. A practical public key cryptosystem provably secure against adaptive chosen ciphertext attack. In *CRYPTO '98*, volume 1462 of *LNCS*, pages 13–25. Springer, 1998.
19. I. B. Damgård. Collision free hash functions and public key signature schemes. In *Advances in Cryptology – EUROCRYPT'87*, volume 304 of *LNCS*, pages 203–216. Springer, 1987.
20. I. B. Damgård. A design principle for hash functions. In G. Brassard, editor, *Advances in Cryptology - CRYPTO '89*, volume 435 of *LNCS*, pages 416–427. Springer-Verlag, 1989.
21. G. Davida. Chosen signature cryptanalysis of the RSA (MIT) public key cryptosystem. *Tech. Rep.*, TR-CS-82-2, 1982.
22. D. E. Denning. Digital signatures with RSA and other public-key cryptosystems. *Commun. ACM*, 27(4):388–392, 1984.
23. A. Desai, A. Hevia, and Y. Yin. A practice-oriented treatment of pseudorandom number generators. In *EUROCRYPT 2002*, pages 368–383. Springer, 2002.
24. Y. G. Desmedt. Private email, 2005.
25. Y. G. Desmedt and A. M. Odlyzko. A chosen text attack on the RSA cryptosystem and some discrete logarithm schemes. In *CRYPTO '85*, pages 516–522. Springer, 1985.
26. W. Diffie and M. E. Hellman. New directions in cryptography. *IEEE Transactions on Information Theory*, IT-22(6):644–654, 1976.
27. Y. Dodis, R. Oliveira, and K. Pietrzak. On the generic insecurity of full-domain hash. In *CRYPTO '05*, volume 3621 of *LNCS*, pages 449–466. Springer, 2005.
28. A. Fiat and A. Shamir. How to prove yourself: practical solutions to identification and signature problems. In *Proceedings on Advances in cryptology—CRYPTO '86*, pages 186–194, London, UK, 1987. Springer-Verlag.

29. FIPS PUB 186-2. Digital signature standard. National Institute of Standards and Technology, 1994.
30. FIPS PUB 186-2 (Change Notice 1). Digital signature standard. National Institute of Standards and Technology, 2001.
31. FIPS PUB 198. The keyed-hash message authentication code (HMAC). National Institute of Standards and Technology, 2002.
32. E. Fujisaki, T. Okamoto, D. Pointcheval, and J. Stern. RSA-OAEP is secure under the RSA assumption. In *Advances in Cryptology– CRYPTO'01*, volume 2139, pages 260–274. Springer, 2001.
33. O. Goldreich, S. Goldwasser, and S. Micali. On the cryptographic applications of random functions. In G. R. Blakley and D. C. Chaum, editors, *CRYPTO '84*, pages 276–288. Springer, 1985. Lecture Notes in Computer Science No. 196.
34. S. Goldwasser, S. Micali, and R. Rivest. A digital signature scheme secure against adaptively chosen message attacks. *SIAM Journal on Computing*, 17(2):281–308, 1988.
35. S. Halevi and H. Krawczyk. Strengthening digital signatures via randomized hashing. In *Advances in Cryptology - CRYPTO'06*, volume 4117 of *LNCS*, pages 41–59. Springer, 2006.
36. S. Halevi and S. Micali. Practical and provably-secure commitment schemes from collision-free hashing. In *Advances in Cryptology - CRYPTO'96*, volume 1109, pages 201–215, 1996.
37. J. Kelsey and B. Schneier. Second preimages on n-bit hash functions for much less than 2^n work. In R. Cramer, editor, *Advances in Cryptology - EUROCRYPT 2005*, volume 3494 of *LNCS*, pages 474–490. Springer-Verlag, 2005.
38. N. Koblitz and A. J. Menezes. Another look at "provable security". *Journal of Cryptology*, 20(1):3–37, January 2007.
39. S. Lucks. Failure-friendly design principle for hash functions. In B. K. Roy, editor, *Advances in Cryptology - ASIACRYPT '05*, volume 3788 of *LNCS*, pages 474–494. Springer, 2005.
40. V. Lyubashevsky, D. Micciancio, C. Peikert, and A. Rosen. Provably secure FFT hashing. Second NIST Workshop on Hash Functions, 2006.
41. A. J. Menezes, P. C. van Oorschot, and S. A. Vanstone. *Handbook of Applied Cryptography*. CRC Press, 1996.
42. R. C. Merkle. *Secrecy, Authentication, and Public Key Systems*. PhD thesis, Stanford University, 1979.
43. R. C. Merkle. One way hash functions and DES. In G. Brassard, editor, *Advances in Cryptology - CRYPTO '89*, volume 435 of *LNCS*, pages 428 – 446. Springer-Verlag, 1989.
44. M. Naor and M. Yung. Universal one-way hash functions and their cryptographic applications. In *Proceedings of the Twenty First Annual ACM Symposium on Theory of Computing*. ACM Press, 1989.
45. D. Pointcheval. The composite discrete logarithm and secure authentication. In *PKC '00*, pages 113–128, London, UK, 2000. Springer-Verlag.
46. B. Preneel. *Analysis and design of cryptographic hash functions*. PhD thesis,

Katholieke Universiteit Leuven, 1993.
47. M. Rabin. Digitalized signatures. In R. A. DeMillo, D. P. Dobkin, A. K. Jones, and R. J. Lipton, editors, *Foundations of Secure Computation*, pages 155–168. Academic Press, 1978.
48. P. Rogaway. Formalizing human ignorance. In *Progress in Cryptology - VIETCRYPT'06*, volume 4341, pages 211–228. Springer, 2006.
49. P. Rogaway and T. Shrimpton. Cryptographic hash-function basics: Definitions, implications, and separations for preimage resistance, second-preimage resistance, and collision resistance. In *Fast Software Encryption – FSE'04*, volume 3017 of *LNCS*, pages 371–388. Springer, 2004.
50. RSA Laboratories. PKCS ♯1 v2.1: RSA Cryptography Standard, June 2002.
51. D. Stinson. Some observations on the theory of cryptographic hash functions. *Des. Codes Cryptography*, 38(2):259–277, 2006.
52. S. Vaudenay. The security of DSA and ECDSA. In *PKC 2003*, volume 2567 of *LNCS*, pages 309–323. Springer-Verlag, 2003.
53. D. Wagner and I. Goldberg. Proofs of security for the unix password hashing algorithm. In D. Gollmann, editor, *ASIACRYPT 2000*, volume 1976 of *LNCS*, pages 560–572. Springer-Verlag, 2000.
54. R. S. Winternitz. Producing a one-way hash function from DES. In *CRYPTO'83*, pages 203–207, 1983.
55. G. Yuval. How to swindle Rabin. *Cryptologia*, 3:187–189, 1979.

Appendix A.

7. Security Requirement Definitions

To make this paper self-contained, we provide here a summary of hash function security requirement definitions referred to in Table **??** in the text. For each requirement listed, we provide: (a) Input and output domains of the hash function h , (b) the computational game between the adversary A and the challenger C, (c) the success condition for the adversary in the game, (d) the definition of adversary's *advantage* in the game, (e) an example of typical required upper bound for adversary advantage for typical adversary resources (here "run-time" t is typically measured as the sum of the run-time in machine instructions plus program length for some fixed computational machine model).

(1) Security Requirement: $h(K,\cdot)$ is a PRF Family[2]

  (a) In/Out Domains: $h : B^b \times B^c \rightarrow B^c$
  (b) Game: (1) C chooses uniformly random bit $b \in B$. If $b = 0$, C chooses uniformly at random a function $h' : B^c \rightarrow B^c$ from the set of all functions from B^c to B^c. If $b = 1$, C chooses uniformly at random a key $K \in B^b$. (2) A runs with no input for time t and makes q queries

to oracle $F : B^c \to B^c$ defined as follows: If $b = 0$, $F(x) \stackrel{\text{def}}{=} h'(x)$, and if $b = 1$, $F(x) \stackrel{\text{def}}{=} h(K, x)$. At the end of the game, A outputs a bit $b' \in B$.

(c) A Success Condition: $b' = b$.

(d) A Advantage: $Adv_{\mathsf{A}} \stackrel{\text{def}}{=} 2|Succ_{\mathsf{A}} - 1/2|$, where $Succ_{\mathsf{A}}$ is A's success probability.

(e) Typical Quantitative Requirement: $Adv_{\mathsf{A}} \leq 2^{-80}$ for any A with resource bounds $t \leq 2^{80}$ and $q \leq 2^{40}$.

(2) Security Requirement: $h_K(\cdot)$ is a CR Family[19]

(a) In/Out Domains: $h_K : B^b \times B^c \to B^c$

(b) Game: (1) C chooses uniformly random hash function key $K \in \mathsf{K}$ (K denotes hash family key space). (2) A runs with input K for time t and outputs $m, m' \in B^b \times B^c$.

(c) A Success Condition: $m \neq m'$ and $h_K(m) = h_K(m')$.

(d) A Advantage: $Adv_{\mathsf{A}} \stackrel{\text{def}}{=} Succ_{\mathsf{A}}$, where $Succ_{\mathsf{A}}$ is A's success probability.

(e) Typical Quantitative Requirement: $Adv_{\mathsf{A}} \leq 2^{-80}$ for any A with resource bounds $t \leq 2^{80}$.

(3) Security Requirement: $H_K(\cdot)$ is a TCR[‡] Family[8,44]

(a) In/Out Domains: $H_K : B^* \to B^c$

(b) Game: (1) A runs on no input and outputs $m \in B^*$, (2) C chooses uniformly random hash function key $K \in \mathsf{K}$ (K denotes hash family key space), (3) A continues to run on input K (for total time t) and outputs $m' \in B^*$.

(c) A Success Condition: $m \neq m'$ and $H_K(m) = H_K(m')$.

(d) A Advantage: $Adv_{\mathsf{A}} \stackrel{\text{def}}{=} Succ_{\mathsf{A}}$, where $Succ_{\mathsf{A}}$ is A's success probability.

(e) Typical Quantitative Requirement: $Adv_{\mathsf{A}} \leq 2^{-80}$ for any A with resource bounds $t \leq 2^{80}$.

(4) Security Requirement: $H_K(\cdot)$ is an eTCR Family[35]

(a) In/Out Domains: $H_K : B^* \to B^c$

(b) Game: (1) A runs on no input and outputs $m \in B^*$, (2) C chooses uniformly random hash function key $K \in \mathsf{K}$ (K denotes hash family

[‡] The term TCR (Target Collision Resistant) was introduced by Bellare and Rogaway[8] as an alternative name for UOWHF (Universal One-Way Hash Function), introduced by Naor and Yung[44]. Both TCR and UOWHF terms are now commonly used in the cryptographic literature to refer to this notion.

key space), (3) A continues to run on input K (for total time t) and outputs $K' \in \mathsf{K}, m' \in B^*$.

(c) A Success Condition: $(m, K) \neq (m', K')$ and $H_K(m) = H_{K'}(m')$.

(d) A Advantage: $Adv_{\mathsf{A}} \stackrel{\text{def}}{=} Succ_{\mathsf{A}}$, where $Succ_{\mathsf{A}}$ is A's success probability.

(e) Typical Quantitative Requirement: $Adv_{\mathsf{A}} \leq 2^{-80}$ for any A with resource bounds $t \leq 2^{80}$.

(5) Security Requirement: $h(\cdot, \cdot)$ is e-SPR[35]

(a) In/Out Domains: $h : B^b \times B^c \to B^c$

(b) Game: (1) A runs on no input and outputs ℓ values $\Delta_1, \ldots, \Delta_\ell$, each in B^b, (2) C chooses uniformly random $r \in B^b$ and computes $m = r \oplus \Delta_\ell$ and $c = H^{c_0}(r \oplus \Delta_1, \ldots, r \oplus \Delta_{\ell-1})$ (here $c_0 \in B^c$ is a fixed IV and $H^{c_0}(x_1, \ldots, x_\ell)$ denotes the output of the ℓ-block Merkle-Damgård (MD) iteration function[19] with ℓ message blocks $(x_1, \ldots, x_\ell)$ and IV c_0), (3) A is given r, continues to run (for total time t) and outputs $c' \in B^c, m' \in B^b$.

(c) A Success Condition: $(m, c) \neq (m', c')$ and $h(m, c) = h(m', c')$.

(d) A Advantage: $Adv_{\mathsf{A}} \stackrel{\text{def}}{=} Succ_{\mathsf{A}}$, where $Succ_{\mathsf{A}}$ is A's success probability.

(e) Typical Quantitative Requirement: $Adv_{\mathsf{A}} \leq 2^{-80}$ for any A with resource bounds $t \leq 2^{80}$ and $\ell \leq 2^{64}$.

(6) Security Requirement: $h(\cdot)$ is preimage resistant (one-way)[53]

(a) In/Out Domains: $h : B^b \to B^c$

(b) Game: (1) C chooses uniformly random $x \in B^b$ and computes $y = f(x)$, (2) A runs on input y for time t and outputs $x' \in B^b$.

(c) A Success Condition: $h(x) = h(x')$.

(d) A Advantage: $Adv_{\mathsf{A}} \stackrel{\text{def}}{=} Succ_{\mathsf{A}}$, where $Succ_{\mathsf{A}}$ is A's success probability.

(e) Typical Quantitative Requirement: $Adv_{\mathsf{A}} \leq 2^{-80}$ for any A with resource bounds $t \leq 2^{80}$.

Scalable Optimal Test Patterns for Crosstalk-induced Faults on Deep Submicron Global Interconnects

Yeow Meng Chee

Division of Mathematical Sciences
School of Physical and Mathematical Sciences
Nanyang Technological University
Singapore 637616
E-mail: ymchee@ntu.edu.sg

Charles J. Colbourn

Department of Computer Science and Engineering
Arizona State University
Tempe, Arizona 85287-8809
USA
Email: charles.colbourn@asu.edu

Capacitance-coupling and mutual inductance between the neighboring wires of global interconnects give rise to crosstalk effects, which are one of the biggest signal integrity problems in DSM circuits today. Previous models for crosstalk-induced faults assume that all the wires of a bus act together to induce crosstalk effects on a single wire. Based on recent simulation results of Sirisaengtaksin and Gupta, we use a more general model of crosstalk-induced faults that allows tradeoffs between efficiency of tests and quality of tests. We construct provably optimal test patterns that covers all crosstalk-induced faults under this general model. Our test patterns admit simple generators for at-speed testing in BIST. The test methodology proposed here can result in huge savings in test cost. In particular, the required linear number of test patterns can be reduced to a constant number of test patterns by accepting only a few percent of error, thus allowing our test methodology to scale with bus width and technology.

Keywords: built-in self-test, crosstalk faults, DSM interconnects, fault model, test patterns

1. Introduction

The explosive growth of the Internet and the amount of data it has to handle have resulted in the need for dramatic performance increase throughout the net. System designers are driven to increase device integration in the form of system-on-chips (SoCs), integrated CPUs, mixed signal and digital VLSI, as well as to decrease feature size so that systems remain deployable. Already, deep submicron (DSM) systems with feature size below 90nm are pervasive. This aggressive scaling of feature size leads to signal integrity problems, as the components are more susceptible to interference due to crosstalk, substrate noise, power bus noise, and distributed delay variations. Design techniques and validation [10,11,13,14] alone cannot deal with all of these errors because noise problems occur also in silicon, which only manifest themselves post-fabrication. Manufacturing testing therefore becomes a very important issue in DSM circuits. Unfortunately, the cost of testing is not scaling: while the cost of silicon manufacturing is decreasing with time, the cost of testing is actually increasing. New test paradigms have to succeed in order to change this situation. For this reason, testing has emerged as one of the most important areas of research in DSM technologies today [6].

The focus of this paper is on testing crosstalk-induced failures on DSM global interconnects, particularly inter-core buses. Crosstalk-induced failures on these interconnects can give rise to logic errors and slowdowns/speedups [2,4,5,8]. Empirical data has shown that crosstalk effects are most significant in long interconnects [7,9]. Due to its timing nature, testing for crosstalk effect needs to be conducted at the operational speed of the circuit under test [3]. The built-in self-test (BIST) framework of testing is considered here as BIST is a more feasible solution for at-speed crosstalk testing [1]. In this framework, models for crosstalk faults are developed and test patterns that are likely to excite crosstalk faults in these models are generated by a processor and applied to the appropriate bus in the normal functional mode of the system. The bus under test is then observed for crosstalk faults under these test patterns. Many models have been considered for crosstalk in long DSM interconnects, but the one that has emerged as being not overly complex to analyze, yet remaining realistic, is the *maximal aggressor fault (MAF) model* of Cuviello *et al.* [7].

In the MAF model, a *victim* is a single wire of a set of interconnects. All of the other interconnects in the set are designated *aggressors*, which act collectively to generate errors on the victim via their coupling capacitances. The MAF model of Cuviello *et al.* [7] considered the four crosstalk-induced faults

(1) positive glitch (g_p),
(2) negative glitch (g_n),
(3) falling delay (d_f), and

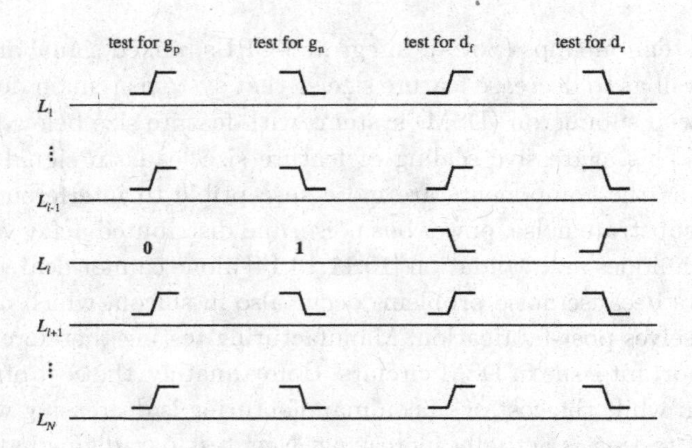

Fig. 1. Required transitions for the MAF model.

(4) rising delay ($\mathrm{d_r}$).

The required transitions on the victim (line L_i) and aggressors to excite the four different possible faults are shown in Fig. 1. For example, to test for $\mathrm{g_p}$ on interconnect L_i, we need the two test vectors $\mathsf{u} = 0 \cdots 000 \cdots 0$ and $\mathsf{v} = 1 \cdots 101 \cdots 1$ (where the i-th component is at static 0) to produce the required transitions. We call $\mathsf{u} \rightarrow \mathsf{v}$ a *test pattern*. Under the MAF model, for a set of N interconnects, there are $4N$ faults to test, and we require $4N$ test patterns.

Our results in this paper generalize the MAF model to allow a tradeoff between efficiency of test and quality of test. We show how to generate the fewest test patterns to cover all crosstalk-induced faults under this generalized model. The number of test patterns required is less than the $4N$ required for the MAF model. If we accept a few percent of error, the required $4N$ test patterns can be reduced to a constant number. Our test methodology is therefore scalable.

2. Testing Under a Generalized MAF Model

Let $L_1, \ldots, L_N$ be a set of N linearly ordered interconnects, so that the neighbors of L_i are L_{i-1} and L_{i+1}, for $2 \leq i \leq N-1$, the neighbor of L_1 is L_2, and the neighbor is L_N is L_{N-1}. For $1 \leq s \leq N$, we define the *MAF model of strength s* (denoted s-MAF) as follows. A victim is a single interconnect as in the MAF model. However, we limit the number of aggressors to the s nearest interconnects on each side of the victim. The faults under consideration are still $\mathrm{g_p}$, $\mathrm{g_n}$, $\mathrm{d_r}$, and $\mathrm{d_f}$. This is a justified model because the neighboring interconnects shield the victim from farther interconnects, and the capacitive coupling effect on the victim decreases with distance, as shown by Sirisaengtaksin and Gupta [12]. The MAF model of Cuviello *et al.* [7] is precisely the $(N-1)$-MAF model. Henceforth, we restrict our attention to $1 \leq s \leq N-2$. It is easy to see that any set of test patterns that covers all faults under the s-MAF model also covers all faults under the s'-MAF model, for all $s' < s$.

We now generate test patterns to cover the crosstalk-induced faults under the s-MAF model. We still have $4N$ faults to test for N interconnects. However, the number of test patterns required can be made smaller than the $4N$ required for the MAF model. In the MAF model, every test patterns covers only one fault, so $4N$ test patterns are necessary. Suppose $N = 8$. In the 2-MAF model, the test pattern $00000000 \to 11011011$ can test for $\mathrm{g_p}$ on victims L_3 and L_6 simultaneously. So fewer than $4 \cdot 8 = 32$ test patterns are required to cover all the 32 possible faults. Let us now formalize this observation.

For a vector u, u_i denotes the i-th coordinate of u. Let $\mathsf{u}, \mathsf{v} \in \{0,1\}^N$. A test pattern $\mathsf{u} \to \mathsf{v}$ can be encoded as a vector x over the set $\Sigma = \{0, 1, +, -\}$ as follows:

$$\mathsf{x}_i = \begin{cases} 0, & \text{if } \mathsf{u}_i = \mathsf{v}_i = 0, \\ 1, & \text{if } \mathsf{u}_i = \mathsf{v}_i = 1, \\ +, & \text{if } \mathsf{u}_i = 0 \text{ and } \mathsf{v}_i = 1, \text{ and} \\ -, & \text{if } \mathsf{u}_i = 1 \text{ and } \mathsf{v}_i = 0. \end{cases} \tag{1}$$

Any given vector over Σ can also be decoded uniquely to a test pattern $\mathsf{u} \to \mathsf{v}$. A vector over Σ is called an *encoded test pattern*. Given this equivalence between test patterns and encoded test patterns, we shall work with encoded test patterns, instead of test patterns, throughout this paper; their representation is more succinct and convenient.

Suppose that T is a set of encoded test patterns that covers all the faults under the s-MAF model. Let A be a $|T|$ by N array so that each row of A is an encoded test pattern of T. Then A has the following properties:

(1) if $N \geq 2s+1$, then in any $2s+1$ consecutive columns of A, each of the four vectors (the *in-between patterns*)

$$\underbrace{\begin{matrix} + & \cdots & + \\ - & \cdots & - \\ + & \cdots & + \\ - & \cdots & - \end{matrix}}_{s} \quad \begin{matrix} 0 \\ 1 \\ - \\ + \end{matrix} \quad \underbrace{\begin{matrix} + & \cdots & + \\ - & \cdots & - \\ + & \cdots & + \\ - & \cdots & - \end{matrix}}_{s}$$

appears in a row at least once;

(2) for $0 \leq t < s$, in the first $\min\{s+t+1, N\}$ columns of A, each of the four vectors (the *start patterns*)

$$\underbrace{\begin{matrix} + & \cdots & + \\ - & \cdots & - \\ + & \cdots & + \\ - & \cdots & - \end{matrix}}_{t} \quad \begin{matrix} 0 \\ 1 \\ - \\ + \end{matrix} \quad \underbrace{\begin{matrix} + & \cdots & + \\ - & \cdots & - \\ + & \cdots & + \\ - & \cdots & - \end{matrix}}_{\min\{s, N-t-1\}}$$

appears in a row at least once (ensuring that each wire with fewer than s neighbours to its left gets tested); and

(3) for $0 \leq t < s$, in the last $\min\{s+t+1, N\}$ columns of A, each of the four vectors (the *end patterns*)

$$\underbrace{\begin{matrix} + & \cdots & + \\ - & \cdots & - \\ + & \cdots & + \\ - & \cdots & - \end{matrix}}_{\min\{s, N-t-1\}} \quad \begin{matrix} 0 \\ 1 \\ - \\ + \end{matrix} \quad \underbrace{\begin{matrix} + & \cdots & + \\ - & \cdots & - \\ + & \cdots & + \\ - & \cdots & - \end{matrix}}_{t}$$

appears in a row at least once (ensuring that each wire with fewer than s neighbours to its right gets tested).

We call an array that satisfies these three properties a *crosstalk test array of length N and strength s*, and denote it by $\mathrm{CTA}(s, N)$. The number of rows of a $\mathrm{CTA}(s, N)$ is called its *size*. Our interest is in determining $\mathrm{CTA}(s, N)$ of small size, since the size corresponds to the number of encoded test patterns required.

The minimum size of a $\mathrm{CTA}(s, N)$ is denoted $\mathsf{C}(s, N)$. A $\mathrm{CTA}(s, N)$ is *optimal* if it has size $\mathsf{C}(s, N)$.

3. Optimal Crosstalk Test Arrays

We begin by establishing a lower bound on $\mathsf{C}(s,N)$.

Lemma 3.1.

$$\mathsf{C}(s,N) \geq \begin{cases} 4N & \textit{if } N \leq s+1; \\ 4(s+1) & \textit{if } N \geq s+2. \end{cases}$$

Proof. There are at exactly $4(s+1)$ distinct start patterns, and any CTA(s,N) must contain each of them as rows. So $\mathsf{C}(s,N) \geq 4(s+1)$.

When $N \leq s+1$, we have $4(s+1) \geq 4N$, giving $\mathsf{C}(s,N) \geq 4N$. □

We now describe a construction for crosstalk test arrays that meet the lower bound of Lemma 3.1.

Define the following vectors in $\{0,1,+,-\}^{s+1}$:

$$\begin{array}{rl} \mathsf{u}_{\mathrm{g_p}}(s) = & 0 \quad \underbrace{\begin{array}{cccc} + & + & \cdots & + \\ - & - & \cdots & - \\ + & + & \cdots & + \\ - & - & \cdots & - \end{array}}_{s} \\ \mathsf{u}_{\mathrm{g_n}}(s) = & 1 \\ \mathsf{u}_{\mathrm{d_f}}(s) = & - \\ \mathsf{u}_{\mathrm{d_r}}(s) = & + \end{array}$$

Let $\mathsf{M}_{\mathrm{g_p}}(s)$, $\mathsf{M}_{\mathrm{g_n}}(s)$, $\mathsf{M}_{\mathrm{d_f}}(s)$, and $\mathsf{M}_{\mathrm{d_r}}(s)$ be $(s+1)\times(s+1)$ circulant matrices whose first rows are given by $\mathsf{u}_{\mathrm{g_p}}(s)$, $\mathsf{u}_{\mathrm{g_n}}(s)$, $\mathsf{u}_{\mathrm{d_f}}(s)$, and $\mathsf{u}_{\mathrm{d_r}}(s)$, respectively.

For nonnegative integers s and t, now let

$$\mathsf{A}(s,t) = \underbrace{\begin{bmatrix} \mathsf{M}_{\mathrm{g_p}}(s) & \cdots & \mathsf{M}_{\mathrm{g_p}}(s) \\ \mathsf{M}_{\mathrm{g_n}}(s) & \cdots & \mathsf{M}_{\mathrm{g_n}}(s) \\ \mathsf{M}_{\mathrm{d_f}}(s) & \cdots & \mathsf{M}_{\mathrm{d_f}}(s) \\ \mathsf{M}_{\mathrm{d_r}}(s) & \cdots & \mathsf{M}_{\mathrm{d_r}}(s) \end{bmatrix}}_{t}.$$

Then $\mathsf{A}(s,t)$ is a $4(s+1)\times t(s+1)$ array over $\{0,1,+,-\}$ (see Fig. 1 for an example of $\mathsf{A}(3,8)$).

It is easy to verify that $\mathsf{A}(s,t)$ is a CTA$(s,t(s+1))$ of size $4(s+1)$ for any positive integer t.

Given $N \geq s+2$, let $t = \lceil N/(s+1) \rceil$. Then $\mathsf{A}(s,t)$ is a CTA$(s,t(s+1))$ of size $4(s+1)$ with at least N columns. Removing the last $t(s+1)-N$ columns from $\mathsf{A}(s,t)$ gives a CTA(s,N) of size $4(s+1)$. When $N \leq s+1$, $\mathsf{A}(N-1,1)$ is a CTA(s,N) of size $4N$. Consequently, we have the following:

Theorem 3.1.

$$\mathsf{C}(s,N)=\begin{cases}4N & \textit{if } N\le s+1;\\ 4(s+1) & \textit{if } N\ge s+2.\end{cases}$$

Corollary 3.1. *Testing for crosstalk-induced faults on an N-bit bus under the s-MAF model can be accomplished with an optimal number of* $\min\{4(s+1),4N\}$ *test patterns. These test patterns are explicitly given by the rows of an appropriately column-truncated* $\mathsf{A}(s,\lceil N/(s+1)\rceil)$.

4. Implementation

Our test methodology follows largely that of Chen *et al.* [3], and focuses on testing for crosstalk-induced faults on inter-core data and address buses, as these are often long and wide and consequently most susceptible to crosstalk defects. Using this methodology, an embedded processor executes a self-test program, with which test patterns $\mathsf{u}\to\mathsf{v}$ (according to the optimal CTA(s,N) constructed in Section 3) are applied to the bus under test. In the presence of crosstalk-induced faults, v becomes distorted at the receiver end of the bus. The processor then stores this error effect to memory as a test response, which can later be unloaded for external analysis. The self-test program for our optimal test patterns under the s-MAF model has very low implementation and timing complexity due to the simple explicit circulant structure of the optimal CTA(s,N) we constructed.

Previous work have all focused on observing error effects on one victim per test pattern. Our test methodology under the s-MAF model allows us to observe multiple victims per test pattern. We describe how this can be done for inter-core data and address buses in the subsections below.

4.1. *Testing Data Buses*

Testing data buses is straightforward. To apply a test pattern $\mathsf{u}\to\mathsf{v}$ to a data bus from a core to the processor, the processor first exchanges u with the core. The processor then requests v from the core. Upon receiving v, the processor writes v to memory for later analysis. Since data buses are bidirectional, crosstalk effects vary as the bus is driven from different directions, so we also need to test the reverse direction. This can be done by having the processor send v to the core after exchanging u.

Identifying bus lines that are victims to crosstalk effects can be done as follows. Suppose that a test pattern $\mathsf{u}\to\mathsf{v}$ is applied and that v' is the final received vector written to memory. Let $x=\mathsf{v}\oplus\mathsf{v}'$, where $\oplus$ is the bitwise

XOR operator. Then the set of values of i for which $\mathsf{x}_i = 1$ indicates the lines that are victims.

4.2. *Testing Address Buses*

To apply a test pattern $\mathsf{u} \rightarrow \mathsf{v}$ to an address bus from the processor to a core, the processor first requests data from address u and v in two consecutive cycles. Since the processor addresses the cores via memory-mapped I/O, the processor will receive data from a different memory address or a different core if v is distorted to v' by crosstalk effects. To be able to identify this fault, the data stored in address v and the data stored in *all possible* v' must be different. Under the s-MAF model, our optimal test patterns can each observe $t = \lceil N/(s+1) \rceil$ victims. So there are altogether 2^t memory addresses, where we must store different values in order for us to identify which lines are victims to crosstalk-induced faults.

5. Significance

We have considered a generalized MAF model and constructed optimal test patterns under this model. We have also seen that our test patterns can be generated with a low-complexity program and results of tests are readily analyzed for victim wires. It remains to show that the s-MAF model, under which we achieved all these results are realistic. This has been done by Sirisaengtaksin and Gupta [12]. They observed that the contribution of lines to crosstalk effects at the victim line decreases as their distances from the victim increases. In particular, they obtained the simulation results in Table I for a 10000 micron long five-bit bus, where a *level j line* is one where there are j lines separating it from the victim line. Wires beyond level two contribute very little to crosstalk effects on the victim wire. In fact, based on these results, Sirisaengtaksin and Gupta suggested that if acceptable error criterion is set to a few percent, then the supporting lines at levels greater than two can be ignored for 0.18 micron CMOS technology.

If a similar error criterion is used, we need only test under the 3-MAF model. This can represent a huge savings in the cost of testing an N-bit bus, since $\mathsf{C}(3, N) = 16$ test patterns suffice to cover all crosstalk-induced faults under the 3-MAF model, regardless of how big N is. For example, under the MAF model, we require $4 \cdot 32 = 128$ test patterns to cover all crosstalk-induced faults in a 32-bit bus, whereas our test methodology requires only 16 test patterns, given by the rows of the array in Fig. 2. This is a 87.5% savings in the number of test patterns required. For a 64-bit bus, our test

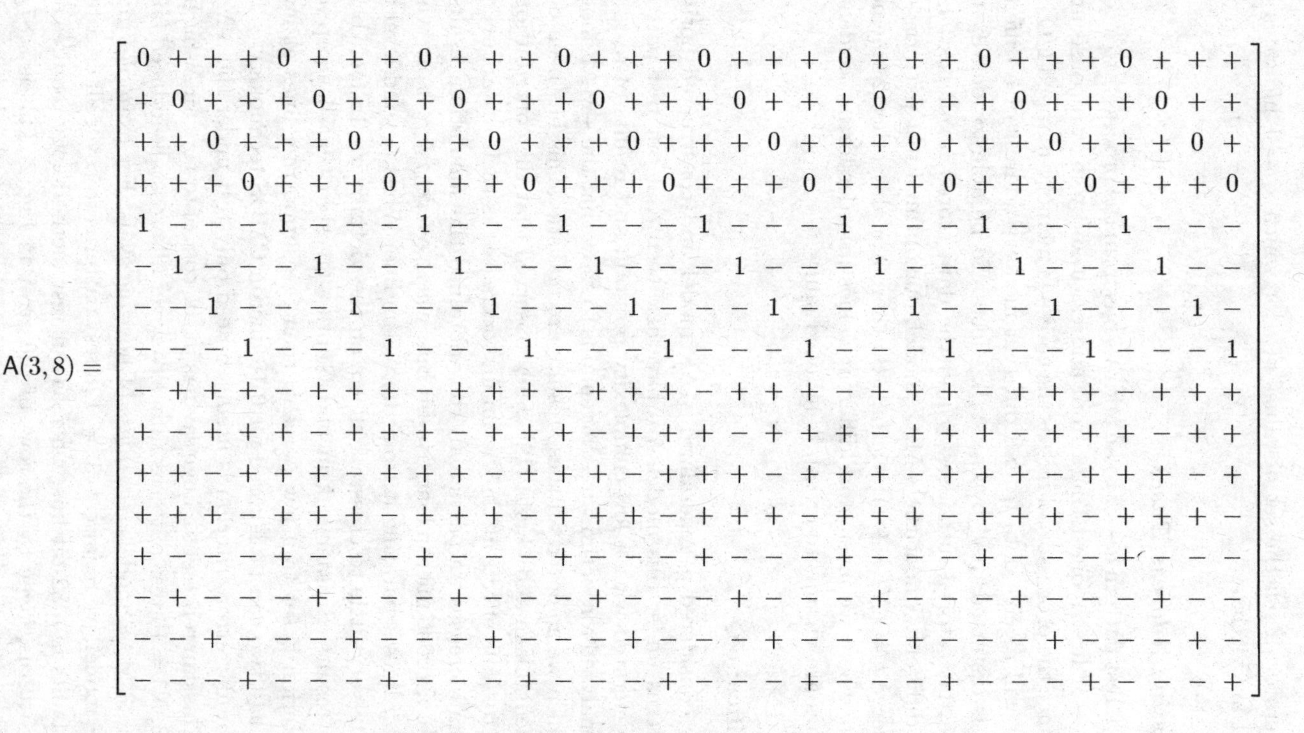

$$\mathsf{A}(3,8) = \left[\begin{array}{cccccccccccccccccccccccccccccccc}
0&+&+&+&0&+&+&+&0&+&+&+&0&+&+&+&0&+&+&+&0&+&+&+&0&+&+&+&0&+&+&+\\
+&0&+&+&+&0&+&+&+&0&+&+&+&0&+&+&+&0&+&+&+&0&+&+&+&0&+&+&+&0&+&+\\
+&+&0&+&+&+&0&+&+&+&0&+&+&+&0&+&+&+&0&+&+&+&0&+&+&+&0&+&+&+&0&+\\
+&+&+&0&+&+&+&0&+&+&+&0&+&+&+&0&+&+&+&0&+&+&+&0&+&+&+&0&+&+&+&0\\
1&-&-&-&1&-&-&-&1&-&-&-&1&-&-&-&1&-&-&-&1&-&-&-&1&-&-&-&1&-&-&-\\
-&1&-&-&-&1&-&-&-&1&-&-&-&1&-&-&-&1&-&-&-&1&-&-&-&1&-&-&-&1&-&-\\
-&-&1&-&-&-&1&-&-&-&1&-&-&-&1&-&-&-&1&-&-&-&1&-&-&-&1&-&-&-&1&-\\
-&-&-&1&-&-&-&1&-&-&-&1&-&-&-&1&-&-&-&1&-&-&-&1&-&-&-&1&-&-&-&1\\
-&+&+&+&-&+&+&+&-&+&+&+&-&+&+&+&-&+&+&+&-&+&+&+&-&+&+&+&-&+&+&+\\
+&-&+&+&+&-&+&+&+&-&+&+&+&-&+&+&+&-&+&+&+&-&+&+&+&-&+&+&+&-&+&+\\
+&+&-&+&+&+&-&+&+&+&-&+&+&+&-&+&+&+&-&+&+&+&-&+&+&+&-&+&+&+&-&+\\
+&+&+&-&+&+&+&-&+&+&+&-&+&+&+&-&+&+&+&-&+&+&+&-&+&+&+&-&+&+&+&-\\
+&-&-&-&+&-&-&-&+&-&-&-&+&-&-&-&+&-&-&-&+&-&-&-&+&-&-&-&+&-&-&-\\
-&+&-&-&-&+&-&-&-&+&-&-&-&+&-&-&-&+&-&-&-&+&-&-&-&+&-&-&-&+&-&-\\
-&-&+&-&-&-&+&-&-&-&+&-&-&-&+&-&-&-&+&-&-&-&+&-&-&-&+&-&-&-&+&-\\
-&-&-&+&-&-&-&+&-&-&-&+&-&-&-&+&-&-&-&+&-&-&-&+&-&-&-&+&-&-&-&+
\end{array}\right]$$

Fig. 2. The array A(3, 8), which is a CTA(3, 32).

methodology under the 3-MAF model results in over 93% savings.

Crosstalk effect	Level 1 line	Level 2 line	Level 3 line
Positive glitch	42.26%	10.87%	0.65%
Negative glitch	36.74%	15.28%	0.99%
Falling delay	25.08%	4.51%	1.55%
Rising delay	96.47%	35.69%	0.88%

The simulation results show that if we are willing to accept a few percent of error in testing, then crosstalk effects are a localized property. This allows our test methodology to scale regardless of the width of the bus under test. We do need, however, to adjust s according to feature size. As feature size reduces, the number of wires in close proximity to a victim increases, so s has to be increased. However, the simulation results of Table I shows that the contribution of these wires to crosstalk effects on the victim decreases rather rapidly.

6. Conclusion

In this paper, we considered a more general model for crosstalk-induced faults that allows for tradeoffs between test efficiency and test quality. Under this model, we constructed provably optimal test patterns for covering all crosstalk-induced faults. Our test patterns admit simple generators for BIST and can be used to obtain multitudes of savings in terms of test cost, by allowing only a few percent of error.

Acknowledgments

The research of Y. M. Chee is supported in part by the Singapore Ministry of Education under Research Grant T206B2204.

References

1. X. Bai, S. Dey, and J. Rajski. Self-test methodology for at-speed test of crosstalk in chip interconnects. In *DAC 2000: Proceedings of the 37th Annual ACM IEEE Design Automation Conference*, pages 619–624, New York, NY, USA, 2000. ACM Press.

2. M. A. Breuer, C. Gleason, and S. Gupta. New validation and test problems for high performance deep sub-micron VLSI circuits. In *Tutorial Notes, VTS 1997: the 15th IEEE VLSI Test Symposium*, 1997.
3. L. Chen, X. Bai, and S. Dey. Testing for interconnect crosstalk defects using on-chip embedded processor cores. *J. Electron. Test.*, 18(4-5):529–538, 2002.
4. W. Chen, M. A. Breuer, and S. K. Gupta. Analytic models for crosstalk delay and pulse analysis under non-ideal inputs. In *Proceedings of the IEEE International Test Conference*, pages 809–818, Washington, DC, USA, 1997. IEEE Computer Society.
5. W. Y. Chen, S. K. Gupta, and M. A. Breuer. Analytical models for crosstalk excitation and propagation in VLSI circuits. *IEEE Trans. Comput. Aided Design Integr. Circuits Syst.*, 21(10):1117–1131, 2002.
6. K.-T. Cheng, M. R. S. Dey, and K. Roy. Test challenges for deep sub-micron technologies. In *DAC 2000: Proceedings of the 37th Annual ACM IEEE Design Automation Conference*, pages 142–149, New York, NY, USA, 2000. ACM Press.
7. M. Cuviello, S. Dey, X. Bai, and Y. Zhao. Fault modeling and simulation for crosstalk in system-on-chip interconnects. In *ICCAD 1999: Proceedings of the 1999 IEEE/ACM International Conference on Computer-Aided Design*, pages 297–303. IEEE Press, 1999.
8. K. T. Lee, C. Nordquist, and J. A. Abraham. Automatic test pattern generation for crosstalk glitches in digital circuits. In *VTS 1998: Proceedings of the 16th IEEE VLSI Test Symposium*, page 34, Washington, DC, USA, 1998. IEEE Computer Society.
9. P. Nordholz, D. Treytnar, J. Otterstedt, H. Grabinski, D. Niggemeyer, and T. W. Williams. Signal integrity problems in deep submicron arising from interconnects between cores. In *VTS 1998: Proceedings of the 16th IEEE VLSI Test Symposium*, pages 28–33, Washington, DC, USA, 1998. IEEE Computer Society.
10. K. Rahmat, J. Neves, and J. Lee. Methods for calculating coupling noise in early design: a comparative analysis. In *ICCD 1998: Proceedings of the IEEE International Conference on Computer Design*, pages 76–81, Los Alamitos, CA, USA, 1998. IEEE Computer Society.
11. K. L. Shepard. Design methodologies for noise in digital integrated circuits. In *DAC 1998: Proceedings of the 35th Annual ACM IEEE Design Automation Conference*, pages 94–99, New York, NY, USA, 1998. ACM Press.
12. W. Sirisaengtaksin and S. K. Gupta. Enhanced crosstalk fault model and methodology to generate tests for arbitrary inter- core interconnect topology. In *ATS 2002: Proceedings of the 11th Asian Test Symposium*, volume 00, page 163, Los Alamitos, CA, USA, 2002. IEEE Computer Society.
13. A. Vittal and M. Marek-Sadowska. Crosstalk reduction for VLSI. *IEEE Trans. Comput. Aided Design Integr. Circuits Syst.*, 16(3):290–298, 1997.
14. H. Zhou and D. F. Wang. Global routing with crosstalk constraints. In *DAC 1998: Proceedings of the 35th Annual ACM IEEE Design Automation Conference*, pages 374–377, New York, NY, USA, 1998. ACM Press.

An Improved Distinguisher for Dragon

Joo Yeon Cho and Josef Pieprzyk

Centre for Advanced Computing – Algorithms and Cryptography,
Department of Computing, Macquarie University,
NSW, Australia, 2109
Email: {jcho,josef}@ics.mq.edu.au

The Dragon stream cipher is one of the focus ciphers which have reached Phase 2 of the eSTREAM project. In this paper, we present a new method of building a linear distinguisher for Dragon. The distinguisher is constructed by exploiting the biases of two S-boxes and the modular addition which are basic components of the nonlinear function F. The bias of the distinguisher is estimated to be around $2^{-75.32}$ which is better than the bias of the distinguisher presented by Englund and Maximov. We have shown that Dragon is distinguishable from a random cipher by using around $2^{150.6}$ keystream words and 2^{64} memory. In addition, we present a very efficient algorithm for computing the bias of linear approximation of modular addition.

Keywords: Stream Ciphers, eSTREAM, Dragon, Distinguishing Attacks, Modular Addition.

1. Introduction

Dragon [1,2] is a word-oriented stream cipher submitted to the eSTREAM project [3]. Dragon is one of the focus ciphers (software category) which are included in Phase 3 of the eSTREAM. During Phase 1, Englund and Maximov presented a distinguishing attack against Dragon [4]. Their distinguisher is constructed using around 2^{155} keystream words and 2^{96} memory.

Unlike Englund and Maximov's work, we use a different approach to find a more efficient distinguisher. In a nut shell, we first derive linear approximations for the basic nonlinear blocks used in the cipher, namely, for the S-boxes and for modular additions. Next we combine those approximations and build a linear approximation for the whole state update function F. While combining these elementary approximations, we use two basic operations that we call cutting and bypassing. The bypassing operation replaces the original component by its approximation. On the other hand, the cutting operation replaces the original component by zero. Then, we design the distinguisher by linking the approximation of the update function F with the observable output keystream for a specific sequence of clocks.

Building the best distinguisher is done in two steps. First, all linear masks for the internal approximations are assumed to be identical. Hence, the mask for distinguisher holding the biggest bias can be found efficiently. Next, the bias of the distinguisher is estimated more precisely by considering the dependencies among internal approximations. This is achieved by allowing the different internal approximation masks that are used in the distinguisher.

In result, the bias of our distinguisher is around $2^{-75.32}$ when 2^{64} bits of internal memory are guessed. Hence, we claim that Dragon is distinguishable from the random cipher after observing around $2^{150.6}$ words with 2^{64} memory for internal state guesses. So our distinguisher is better than the one presented in the paper [4]. Our distinguisher is also described explicitly by showing the best approximations of the nonlinear components of the cipher. In contrast, the previous best distinguishing attack by Englund and Maximov used a statistical argument to evaluate a bias of the function F.

This paper is organized as follows. Section 2 presents a brief description of Dragon. In Section 3, a collection of linear approximations of nonlinear components for Dragon is presented. Next, a distinguisher is built by combining the approximations. In Section 4, the distinguisher is improved by considering the dependencies of intermediate approximations. Section 5 concludes the work.

2. A brief description of Dragon

Dragon consists of a 1024-bit nonlinear feedback register, a nonlinear state update function, and a 64-bit internal memory M. Dragon uses two sizes of key and initialization vector that is either 128 or 256 bits and produces a 64-bit (two words) output per clock. The nonlinear state update function (the function F) takes six words (192 bits) as the input and produces six words (192 bits) as the output. Among the output words of the function F, two words are used as new state words and two words are produced as a keystream. The detail structure of the function F is displayed in Figure 1. Suppose that the 32-bit input x is split into four bytes, i.e. $x = x_0||x_1||x_2||x_3$, where x_i stands for a single byte and $||$ denotes a concatenation. The byte x_0 denotes the most significant byte and x_3 denotes the least significant byte. The functions G and H are components of the function F and are constructed from the two basic 8×32 S-boxes S_1 and

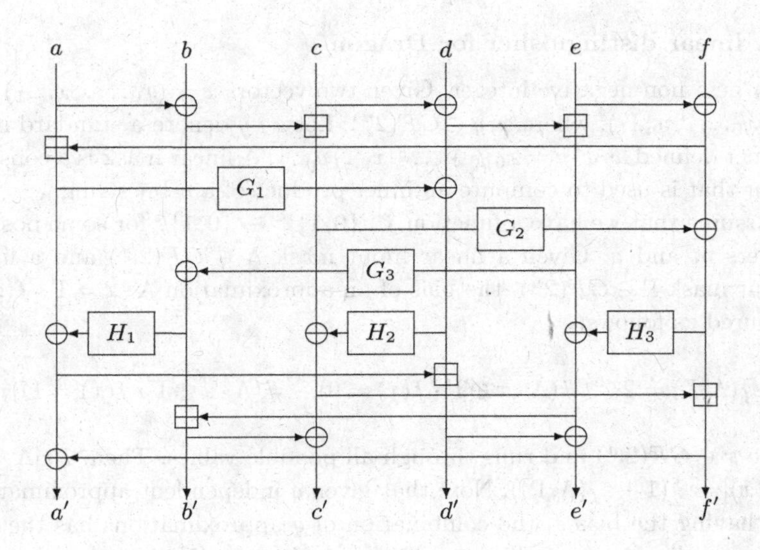

Fig. 1. *F* function

S_2 in the following way.

$$G_1(x) = S_1(x_0) \oplus S_1(x_1) \oplus S_1(x_2) \oplus S_2(x_3)$$
$$G_2(x) = S_1(x_0) \oplus S_1(x_1) \oplus S_2(x_2) \oplus S_1(x_3)$$
$$G_3(x) = S_1(x_0) \oplus S_2(x_1) \oplus S_1(x_2) \oplus S_1(x_3)$$
$$H_1(x) = S_2(x_0) \oplus S_2(x_1) \oplus S_2(x_2) \oplus S_1(x_3)$$
$$H_2(x) = S_2(x_0) \oplus S_2(x_1) \oplus S_1(x_2) \oplus S_2(x_3)$$
$$H_3(x) = S_2(x_0) \oplus S_1(x_1) \oplus S_2(x_2) \oplus S_2(x_3)$$

The keystream is generated as follows.

(1) Input : $\{B_0, B_1, \ldots, B_{31}\}$ and $M = (M_L||M_R)$, where M_L is a upper word and M_R is a lower word of M.
(2) Assume that $a = B_0, b = B_9, c = B_{16}, d = B_{19}, e = B_{30} \oplus M_L, f = B_{31} \oplus M_R$, where $M = M_R||M_L$.
(3) Compute $(a', b', c', d', e', f') = F(a, b, c, d, e, f)$.
(4) Update the state $B_0 = b', B_1 = c'$ and $B_i = B_{i-2}, 2 \leq i \leq 31, M = M + 1$.
(5) Output the keystream : $k = (a'||e')$.

For a detailed description of Dragon, we refer the reader to the paper [1].

3. A linear distinguisher for Dragon

Let n be a non-negative integer. Given two vectors $x = (x_0, \ldots, x_{n-1})$ and $y = (y_0, \ldots, y_{n-1})$, where $x, y \in GF(2^n)$. Let $x \cdot y$ denote a standard inner product defined as $x \cdot y = x_0 y_0 \oplus \ldots \oplus x_{n-1} y_{n-1}$. A linear mask is a constant vector that is used to compute an inner product of a n-bit string.

Assume that we have a function $f : \{0,1\}^m \rightarrow \{0,1\}^n$ for some positive integers m and n. Given a linear input mask $\Lambda \in GF(2^m)$ and a linear output mask $\Gamma \in GF(2^n)$, the bias of an approximation $\Lambda \cdot x = \Gamma \cdot f(x)$ is measured as follows:

$$\varepsilon_f(\Lambda, \Gamma) = 2^{-n}(\#(\Lambda \cdot x \oplus \Gamma \cdot f(x) = 0) - \#(\Lambda \cdot x \oplus \Gamma \cdot f(x) = 1)),$$

where $x \in GF(2^m)$ and runs through all possible values. Then, $Pr[\Lambda \cdot x = \Gamma \cdot f(x)] = \frac{1}{2}(1 + \varepsilon_f(\Lambda, \Gamma))$. Note that given q independent approximations each having the bias ε, the combination of q approximations has the bias of ε^q according to the well-known Piling-up Lemma [5].

3.1. *Approximations of functions G and H*

According to the structure of the functions G and H, the essential components of the functions G and H are the two S-boxes: S_1 and S_2. Hence, the linear approximations of the functions G and H can be constructed by combining approximations of S_1 and S_2 appropriately. In particular, for our distinguisher which will be described in the next subsection, we need special forms of approximations as displayed in Table 1. Note that the approximations of the function G use identical masks for both the input and output, while the function H uses an output mask only. The reason for this will be given in Subsection 3.3. The approximations of the form $\Gamma \cdot G(x) = \Gamma \cdot x$ are called **bypassing** ones, whereas the approximations of the form $\Gamma \cdot H(x) = 0$ are named **cutting** ones. Table 1 shows the examples of such approximations with high biases.

Table 1. Cutting and bypassing approximations of the function G and H

approximation	bias	example
$\Gamma \cdot H(x) = 0$	$\varepsilon_H(0, \Gamma)$	$\varepsilon_H(0, \text{0x4810812B}) = -2^{-7.16}$
$\Gamma \cdot x = \Gamma \cdot G_1(x)$	$\varepsilon_{G_1}(\Gamma, \Gamma)$	$\varepsilon_{G_1}(\text{0x09094102}, \text{0x09094102}) = -2^{-9.33}$
$\Gamma \cdot x = \Gamma \cdot G_2(x)$	$\varepsilon_{G_2}(\Gamma, \Gamma)$	$\varepsilon_{G_2}(\text{0x90904013}, \text{0x90904013}) = -2^{-9.81}$

3.1.1. *Approximations of the function H*

Assume that a 32-bit word x is a uniformly distributed random variable. If the word x is divided into four bytes so $x = x_0||x_1||x_2||x_3$, where x_i denotes the i-th byte of x, then the approximation $\Gamma \cdot H_1(x) = 0$ can be represented as

$$\Gamma \cdot H_1(x) = \Gamma \cdot S_2(x_0) \oplus \Gamma \cdot S_2(x_1) \oplus \Gamma \cdot S_2(x_2) \oplus \Gamma \cdot S_1(x_3) = 0.$$

Hence, the bias $\varepsilon_{H_1}(0, \Gamma)$ can be computed as

$$\varepsilon_{H_1}(0, \Gamma) = \varepsilon_{S_2}(0, \Gamma)^3 \times \varepsilon_{S_1}(0, \Gamma),$$

where $\varepsilon_{S_i}(0, \Gamma)$ denotes the bias of the approximation $\Gamma \cdot S_i(x_j) = 0$. Due to the structure of the function H, the approximations $\Gamma \cdot H_1(x) = 0, \Gamma \cdot H_2(x) = 0$ and $\Gamma \cdot H_3(x) = 0$ have identical biases when the input x is an independent random variable. Hence, $\varepsilon_{H_1}(0, \Gamma) = \varepsilon_{H_2}(0, \Gamma) = \varepsilon_{H_3}(0, \Gamma)$.

3.1.2. *Approximations of the function G*

A 32-bit word x is assumed to be a uniformly distributed random variable. If the word x is divided into four bytes such as $x = x_0||x_1||x_2||x_3$, and a mask Γ is divided into four submasks such that $\Gamma = \Gamma_0||\Gamma_1||\Gamma_2||\Gamma_3$, where $\Gamma_i \in \{0,1\}^8$, then the approximation $\Gamma \cdot x = \Gamma \cdot G(x)$ can be split into

$$\begin{aligned}\Gamma \cdot (x \oplus G_1(x)) = (\Gamma_0 \cdot x_0 \oplus \Gamma \cdot S_1(x_0)) \oplus (\Gamma_1 \cdot x_1 \oplus \Gamma \cdot S_1(x_1)) \\ \oplus (\Gamma_2 \cdot x_2 \oplus \Gamma \cdot S_1(x_2)) \oplus (\Gamma_3 \cdot x_3 \oplus \Gamma \cdot S_2(x_3)) = 0\end{aligned}$$

Hence, the bias $\varepsilon_G(\Gamma, \Gamma)$ can be computed as follows

$$\varepsilon_G(\Gamma, \Gamma) = \varepsilon_{S_1(x_0)}(\Gamma_0, \Gamma) \times \varepsilon_{S_1(x_1)}(\Gamma_1, \Gamma) \times \varepsilon_{S_1(x_2)}(\Gamma_2, \Gamma) \times \varepsilon_{S_2(x_3)}(\Gamma_3, \Gamma),$$

where $\varepsilon_{S_i(x_j)}(\Gamma_j, \Gamma)$ denotes the bias of the approximation $\Gamma_j \cdot x_j \oplus \Gamma \cdot S_i(x_j) = 0$.

3.2. *Linear approximations of modular addition*

Let x and y be uniformly distributed random vectors, where $x, y \in GF(2^n)$ for a positive n. Given a mask $\Gamma \in GF(2^n)$ that is used for both the input and output, a linear approximation of modular addition where an input and an output masks are Γ is defined as follows:

$$Pr[\Gamma \cdot (x \boxplus y) = \Gamma \cdot (x \oplus y)] = \frac{1}{2}(1 + \varepsilon_+(\Gamma, \Gamma)), \quad (1)$$

where the bias of the approximation is denoted by $\varepsilon_+(\Gamma, \Gamma)$. Also, given a vector x, the Hamming weight of x is defined as the number of nonzero coordinates of x.

Theorem 3.1. *Let n and m be positive integers. Given a linear mask $\Gamma = (\gamma_{n-1}, \cdots, \gamma_0)$, where $\gamma_i \in \{0,1\}$, we assume that the Hamming weight of Γ is m. If a vector $W_\Gamma = (w_{m-1}, w_{m-2} \cdots, w_0)$ denotes the bit positions of Γ, where $\gamma_i = 1$ and $w_{m-1} > \cdots > w_0$, then a bias $\varepsilon_+(\Gamma, \Gamma)$ is determined as follows.*

If m is even, then,

$$\varepsilon_+(\Gamma, \Gamma) = 2^{-d_1}, \quad \textit{where } d_1 = \sum_{i=0}^{m/2-1} (w_{2i+1} - w_{2i}), \tag{2}$$

If m is odd, then

$$\varepsilon_+(\Gamma, \Gamma) = 2^{-d_2}, \quad \textit{where } d_2 = \sum_{i=1}^{(m-1)/2} (w_{2i} - w_{2i-1}) + w_0. \tag{3}$$

Proof. See Appendix A. □

For example, if Γ = 0x0600018D, the Hamming weight of the mask Γ is 7 and $W_\Gamma = (26, 25, 8, 7, 3, 2, 0)$. Hence, the bias $\varepsilon_+(\Gamma, \Gamma) = 2^{-[(26-25)+(8-7)+(3-2)]} = 2^{-3}$.

Corollary 3.1. *Let m be a positive integer. Given a mask Γ whose Hamming weight is m, the approximation $\Gamma \cdot (x \boxplus y) = \Gamma \cdot (x \oplus y)$ has at most a bias of $2^{-(m-1)/2}$.*

Proof. See Appendix B. □

3.3. *Linear approximation of the function F*

According to the state update rule of Dragon, the following relation between two state words at the clocks t and $t+15$ holds [a]

$$B_0[t] = B_{30}[t+15], \quad t = 0, 1, \ldots. \tag{4}$$

We know that $a = B_0$ and $e = B_{30} \oplus M_L$, where a and e are two words of the function F. Then, we try to find the linear approximations $\Gamma \cdot a' = \Gamma \cdot a$

[a] This relation was also observed in [4].

and $\Gamma \cdot e' = \Gamma \cdot e$, where a' and e' are two output words of the function F that are produced as keystream.

We regard the outputs of the functions G and H as independent and uniformly distributed random variables. This assumption is reasonable since each G and H functions have unique input parameters so that the output of the functions G and H are mutually independent. Hence, the functions G and H can be described without input parameters as shown below.

3.3.1. *The approximation of* a'

As illustrated in Figure 1, an output word a' is expressed by the following relation

$$a' = [(a \boxplus (e \oplus f)) \oplus H_1] \oplus [(e \oplus f \oplus G_2) \boxplus (H_2 \oplus ((a \oplus b) \boxplus c))]. \quad (5)$$

Due to the linear property of Γ, we know that

$$\Gamma \cdot a' = \Gamma \cdot [(a \boxplus (e \oplus f)) \oplus H_1] \oplus \Gamma \cdot [(e \oplus f \oplus G_2) \boxplus (H_2 \oplus ((a \oplus b) \boxplus c))].$$

By applying Approximation (1), we get

$$\Gamma \cdot [(e \oplus f \oplus G_2) \boxplus (H_2 \oplus ((a \oplus b) \boxplus c))] = \Gamma \cdot (e \oplus f \oplus G_2) \oplus \Gamma \cdot [(H_2 \oplus ((a \oplus b) \boxplus c))],$$

which holds with the bias of $\varepsilon_+(\Gamma, \Gamma)$. Hence, we have

$$\Gamma \cdot a' = \Gamma \cdot [(a \boxplus (e \oplus f)) \oplus H_1] \oplus \Gamma \cdot (e \oplus f \oplus G_2) \oplus \Gamma \cdot [H_2 \oplus ((a \oplus b) \boxplus c)].$$

Next, the two types of approximations are used in our analysis. First, cutting approximations are used for the functions H_1 and H_2. That is, we use $\Gamma \cdot H_1 = 0$ and $\Gamma \cdot H_2 = 0$, which hold with the biases of $\varepsilon_{H_1}(0, \Gamma)$ and $\varepsilon_{H_2}(0, \Gamma)$, respectively. Intuitively, these approximations allow to simplify the form of the final approximation of the function F by replacing the output variables of a nonlinear component by zeros.

Second, bypassing approximations are used for the function G_2. That is, we use $\Gamma \cdot G_2 = \Gamma \cdot [(a \oplus b) \boxplus c]$ that has a bias $\varepsilon_{G_2}(\Gamma, \Gamma)$. In this category of approximations we are able to replace a combination of output variables by a combination of input variables. Then, we can write that

$$\begin{aligned} \Gamma \cdot a' &= \Gamma \cdot [(a \boxplus (e \oplus f))] \oplus \Gamma \cdot (e \oplus f \oplus [(a \oplus b) \boxplus c]) \oplus \Gamma \cdot [(a \oplus b) \boxplus c] \\ &= \Gamma \cdot [(a \boxplus (e \oplus f))] \oplus \Gamma \cdot (e \oplus f). \end{aligned}$$

Finally, by applying Approximation (1) for the modular addition, we obtain

$$\Gamma \cdot a' = \Gamma \cdot a. \quad (6)$$

We know that $\Gamma \cdot [(a \boxplus (e \oplus f))] = \Gamma \cdot a \oplus \Gamma \cdot (e \oplus f)$ holds with the bias of $\varepsilon_+(\Gamma, \Gamma)$. Therefore, the bias of Approximation (6) can be computed from the biases of the component approximations as follows:

$$\varepsilon_{a'}(\Gamma, \Gamma) = \varepsilon_+(\Gamma, \Gamma)^2 \times \varepsilon_{H_1}(0, \Gamma) \times \varepsilon_{H_2}(0, \Gamma) \times \varepsilon_{G_2}(\Gamma, \Gamma).$$

Since the 32-bit word a' is an upper part of a 64-bit keystream output at each clock, Approximation (6) is equivalent to the following expression

$$\Gamma \cdot k_0[t] = \Gamma \cdot B_0[t], \tag{7}$$

where $k_0[t]$ denotes the upper part of a 64-bit k at clock t.

3.3.2. *The approximation of* e'

As depicted in Figure 1, an output word e' can be described as

$$e' = [((a \boxplus (e \oplus f)) \oplus H_1) \boxplus (c \oplus d \oplus G_1)] \oplus [H_3 \oplus ((c \oplus d) \boxplus e)]. \tag{8}$$

Similarly to the case of a', we would like to obtain an approximation $\Gamma \cdot e' = \Gamma \cdot e$. To do this, we first apply Approximation (1) for modular addition and as the result we get

$$\Gamma \cdot e' = \Gamma \cdot [(a \boxplus (e \oplus f)) \oplus H_1] \oplus \Gamma \cdot (c \oplus d \oplus G_1) \oplus \Gamma \cdot [H_3 \oplus ((c \oplus d) \boxplus e)].$$

Next, we apply the cutting approximations for functions H_1, H_3 and the bypassing approximation for the function G_1. That is, we use the following approximations

$$\Gamma \cdot H_1 = 0, \quad \Gamma \cdot H_3 = 0, \quad \Gamma \cdot G_1 = \Gamma \cdot [a \boxplus (e \oplus f)]$$

that hold with the biases $\varepsilon_{H_1}(0, \Gamma), \varepsilon_{H_3}(0, \Gamma)$ and $\varepsilon_{G_1}(\Gamma, \Gamma)$, respectively. These approximations are plugged into the above relation and we obtain the following result

$$\begin{aligned}\Gamma \cdot e' &= \Gamma \cdot [(a \boxplus (e \oplus f))] \oplus \Gamma \cdot (c \oplus d \oplus [a \boxplus (e \oplus f)]) \oplus \Gamma \cdot [(c \oplus d) \boxplus e] \\ &= \Gamma \cdot (c \oplus d) \oplus \Gamma \cdot [(c \oplus d) \boxplus e].\end{aligned}$$

Finally, by applying Approximation (1) for modular addition, we can conclude that output e' and input e satisfy the following approximation

$$\Gamma \cdot e' = \Gamma \cdot e \tag{9}$$

with the bias $\varepsilon_{e'}(\Gamma, \Gamma) = \varepsilon_+(\Gamma, \Gamma)^2 \times \varepsilon_{H_1}(0, \Gamma) \times \varepsilon_{H_3}(0, \Gamma) \times \varepsilon_{G_1}(\Gamma, \Gamma)$. Since the 32-bit word e' is a lower part of a 64-bit keystream output k at each clock, Approximation (9) is equivalent to the following expression

$$\Gamma \cdot k_1[t] = \Gamma \cdot (B_{30}[t] \oplus M_L[t]), \tag{10}$$

where $k_1[t]$ and $M_L[t]$ denote the lower part of a 64-bit k and the upper part of a 64-bit memory word M at clock t, respectively.

3.4. *Building the distinguisher*

According to Equation (4), Approximations (7) and (10) can be combined in such a way that

$$\Gamma \cdot k_0[t] = \Gamma \cdot B_0[t] = \Gamma \cdot B_{30}[t+15] = \Gamma \cdot (k_1[t+15] \oplus M_L[t+15]).$$

By guessing (partially) the initial value of M, we can build the following distinguisher

$$\Gamma \cdot (k_0[t] \oplus k_1[t+15]) = \Gamma \cdot M_L[t+15]. \tag{11}$$

For the correctly guessed initial value of M, the distinguisher (11) shows the bias

$$\begin{aligned}\varepsilon_D(\Gamma,\Gamma) &= \varepsilon_{a'}(\Gamma,\Gamma) \times \varepsilon_{e'}(\Gamma,\Gamma) \\ &= \varepsilon_+(\Gamma,\Gamma)^4 \times \varepsilon_{H_1}(0,\Gamma)^2 \times \varepsilon_{H_2}(0,\Gamma) \times \varepsilon_{H_3}(0,\Gamma) \times \varepsilon_{G_1}(\Gamma,\Gamma) \times \varepsilon_{G_2}(\Gamma,\Gamma)\end{aligned} \tag{12}$$

We implemented a mask search for the function F to achieve the distinguisher with the biggest bias. The space of a linear mask Γ contains $2^{32}-1$ elements. For each mask Γ, the following procedure is performed to compute the bias given by Expression (12).

Step 1. For an input x that varies from 0 to 255, measure the biases of $\Gamma \cdot S_1(x) = 0$ and $\Gamma \cdot S_2(x) = 0$, respectively. Then, compute $\varepsilon_{H_1}(0,\Gamma)$, $\varepsilon_{H_2}(0,\Gamma)$ and $\varepsilon_{H_3}(0,\Gamma)$.
Step 2. The mask Γ is divided into four submasks $\Gamma = \Gamma_0||\Gamma_1||\Gamma_2||\Gamma_3$. For an input x that varies from 0 to 255, measure the bias of $\Gamma \cdot S_1(x) = \Gamma_i \cdot x$ and $\Gamma \cdot S_2(x) = \Gamma_i \cdot x$ for some $0 \le i \le 3$. Then, compute the biases $\varepsilon_{G_1}(\Gamma,\Gamma)$ and $\varepsilon_{G_2}(\Gamma,\Gamma)$.
Step 3. Determine the bias $\varepsilon_+(\Gamma,\Gamma)$ using Theorem 1.
Step 4. Finally, compute $\varepsilon_D(\Gamma,\Gamma)$.

3.5. *Our results*

We searched for a linear mask that maximizes the bias (12). Due to Corollary 3.1, the bias $\varepsilon_+(\Gamma,\Gamma)$ decreases exponentially as long as the Hamming weight of a linear mask increases. Hence, there is a better chance to achieve higher bias when the Hamming weight is smaller.

We found that the best linear approximation of the function F is using Equation (11) with the mask $\Gamma = 0x0600018D$. The bias of the distinguisher in this case is $2^{-75.8}$ as listed in Table 2. In order to remove the impact of the unknown state of the internal memory on the bias, we need to guess the first 27 bits of initial value of M_L and 32 bits of M_R. Hence, we need to store all possible values of the internal state which takes $2^{27+32} = 2^{59}$ bits.

Table 2. The bias of distinguisher

Γ	$\varepsilon_+(0,\Gamma)$	$\varepsilon_H(\Gamma,\Gamma)$	$\varepsilon_{G_1}(\Gamma,\Gamma)$	$\varepsilon_{G_2}(\Gamma,\Gamma)$	$\varepsilon_{a'}(\Gamma,\Gamma)$	$\varepsilon_{e'}(\Gamma,\Gamma)$	$\varepsilon_D(\Gamma,\Gamma)$
0x0600018D	2^{-3}	$-2^{-8.58}$	$2^{-13.59}$	$2^{-15.91}$	$-2^{-39.1}$	$-2^{-36.7}$	$2^{-75.8}$

4. Improving the distinguisher

In this section, we generalize a method presented in Section 3. [b] First, we apply different linear masks for each component of the function F and combine them to build the distinguisher. Figure 2 illustrates how different linear masks can be applied for each component of the function F. Second, we consider the internal dependencies for the approximations of the function F. Since the approximations of the components of the F cancel each other, the bias of the distinguisher can be accurately computed by trying all possible internal approximations induced by different linear masks. Based on these two observations, we searched extensively for a new distinguisher that could improve the efficiency of our attack.

A new distinguisher can be built from the relation of (a, a') and (e, e') presented in Equations (5) and (8). A basic requirement for establishing a distinguisher is to apply the identical mask Φ to the state a at clock t and to the state e at clock $t + 15$, as stated in Section 3. However, this time, the other internal masks can be different, as shown in Figure 2. We set up the six masks, $\{\Lambda_1, \cdots, \Lambda_6\}$, for the components of the F and build the following approximations:

$$\begin{aligned}\Lambda_2 \cdot a' &= \Lambda_2 \cdot [(a \boxplus (e \oplus f)) \oplus H_1] \oplus \Lambda_2 \cdot [(e \oplus f \oplus G_2) \boxplus (H_2 \oplus ((a \oplus b) \boxplus c))] \\ &= \Phi \cdot a \oplus \Lambda_1 \cdot (e \oplus f) \oplus \Lambda_2 \cdot H_1 \oplus \Lambda_1 \cdot (e \oplus f \oplus G_2) \oplus \Lambda_3 \\ &\quad \cdot [H_2 \oplus ((a \oplus b) \boxplus c)] \\ &= \Phi \cdot a \end{aligned} \tag{13}$$

[b]This section was inspired by the distinguishing attack on SNOW 2.0 presented by Nyberg and Wallen [6].

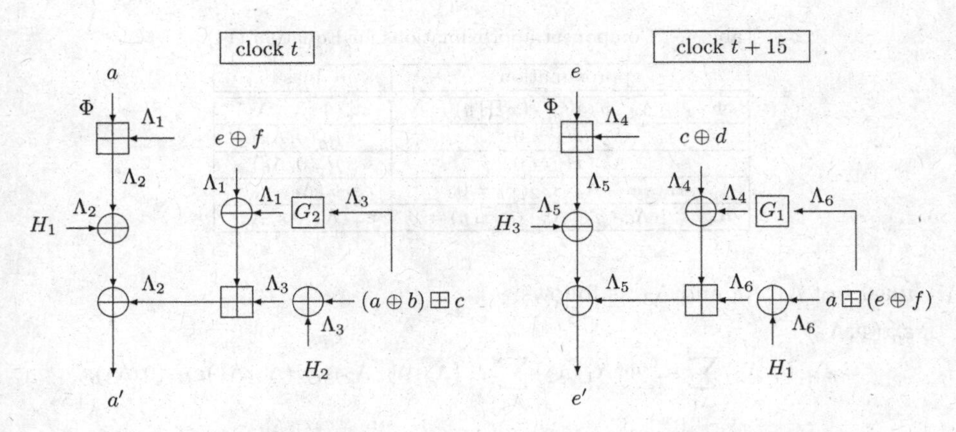

Fig. 2. Generalized linear masks for approximations of the function F

$$\begin{aligned}\Lambda_5 \cdot e' &= \Lambda_5 \cdot [((c \oplus d) \boxplus e) \oplus H_3] \oplus \Lambda_5 \cdot [((a \boxplus (e \oplus f)) \oplus H_1) \boxplus (c \oplus d \oplus G_1)] \\ &= \Lambda_4 \cdot (c \oplus d) \oplus \Phi \cdot e \oplus \Lambda_5 \cdot H_3 \oplus \Lambda_6 \cdot (a \boxplus (e \oplus f)) \oplus \Lambda_6 \\ &\quad \cdot H_1 \oplus \Lambda_4 \cdot (c \oplus d \oplus G_1) \\ &= \Phi \cdot e \end{aligned} \tag{14}$$

The component-wise approximations required for Approximations (13) and (14) are listed in Tables 3 and 4. According to the well-known theorem

Table 3. Component approximations for Equation (13)

approximation	bias
$\Phi \cdot x \oplus \Lambda_1 \cdot y \oplus \Lambda_2 \cdot (x \boxplus y) = 0$	$\varepsilon_+(\Phi, \Lambda_1, \Lambda_2)$
$\Lambda_2 \cdot H_1 = 0$	$\varepsilon_{H_1}(0, \Lambda_2)$
$\Lambda_3 \cdot H_2 = 0$	$\varepsilon_{H_2}(0, \Lambda_3)$
$\Lambda_3 \cdot x \oplus \Lambda_1 \cdot G_1(x) = 0$	$\varepsilon_{G_1}(\Lambda_3, \Lambda_1)$
$\Lambda_1 \cdot x \oplus \Lambda_3 \cdot y \oplus \Lambda_2 \cdot (x \boxplus y) = 0$	$\varepsilon_+(\Lambda_1, \Lambda_3, \Lambda_2)$

[7] the correlation of approximations can be computed as a sum of partial correlations over all intermediate linear masks. For theoretical analysis of the theorem, we refer the reader to the paper of [7]. Hence, the bias of Approximation (13) is computed as a sum of partial biases induced by the

Table 4. Component approximations for Equation (14)

approximation	bias
$\Phi \cdot x \oplus \Lambda_4 \cdot y \oplus \Lambda_5 \cdot (x \boxplus y) = 0$	$\varepsilon_+(\Phi, \Lambda_4, \Lambda_5)$
$\Lambda_5 \cdot H_3 = 0$	$\varepsilon_{H_3}(0, \Lambda_5)$
$\Lambda_6 \cdot H_1 = 0$	$\varepsilon_{H_1}(0, \Lambda_6)$
$\Lambda_6 \cdot x \oplus \Lambda_4 \cdot G_2(x) = 0$	$\varepsilon_{G_2}(\Lambda_6, \Lambda_4)$
$\Lambda_4 \cdot x \oplus \Lambda_6 \cdot y \oplus \Lambda_5 \cdot (x \boxplus y) = 0$	$\varepsilon_+(\Lambda_4, \Lambda_6, \Lambda_5)$

masks of Λ_1, Λ_2 and Λ_3 as follows:

$$\begin{aligned}&\varepsilon_{a'}(\Phi, \Lambda_2)\\ &\quad = \varepsilon_{H_1}(0, \Lambda_2) \sum_{\Lambda_1} \varepsilon_+(\Phi, \Lambda_1, \Lambda_2) \sum_{\Lambda_3} \varepsilon_+(\Lambda_3, \Lambda_1, \Lambda_2) \varepsilon_{G_2}(\Lambda_3, \Lambda_1) \varepsilon_{H_2}(0, \Lambda_3).\end{aligned} \tag{15}$$

Similarly, the bias of Approximation (14) using the masks of Λ_4, Λ_5 and Λ_6 can be computed as follows:

$$\begin{aligned}&\varepsilon_{e'}(\Phi, \Lambda_5)\\ &\quad = \varepsilon_{H_3}(0, \Lambda_5) \sum_{\Lambda_4} \varepsilon_+(\Phi, \Lambda_4, \Lambda_5) \sum_{\Lambda_6} \varepsilon_+(\Lambda_4, \Lambda_6, \Lambda_5) \varepsilon_{G_1}(\Lambda_6, \Lambda_4) \varepsilon_{H_1}(0, \Lambda_6).\end{aligned} \tag{16}$$

Hence, according to Subsection 3.4, a new distinguisher can be derived from Approximations (13) and (14) as follows:

$$\Lambda_2 \cdot k_0[t] \oplus \Lambda_5 \cdot k_1[t+15] = \Phi \cdot M_L[t+15] \tag{18}$$

with the bias of

$$\varepsilon_D(\Lambda_2, \Lambda_5) = \sum_{\Phi} \varepsilon_{a'}(\Phi, \Lambda_2) \varepsilon_{e'}(\Phi, \Lambda_5). \tag{19}$$

Note that we need to guess 64 memory bits this time since the linear mask Φ can be an arbitrary value among $[1, 2^{32} - 1]$.

4.1. *Experiments*

In order to find the distinguisher holding the biggest bias, we need to search all possible combinations of Γ_2 and Γ_5 and test their biases by Equation (19). Furthermore, for each Γ_2 and Γ_5, the computation of Equation (19) requires a large number of iterations due to large sizes of the intermediate masks. Hence, our experiments focus on reducing the overall size of the masks that are required for the computation of the bias. To achieve this goal, we implemented two techniques that can remove a large portion of terms from the summation in Equation (19).

In the first technique, we remove the unnecessary terms from Equations (15) and (16). Note that these terms are generated by the condition $\varepsilon = 0$.

The approximations of the modular addition have non-trivial biases only in a portion of bits which is determined by the values of an input and an output masks.

Lemma 4.1. *Assume that the bias of the approximation* $\Lambda_1 \cdot x \oplus \Lambda_3 \cdot y \oplus \Lambda_2 \cdot (x \boxplus y) = 0$ *is represented by* $\varepsilon_+(\Lambda_1, \Lambda_3, \Lambda_2)$. *Given* $\Lambda_2 = b_{31}b_{30}\cdots b_0$ *where* b_i *stands for the i-th bit of* Λ_2, *we assume that the most significant non-zero bit of* Λ_2 *is located in the bit position of* b_t *where* $0 \leq t \leq 31$. *Then, the bias* $\varepsilon_+(\Lambda_1, \Lambda_3, \Lambda_2)$ *is zero when* $\Lambda_1, \Lambda_3 < 2^t$ *or* $\Lambda_1, \Lambda_3 \geq 2^{t+1}$. *In other words, we conclude that*

$$\sum_{\Lambda_1=1}^{2^{32}-1} \sum_{\Lambda_3=1}^{2^{32}-1} \varepsilon_+(\Lambda_1, \Lambda_3, \Lambda_2) = \sum_{\Lambda_1=2^t}^{2^{t+1}-1} \sum_{\Lambda_3=2^t}^{2^{t+1}-1} \varepsilon_+(\Lambda_1, \Lambda_3, \Lambda_2),$$

Proof. See Appendix C. □

According to Lemma 4.1, the value of the biases (15) and (16) depends on the size of the output mask of the modular addition. For example, if Λ_2 = 0x0600018D, then, $\varepsilon_+(\Lambda_1, \Lambda_3, \Lambda_2)$ becomes zero when $\Lambda_1, \Lambda_3 <$ 0x04000000 or $\Lambda_1, \Lambda_3 \geq$ 0x08000000.

In the second technique, we restrict the biases of the modular additions in Equations (15) and (16) and reduce the number of iterations. Instead of iterating the full space of the intermediate masks, we use only relatively highly biased approximations of the modular additions. In the paper of [6], authors proposed an efficient algorithm for finding all input and output masks for addition with a given correlation. This algorithm enables us to reduce the number of iteration for Equations (15) and (16) significantly. We restricted the effective correlation of the modular addition up to $\pm 2^{-24}$, as suggested in the paper of [6].

Based on these techniques, we re-calculated the bias of Distinguisher (18) and found that the bias is estimated to be $2^{-75.32}$. It is interesting to observe how the biases can be improved by considering the dependencies of the combinations of the approximation. Table 5 shows that the bias measurement without considering the dependencies can underestimate the real bias of the approximation.

Due to the restrictions on computing resources, we searched for the best distinguisher under the condition that $\Lambda_2 = \Lambda_5$ and we could not perform the experiment for the cases when the different values of Λ_2 and Λ_5 are allowed. Even though Bias (19) tends to be high when $\Lambda_2 = \Lambda_5$, there is a possibility that two different values of Λ_2 and Λ_5 may lead to a distinguisher with a bigger bias. We leave this issue as an open problem.

Table 5. Comparison of the bias of the distinguisher computed by two methods

$\Lambda_2 = \Lambda_5$	ε_D without dependencies	ε_D with dependencies
0x0600018D	$2^{-75.81}$	$2^{-75.32}$
0x002C0039	$-2^{-129.55}$	$2^{-81.77}$
0x00001809	$2^{-84.13}$	$2^{-79.51}$

5. Conclusion

In this paper, we presented a new distinguisher for Dragon. Since the amount of observations for the distinguishing attack is by far larger than the limit of keystream available from a single key, our distinguisher leads only to a theoretical attack on Dragon. However, our analysis shows that some approximations of the functions G and H have larger biases than the ones expected by the designers. As far as we know, our distinguisher is the best one for Dragon published so far in open literature. In addition, we present an efficient algorithm to compute the bias of approximation of modular addition, which is expected to be useful for other attacks against ciphers using modular additions.

Acknowledgment

We wish to thank Matt Henricksen for invaluable comments. The authors were supported by ARC grants DP0451484, DP0663452 and Macquarie University ARC Safety Net Grant.

References

1. E. Dawson, K. Chen, M. Henricksen, W. Millan, L. Simpson, H. Lee and S. Moon, Dragon: A fast word based stream cipher eSTREAM, ECRYPT Stream Cipher Project, Report 2005/006, (2005), http://www.ecrypt.eu.org/stream.
2. K. Chen, M. Henricksen, W. Millan, J. Fuller, L. Simpson, E. Dawson, H. Lee and S. Moon, Dragon: A fast word based stream cipher, in *Information Security and Cryptology - ICISC 2004*, eds. C. Park and S. Chee, Lecture Notes in Computer Science, Vol. 3506 (Springer, 2004).
3. E. NoE, eSTREAM - the ECRYPT stream cipher project Available at http://www.ecrypt.eu.org/stream/, (2005).

4. H. Englund and A. Maximov, Attack the Dragon, in *Progress in Cryptology - INDOCRYPT 2005*, eds. S. Maitra, C. E. V. Madhavan and R. Venkatesan, Lecture Notes in Computer Science, Vol. 3797 (Springer, 2005).
5. M. Matsui, Linear cryptoanalysis method for des cipher, in *EUROCRYPT*, 1993.
6. K. Nyberg and J. Wallen, Improved linear distinguishers for SNOW 2.0., in *Fast Software Encryption - FSE 2006*, ed. M. J. B. Robshaw, Lecture Notes in Computer Science, Vol. 4047 (Springer, 2006).
7. K. Nyberg, *Discrete Applied Mathematics* **111**, 177 (2001).

Appendix A. Proof of Theorem 3.1

Suppose that $z = x \boxplus y$ where $x = (x_{n-1}, \cdots, x_0), y = (y_{n-1}, \cdots, y_0)$ and $z = (z_{n-1}, \cdots, z_0)$. Then, each z_i bit is expressed a function of $x_i, \cdots, x_0$ and $y_i, \cdots, y_0$ bits as follows

$$z_0 = x_0 \oplus y_0, \quad z_i = x_i \oplus y_i \oplus x_{i-1} y_{i-1} \oplus \sum_{j=0}^{i-2} x_j y_j \prod_{k=j+1}^{i-1} (x_k \oplus y_k), \quad i = 1, \cdots, n.$$

If we define the carry $R(x, y)$ as

$$R(x,y)_0 = x_0 y_0, \quad R(x,y)_i = x_i y_i \oplus \sum_{j=0}^{(i-1)} x_j y_j \prod_{k=j+1}^{i} (x_k \oplus y_k), \quad i = 1, 2, \ldots,$$

then, it is clear that $z_i = x_i \oplus y_i \oplus R(x,y)_{i-1}$ for $i > 0$. By the definition, $R(x,y)_i$ has the following recursive relation

$$R(x,y)_i = x_i y_i \oplus (x_i \oplus y_i) R(x,y)_{i-1}. \tag{A.1}$$

First, we examine the bias of the Γ of which the Hamming weight is 2, i.e. $m = 2$. Without loss of generality, we assume that $\gamma_i = 1$ and $\gamma_j = 1$ where $0 \leq j < i < n$. Then, by Relation (A.1), Approximation (1) is expressed as

$$\begin{aligned} \Gamma \cdot (x \boxplus y) \oplus \Gamma \cdot (x \oplus y) &= z_i \oplus z_j \oplus (x_i \oplus y_i) \oplus (x_j \oplus y_j) \\ &= R(x,y)_{i-1} \oplus R(x,y)_{j-1} \\ &= x_{i-1} y_{i-1} \oplus (x_{i-1} \oplus y_{i-1}) R(x,y)_{i-2} \oplus R(x,y)_{j-1}. \end{aligned}$$

Let us denote $p_{i-1} = Pr[R(x,y)_{i-1} \oplus R(x,y)_{j-1} = 0]$. Since x_i and y_i are assumed as uniformly distributed random variables, the probability p_{i-1} is split into the three cases as follows.

$$p_{i-1} = \begin{cases} Pr[R(x,y)_{j-1} = 0], & \text{if } (x_{i-1}, y_{i-1}) = (0,0) \\ Pr[1 \oplus R(x,y)_{j-1} = 0], & \text{if } (x_{i-1}, y_{i-1}) = (1,1) \\ Pr[R(x,y)_{i-2} \oplus R(x,y)_{j-1} = 0], & \text{if } (x_{i-1}, y_{i-1}) = (0,1), (1,0) \end{cases}$$

Clearly, $Pr[R(x,y)_{j-1} = 0] = 1 - Pr[1 \oplus R(x,y)_{j-1} = 0]$. Hence, we get

$$p_{i-1} = \frac{1}{4} + \frac{1}{2} Pr[R(x,y)_{i-2} \oplus R(x,y)_{j-1} = 0] = \frac{1}{4} + \frac{1}{2} p_{i-2}.$$

If $j = i - 1$, then $Pr[R(x,y)_{i-2} \oplus R(x,y)_{j-1} = 0] = 1$. Hence, $p_{i-1} = \frac{1}{4} + \frac{1}{2} = \frac{3}{4}$. Otherwise, p_{i-2} is determined recursively by the same technique used as above until p_{j-1} is reached.

Hence, we obtain the following result

$$p_{i-1} = \frac{1}{4}(1 + \cdots + 2^{-(i-j-1)}) + 2^{-(i-j)} = \frac{1}{2}(1 + 2^{-(i-j)}). \qquad \text{(A.2)}$$

Therefore, the bias $\varepsilon_+(\Gamma, \Gamma)$ is determined by the difference between two position i and j of Γ only.

Next, we consider the case that Γ has an arbitrary Hamming weight, which is denoted m. Assume that we convert m into an even number m' by using the following technique.

- If m is even, then set $m' = m$.
- If m is odd and $\gamma_0 = 0$, then set $\gamma_0 = 1$ and $m' = m + 1$.
- If m is odd and $\gamma_0 = 1$, then set $\gamma_0 = 0$ and $m' = m - 1$.

In result, the Γ is transformed to Γ' which has the Hamming weight of m'. Since the modular addition is linear for the least significant bit, $\varepsilon_+(\Gamma, \Gamma) = \varepsilon_+(\Gamma', \Gamma')$. Hence, a new position vector for Γ' is defined as $W_{\Gamma'} = (w_{m'-1}, \cdots, w_0)$, where $0 \leq w_j < n$.

Now, we decompose Γ' into a combination of sub-masks which have the Hamming weight of 2. That is, Γ is expressed as

$$\Gamma = \Omega_{m'/2-1} \oplus \cdots \oplus \Omega_0,$$

where Ω_k is a sub-mask which has the nonzero coordinates only at position w_{2k} and w_{2k+1} for $k = 0, 1, \cdots, \frac{m'}{2} - 1$. Clearly, the number of such sub-masks is $\frac{m'}{2}$. For example, if $\Gamma = (0,0,1,1,0,1,1)$, then $\Gamma = \Omega_1 \oplus \Omega_0 = (0,0,1,1,0,0,0) \oplus (0,0,0,0,0,1,1)$.

From (A.2), we know that the bias of $\Omega_k \cdot (x \boxplus y) \oplus \Omega_k \cdot (x \oplus y)$ is only determined by the difference $w_{2k+1} - w_{2k}$. Hence, according to Piling-up Lemma [5], the bias of $\Gamma \cdot (x \boxplus y) \oplus \Gamma \cdot (x \oplus y)$ is obtained by combining the $\frac{m'}{2}$ approximations. Note that the there are no inter-dependencies among sub-masks. Therefore, the claimed bias is computed as

$$\varepsilon_+(\Gamma, \Gamma) = 2^{-[w_{m'-1} - w_{m'-2}] + \cdots + [w_1 - w_0]}.$$

If m' is replaced by m, we obtain the claimed bias. $\square$

Appendix B. Proof of Corollary 3.1

Recall Theorem 3.1. If m is even, then,

$$d_1 = \sum_{i=0}^{m/2-1} (w_{2i+1} - w_{2i}) \geq \sum_{i=0}^{m/2-1} 1 = m/2.$$

If m is odd, then,

$$d_2 = \sum_{i=1}^{(m-1)/2} (w_{2i} - w_{2i-1}) + w_0 \geq \sum_{i=1}^{(m-1)/2} 1 = (m-1)/2.$$

Hence, the bias $\varepsilon_+(\Gamma, \Gamma) \leq 2^{-(m-1)/2}$. □

Appendix C. Proof of Lemma 4.1

Let x_i and y_i denote the i-th bits of 32-bit words x and y. According to the notation used in Appendix A, the approximation using the output mask Λ_2 can be expressed as

$$\Lambda_2 \cdot (x \boxplus y) = x_t \oplus y_t \oplus R(x,y)_{t-1} \oplus A(x,y)_{t-1},$$

where $A(x,y)_{t-1}$ is a function which does not contain x_t and y_t bits as variables.

When $\Lambda_1 < 2^t$ or $\Lambda_3 < 2^t$, the input approximation $\Lambda_1 \cdot x \oplus \Lambda_3 \cdot y$ does not contain x_t or y_t bit as a variable. Thus, $\Lambda_1 \cdot x \oplus \Lambda_3 \cdot y \oplus \Lambda_2 \cdot (x \boxplus y)$ retains a linear term x_t or y_t so that the bias of the approximation becomes zero.

On the other hand, given $\Lambda_1 \geq 2^{t+1}$ or $\Lambda_3 \geq 2^{t+1}$, the input approximation $\Lambda_1 \cdot x \oplus \Lambda_3 \cdot y$ contain x_u or y_v bit as a variable where $u, v > t$. Thus, $\Lambda_1 \cdot x \oplus \Lambda_3 \cdot y \oplus \Lambda_2 \cdot (x \boxplus y)$ retains a linear term x_u or y_v so that the bias of the approximation becomes zero. □

Constructing Perfect Hash Families Using a Greedy Algorithm

Charles J. Colbourn
Computer Science and Engineering
Arizona State University
P.O. Box 878809,
Tempe, AZ 85287, U.S.A.
charles.colbourn@asu.edu

A *perfect hash family* $\mathsf{PHF}(N; k, v, t)$ is an $N \times k$ array on v symbols with $v \geq t$, in which in every $N \times t$ subarray, at least one row is comprised of distinct symbols. Perfect hash families have a wide range of applications in cryptography, particularly to secure frameproof codes, in database management, and are indirectly used in software interaction testing. A simple one-row-at-a-time greedy algorithm for constructing small perfect hash families is described. The algorithm is deterministic, and its worst-case runtime is polynomial in k and v but exponential in t; consequently, when t is fixed, the algorithm runs in polynomial time in the worst case. It provides a deterministic guarantee that the number N of rows produced is $O(\log k)$. In addition to these strong asymptotic guarantees, the method is shown to be practical for the computation of perfect hash families for "small" values of k and t.

1. Introduction

A *perfect hash family* $\mathsf{PHF}(N; t, k, v)$ is an $N \times k$ array, in which in every $N \times t$ subarray, at least one row is comprised of distinct symbols. Figure 1 shows a $\mathsf{PHF}(6; 12, 3, 3)$. For instance, in columns 1, 3, and 5, the first row contains `1 0 2`. The smallest N for which a $\mathsf{PHF}(N; k, v, t)$ exists is the *perfect hash family number*, denoted $\mathsf{PHFN}(k, v, t)$. An older [6] and a newer [10] survey on PHFs can be consulted for more extensive background.

Mehlhorn [8] defined perfect hash families as follows: A (k, v)-*hash function* is a function $h : A \to B$, where $|A| = k$ and $|B| = v$. For any given subset $X \subseteq A$, the function h is *perfect* if h is injective on X, i.e., if $h|_X$ is one-to-one. Given integers k, v, t so that $k \geq v \geq t \geq 2$, define a $\mathsf{PHF}(N; k, v, t)$ as a set $\mathcal{H}$ with $|\mathcal{H}| = N$ of (k, v)-hash functions such that $h : A \to B$ for each $h \in \mathcal{H}$ where $|A| = k$ and $|B| = v$ with the property that for any $X \subseteq A$ with $|X| = t$, there exists at least one $h \in \mathcal{H}$ such that $h|_X$ is one-to-one. This definition is equivalent to the array definition. Con-

$$\begin{bmatrix} 1&0&0&2&2&2&1&1&2&1&0&2 \\ 2&0&1&1&2&0&2&0&1&1&2&1 \\ 2&0&2&1&2&1&0&2&2&1&1&0 \\ 0&1&2&2&1&2&2&0&1&1&0&0 \\ 2&0&1&2&1&1&2&2&0&1&2&1 \\ 0&2&1&0&2&2&2&1&0&1&2&1 \end{bmatrix}$$

Fig. 1. A $\mathsf{PHF}(6; 12, 3, 3)$

sider each row of the array to be a function h, and take $A = \{1, 2, \ldots, k\}$. Then the value in column i of the row for h is the value of $h(i)$.

Mehlhorn [8] introduced perfect hash families as an efficient tool for compact storage and fast retrieval of frequently used information, such as reserved words in programming languages or command names in interactive systems. Stinson, Trung, and Wei [9] establish that perfect hash families, and a variation known as "separating hash families", can be used to construct separating systems, key distribution patterns, group testing algorithms, cover-free families, and secure frameproof codes. Perfect hash families have also recently found applications in broadcast encryption [7] and threshold cryptography [1]. Finally, perfect hash familes arise as ingredients in some recursive constructions for covering arrays [5]. Covering arrays have a wide range of applications, most prominently in software interaction testing.

Walker and Colbourn [10] computed upper bounds for $3 \leq t \leq 6$, $t \leq v \leq 50$, $v \leq k \leq 500{,}000$. Many tables are online at `www.phftables.com`. Available direct constructions, while yielding small PHFs, are only occasionally effective. Recursions, while more generally applicable, necessitate knowledge of appropriate small ingredients. Hence one resorts to computation. Metaheuristic search, such as tabu search [10], has been applied to the search for perfect hash families. Its primary drawback is that it is computationally intensive. This can limit the range of parameters for which solutions can be found within reasonable time.

In this paper we therefore develop a simpler heuristic method, based on selecting one row at a time in a greedy manner. This type of method is not new, having been used for the construction of different types of combinatorial arrays [2–4]. Our interest is in obtaining a method that is deterministic and efficient. Hence we construct each new row by repeatedly selecting a column and a symbol to place in that column. At each step, a value to be placed in the next row in a specific column is selected based on the number of t-subsets of columns that the selection will guarantee to separate as well as those that it can be *expected* to distinguish. Hence while the method is greedy, it incorporates an implicit look-ahead to future selections.

The essential problem with greedy methods of this type is that one usually obtains no guarantee about the number of rows needed. While it is known that $\mathsf{PHFN}(k, v, t) = O(\log k)$ for fixed v and t with $v \geq t$, naive greedy methods provide no guarantee that they produce arrays meeting this bound. Cohen, Litsyn, and Zémor [4] show that a greedy method that considers all possible rows guarantee such logarithmic growth in a wide variety of problems, but their methods lead to efficient randomized techniques, or to exponential time deterministic techniques. Hence in Section 2 we develop a deterministic greedy algorithm whose runtime is polynomial when t is fixed, and that guarantees to produce a PHF whose number of rows is bounded above by $c_{t,v} \log k$, where $c_{t,v}$ is a constant depending only on t and v.

Perhaps more surprisingly, the greedy method developed does not just behave well asymptotically. Rather it produces useful new perfect hash families even when k, t, and v are "small". In Section 3 we report a few computational results from the method.

2. Density and a Greedy Algorithm

Choose an alphabet size v and a number of columns k, to remain fixed throughout. Index the columns by $C = \{c_1, \ldots, c_k\}$. In general we suppose that some rows of a v-ary array have already been selected, and we are in the process of building a further row. In the rows already built, any t-subset $T \subseteq C$ is *separated* if some row contains distinct symbols in every column in T, in which case we write $\lambda(T) = 0$; if not separated in any row, we write $\lambda(T) = 1$.

Let $\rho = (\rho_1, \ldots, \rho_k)$ denote a *partial row*, i.e. a k-tuple in which $\rho_i \in \{1, \ldots, v\} \cup \{\star\}$ for $1 \leq i \leq k$. We interpret a numerical value to mean that the value in the corresponding column has been *fixed* to the indicated value, while $\star$ indicates that the value in the column has not yet been selected (and the column is *free*).

We ask a simple question. If we complete ρ so that all free columns are fixed randomly, how many further t-subsets of columns do we expect to separate? To answer this question, we first define the *density*, $\delta(\varphi)$, of the partial row φ having all columns free, as follows:

$$\delta(\varphi) = \sum_{\{f_1,\ldots,f_t\}=T\subseteq C} \lambda(T)\frac{v\cdot(v-1)\cdot\cdots\cdot(v-t+1)}{v^t}.$$

The number of t-subsets yet to be separated is $\Lambda = \sum_{\{f_1,\ldots,f_t\}=T\subseteq C}\lambda(T)$. Hence $\delta(\varphi)$ is a fixed fraction of the number of t-subsets remaining to be separated. Before proceeding, let us observe that $\delta(\varphi)$ is indeed the expected number of additional t-subsets separated. Among the v^k possible rows, there are $v\cdot(v-1)\cdot\cdots\cdot(v-t+1)\cdot v^{k-t}$ that separate a specific t-subset T. Summing over all t-subsets yet to be separated, the *total* number separated by all possible choices of rows is $\Lambda v\cdot(v-1)\cdot\cdots\cdot(v-t+1)\cdot v^{k-t}$, and hence on average each row separates $\Lambda\frac{v\cdot(v-1)\cdot\cdots\cdot(v-t+1)}{v^t}$ additional t-subsets. This is $\delta(\varphi)$.

As a consequence, if we repeatedly choose a row that separates at least the average number ($\delta(\varphi)$) of additional t-subsets, the number of rows chosen in total will not exceed $\frac{\log\binom{k}{t}}{\log\frac{v^t}{v^t-v\cdot(v-1)\cdot\cdots\cdot(v-t+1)}}$ rows. Of course the challenge is to select a row that separates at least as many additional t-subsets as the average, and to do so in *deterministic polynomial time*. To do this, we extend the notion of density.

Consider a set $T\subseteq C$, $|T|=t$, in a partial row ρ. Suppose that s columns of T are fixed and that the remaining $t-s$ are free. Then set $\chi(T,\rho)=0$ if any two fixed columns contain the same symbol, and $\chi(T,\rho)=\frac{(v-s)\cdot(v-s-1)\cdot\cdots\cdot(v-t+1)}{v^{t-s}}$ otherwise. Then $\chi(T,\rho)$ is the ratio of the number of ways in which these t columns could be fixed to separate T to all ways to fix these columns.

We extend the definition of density as follows:

$$\delta(\rho) = \sum_{T\subseteq C,\ |T|=t} \lambda(T)\chi(T,\rho).$$

Now let c be a free column in ρ, and let $\mu(\rho,c,i)$ be the partial row obtained from ρ by setting $\rho_c = i$. The key observation is:

Lemma 2.1.

$$\delta(\rho) = \frac{1}{v}\sum_{i=1}^{v}\delta(\mu(\rho,c,i)).$$

Proof. The contribution to $\delta(\rho)$ from every t-subset T with $c\notin T$ remains unchanged in $\delta(\mu(\rho,c,i))$ and hence is added v times, and the contribution divided by v; hence the contribution is the same. When $c\in T$, we claim that $\chi(T,\rho) = \sum_{i=1}^{v} \chi(T,\mu(\rho,c,i))$. Indeed consider $\chi(T,\rho)$; suppose that T had s fixed columns and $t-s$ free ones. If the fixed ones contain two symbols the same, then $\chi(T,\rho) = \chi(T,\mu(\rho,c,i)) = 0$ for all $1 \le i \le v$. Otherwise $\chi(T,\rho) = \frac{(v-s)\cdot(v-s-1)\cdot\dots\cdot(v-t+1)}{v^{t-s}}$. For s values of i we find that $\chi(T,\mu(\rho,c,i)) = 0$ because the choice repeats a symbol in an already fixed column. And if choice i does not, then $\chi(T,\mu(\rho,c,i)) = \frac{(v-s-1)\cdot(v-s-2)\cdot\dots\cdot(v-t+1)}{v^{t-s-1}}$. But then $(v-s)\frac{(v-s-1)\cdot(v-s-2)\cdot\dots\cdot(v-t+1)}{v^{t-s-1}} = v\frac{(v-s)\cdot(v-s-1)\cdot\dots\cdot(v-t+1)}{v^{t-s}}$, and the equality holds. □

Lemma 2.1 says that, in any column c, there is at least one selection of a value that, if chosen, does not reduce the expected number of additional t-subsets separated.

This provides the key to a deterministic greedy algorithm, which we call the *density algorithm*:

Start with an empty set of rows R.
Mark all t-subsets of columns as not separated.
while some t-subset of columns is not separated by a row of R
    Let ρ be the empty partial row
    while a free column c remains in ρ
        Select the value of i that maximizes $\delta(\mu(\rho,c,i))$
        Replace ρ by $\mu(\rho,c,i)$ (fix column i and set $\rho_c = i$)
    Add row ρ to R
Output R.

Lemma 2.2. *For fixed v, t with $v \ge t$, there is a fixed constant $c_{t,v}$ such that the density algorithm produces a perfect hash family* $\mathsf{PHF}(N;k,v,t)$ *with* $N \le c_{t,v}\log_2\binom{k}{t} \le tc_{t,v}\log_2 k$, *where* $c_{t,v} = \left(\log_2 \frac{v^t}{v^t - v\cdot(v-1)\cdot\dots\cdot(v-t+1)}\right)^{-1}$.

Proof. In selecting a row, the initial partial row has density $\delta(\varphi) = \Lambda\frac{v\cdot(v-1)\cdot\dots\cdot(v-t+1)}{v^t}$ where Λ is the remaining number of t-subsets to be separated. By Lemma 2.1, every fixing of a column does not decrease density. When no columns are free, $\delta(\rho)$ is the actual number of additional t-subsets separated. Thus after each row is selected, at least the fraction $\frac{v\cdot(v-1)\cdot\dots\cdot(v-t+1)}{v^t}$ of the remaining t-subsets of columns are separated. After N rows are selected, at most $\binom{k}{t}\left(\frac{v^t - v\cdot(v-1)\cdot\dots\cdot(v-t+1)}{v^t}\right)^N$ t-subsets remain to be separated. So choosing N so that this quantity is less than 1 ensures that all are separated. Solve for N. □

Lemma 2.3. *The density algorithm runs in time that is polynomial in v*

and k, but exponential in t. Hence its running time is polynomial for fixed t.

Proof. Consider the constants $c_{t,v}$ in Lemma 2.2. When $v \geq t \geq 2$, $c_{t,v} \leq 2$. Hence the number of rows produced is bounded by $2t \log_2 k$ by Lemma 2.2. In generating each row, k columns must be fixed. In the fixing of a single column, the change in density involves $\binom{k-1}{t-1}$ t-subsets, each of which requires a computation that is linear in v. Therefore the update can be done in $O(v \cdot k^{t-1})$ time for each column, to obtain an $O(v \cdot k^t)$ running time for selecting each row, and hence an $O((t \log_2 k)(v \cdot k^t))$ running time for the algorithm. □

3. Computational Results

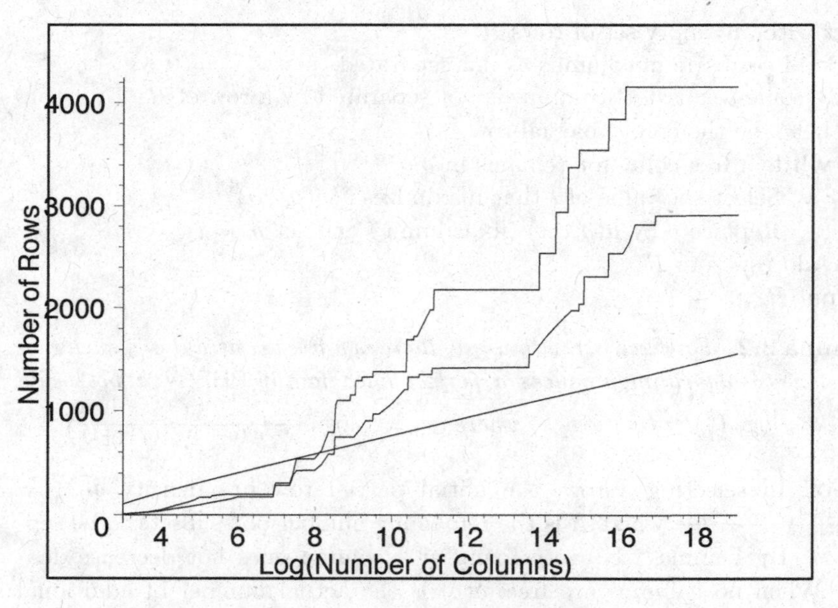

Fig. 2. Upper Bounds for PHFN$(k, 5, 5)$, $2^3 \leq k \leq 2^{18}$

In Figure 2 three bounds are shown on PHFN$(k, 5, 5)$ for $3 \leq \log_2 k \leq 18$. The irregular graph yielding the largest bound is reports the last published bounds [10]. The irregular graph in the middle is the updated bound at `www.phftables.com` using the small perfect hash families produced here. Remarkably, the best bound for $k \geq 500$ is given in the third graph, which is the bound from the theoretical analysis of the density method. As expected, density does not perform as well for smaller values of k; indeed it is surprising that it outperforms the known recursive and direct techniques at all! In a sense, the comparison is misleading. The perfect hash families resulting from direct and recursive constructions can all be easily constructed, even for hundreds of thousands of columns. On the other hand, although density is an efficient technique, it is not realistic to apply the method for hundreds, let alone thousands, of columns. Thus the explicit construction of perfect hash families for large k is not well solved by density.

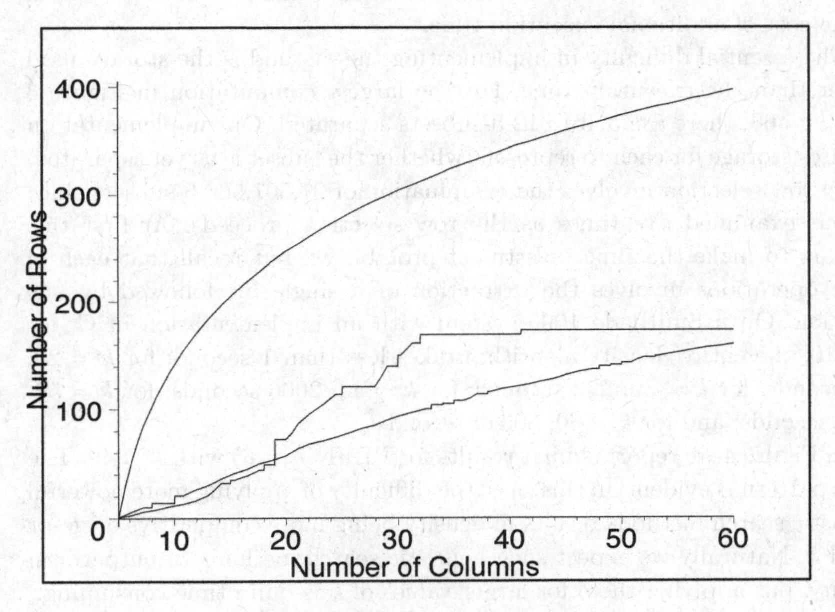

Fig. 3. Upper Bounds for PHFN$(k, 5, 5)$, $5 \leq k \leq 60$

We focus now on smaller values of k. In Figure 3, three graphs are shown for upper bounds on PHFN$(k, 5, 5)$ with $k \leq 60$. The irregular graph, generally in the "middle", plots the best upper bound known from all methods [10] prior to the application of the density method developed here. The largest upper bound plots the theoretical upper bound produced by Lemma 2.2. As one might expect, the theoretical guarantee is not at all competitive.

However, a surprise is in store. Rather than simply selecting a new row that is at least as good as the average, at every step in building the row, the expected number of additional t-subsets separated is maximized. As we have seen, this ensures that at least the average number are separated, but in practice one often separates more. One might expect this to have a negligible effect, but the effect is rather dramatic. The third graph plotted in Figure 3, yielding the best bound for $19 \leq k \leq 60$, shows the number of rows produced by the density method. These results are obtained by executing the method 10 times (*repetitions*), selecting the best size. In addition, since ties occur among the densities in the selection of a value for a column in a row, we break ties randomly; for every row we build 10 *candidates* and select one that separates the most additional t-subsets. Increasing numbers of candidates and/or repetitions often reduces the number of rows produced, at the cost of additional execution time.

The essential difficulty in implementing the method is the storage used rather than the execution time. For the largest computation in Figure 3 with $k = 60$, there are 5,461,512 5-subsets separated. Our implementation requires storage for each to represent whether the subset is as yet separated. Every row selection involves the examination of 27,307,560 5-subsets since each is examined five times as the row selection proceeds. At first this appears to make the time investment prohibitive, but recall that each of these operations involves the extraction of a single bit followed by one addition. On a SunBlade 1000 system with an implementation in C, for $k = 10$, the entire density algorithm takes less than 1 second; for $k = 20$, 15 seconds; for $k = 30$, 273 seconds; for $k = 40$, 2006 seconds; for $k = 50$, 9028 seconds; and for $k = 60$, 30349 seconds.

In Figure 4 we report similar results for PHFN$(k, 6, 6)$ with $k \leq 40$. The same pattern is evident; in this case, the difficulty of applying more powerful heuristic search methods results in density being more competitive even for small k. Naturally we expect such heuristic search methods to outperform density, but applying them for larger values of k is quite time-consuming.

At `www.phftables.com`, results are reported for a wide variety of values of t and v, restricted to cases when $t \leq 6$. While many improvements of previous results are obtained by this density technique, the two parameter situations discussed here illustrate the general situation well so we do not report each individual improvement obtained.

4. Conclusions

The density algorithm gives a deterministic algorithm for generating perfect hash families; it both guarantees to generate an array with k columns having $O(t \log k)$ rows, and runs in worst-case time that is polynomial in k and v, although exponential in t. It is perhaps surprising that for fixed t one can

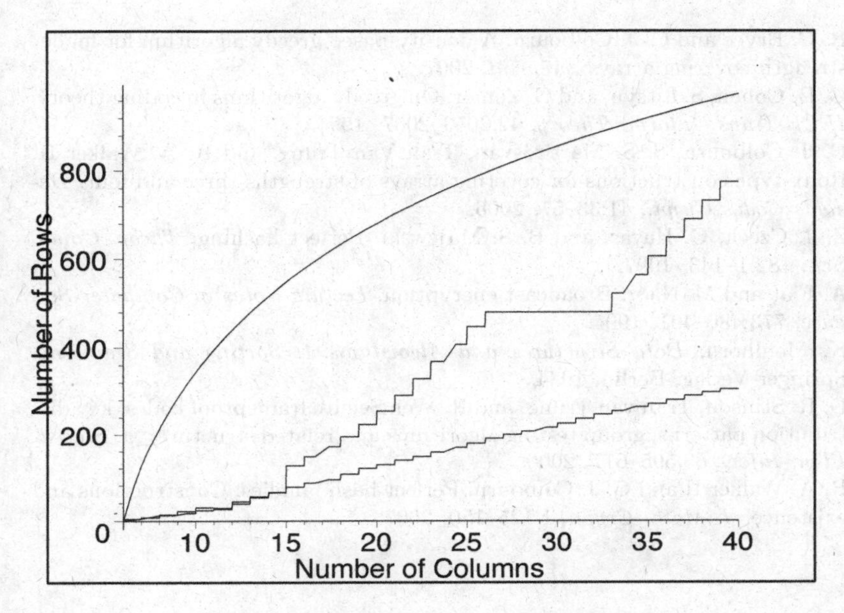

Fig. 4. Upper Bounds for PHFN$(k, 6, 6)$, $6 \leq k \leq 40$

get worst-case logarithmic growth in the number of rows from a polynomial time algorithm.

However the bigger surprise is that the method leads to an explicit logarithmic bound that improves upon known direct and recursive explicit constructions for perfect hash families. Indeed when implemented, for numbers of columns in the range $20 \leq k \leq 60$, the density algorithm gives a practical technique for the construction of perfect hash families with the fewest rows of any that are presently known.

Acknowledgements

Thanks to Renée Bryce for helpful discussions concerning density, and to Robby Walker for existence tables to determine the effectiveness of the method.

References

1. S. R. Blackburn, M. Burmester, Y. Desmedt, and P. R. Wild. Efficient multiplicative sharing schemes. *Lecture Notes in Computer Science*, 1070:107–118, 1996.
2. R. C. Bryce and C. J. Colbourn. The density algorithm for pairwise interaction testing. *Software Testing, Verification, and Reliability*, 17:159–182, 2007.

3. R. C. Bryce and C. J. Colbourn. A density-based greedy algorithm for higher strength covering arrays. *preprint*, 2007.
4. G. D. Cohen, S. Litsyn, and G. Zémor. On greedy algorithms in coding theory. *IEEE Trans. Inform. Theory*, 42:2053–2057, 1996.
5. C. J. Colbourn, S. S. Martirosyan, Tran Van Trung, and R. A. Walker II. Roux-type constructions for covering arrays of strengths three and four. *Designs Codes Crypt.*, 41:33–57, 2006.
6. Z. J. Czech, G. Havas, and B. S. Majewski. Perfect hashing. *Theor. Comp. Sci.*, 182:1–143, 1997.
7. A. Fiat and M. Naor. Broadcast encryption. *Lecture Notes in Computer Science*, 773:480–491, 1994.
8. K. Mehlhorn. *Data Structures and Algorithms 1: Sorting and Searching.* Springer-Verlag, Berlin, 1984.
9. D. R. Stinson, Tran van Trung, and R. Wei. Secure frameproof codes, key distribution patterns, group testing algorithms and related structures. *J. Statist. Plan. Infer.*, 86:595–617, 2000.
10. R. A. Walker II and C. J. Colbourn. Perfect hash families: Constructions and existence. *J. Math. Crypt.*, 1:125–150, 2007.

Two-Weight Codes Punctured from Irreducible Cyclic Codes

Cunsheng Ding
Department of Computer Science and Engineering
The Hong Kong University of Science and Technology
Clearwater Bay, Kowloon, Hong Kong
E-mail: `cding@cse.ust.hk`
`www.cse.ust.hk/faculty/cding/`

Jinquan Luo
Department of Mathematics
Yangzhou University
Yangzhou, Jiangsu Province 225009, China
E-mail: `luojq01@gmail.com`

Harald Niederreiter
Department of Mathematics
National University of Singapore
2 Science Drive 2, Singapore 117543, Republic of Singapore
E-mail: `nied@math.nus.edu.sg`
`www.math.nus.edu.sg/~nied`

In this paper, a class of two-weight linear codes over $\mathbb{F}_q$ is presented. This family of codes is punctured from irreducible cyclic codes and contains a subclass of MDS codes and other optimal codes. Their dual codes are also optimal in some cases. The punctured codes are in general better than the original irreducible cyclic codes in terms of information rate and error correcting capability.

Keywords: Cyclotomy; linear codes; two-weight codes.

1. Introduction

Let q be an arbitrary prime power. An $[n, k, d]$ linear code over $\mathbb{F}_q$ is a k-dimensional subspace of $\mathbb{F}_q^n$ with minimum Hamming distance d. Let $r = q^m$ with a positive integer m and let N be a positive integer dividing $r-1$. Put $n = (r-1)/N$. Let g be a primitive element of $\mathbb{F}_r$ and let $\theta = g^N$. The set

$$\mathsf{C}(N, q^m) = \{\mathbf{c}(\beta) = (\mathrm{Tr}(\beta), \mathrm{Tr}(\beta\theta), \ldots, \mathrm{Tr}(\beta\theta^{n-1})) : \beta \in \mathbb{F}_r\} \quad (1)$$

is called an *irreducible cyclic* $[n, m_0]$ *code* over $\mathbb{F}_q$, where Tr is the trace function from $\mathbb{F}_r$ onto $\mathbb{F}_q$ and m_0 divides m.

The weight distribution of irreducible cyclic codes is quite complicated in general [11]. However, in certain special cases the weight distribution is known. We summarize these cases below.

(1) When $N|(q^j+1)$ for some j being a divisor of $m/2$, which is called the *semi-primitive case*, the codes have two weights. These codes were studied by Delsarte and Goethals [4], McEliece [12], and Baumert and McEliece [1].
(2) When $N = 2$, the weight distribution was found by Baumert and McEliece [1].
(3) When N is a prime with $N \equiv 3 \pmod 4$ and $\mathrm{ord}_q(N) = (N-1)/2$, the weight distribution was determined by Baumert and Mykkeltveit [2].

McEliece also generalized these results [12] and showed that the weights of an irreducible cyclic code can be expressed as a linear combination of Gauss sums via the Fourier transform [13] (see also McEliece and Rumsey [14], Fitzgerald and Yucas [6], van der Vlugt [17], and the references therein).

Two-weight codes are a class of interesting codes which are closely related to combinatorial designs, finite geometry, and graph theory. Information on them can be found in Baumert and McEliece [1], Calderbank and Kantor [3], Wolfmann [19], Delsarte and Goethals [4], Langevin [9], Schmidt and White [16], Wolfmann [20,21], and Vega and Wolfmann [18].

The objective of this paper is to improve the irreducible cyclic codes in certain cases. Our idea of the improvement is to puncture these codes in a proper way. The codes reported in this paper have only two nonzero weights and are better than the original irreducible cyclic codes in terms of error-correcting capability and information rate. They contain a subclass of MDS codes and other optimal codes. Their duals are also very good in general, and contain optimal codes.

2. The class of two-weight codes

Let p be a prime, $q = p^s$, $m = 2lk$, $r = q^m$, where s, l, and k are positive integers. Let h be a positive divisor of q^k+1, and so of $(r-1)/(q-1)$, and let g be a primitive element of $\mathbb{F}_r$. Throughout this paper we always assume $h < \sqrt{r}+1$. Define the additive character χ on $\mathbb{F}_r$ by $\chi(x) = \zeta_p^{\mathrm{Tr}_{r/p}(x)}$, where $\zeta_p = \exp(2\pi\sqrt{-1}/p)$ and $\mathrm{Tr}_{r/p} : \mathbb{F}_r \to \mathbb{F}_p$ is the absolute trace mapping.

Let $\mathrm{Tr}_{r/q} : \mathbb{F}_r \to \mathbb{F}_q$ and $\mathrm{Tr}_{q/p} : \mathbb{F}_q \to \mathbb{F}_p$ be further trace mappings. Define

$$C_i = \left\{ g^{i+jh} : 0 \leq j < \frac{r-1}{h} \right\} \tag{2}$$

for each i with $0 \leq i \leq h-1$. These C_i $(0 \leq i \leq h-1)$ are called the *cyclotomic classes* of order h in $\mathbb{F}_r^*$.

For any $\alpha \in \mathbb{F}_r$, define the exponential sum

$$S(\alpha) = \sum_{x \in \mathbb{F}_r} \chi(\alpha x^h).$$

Lemma 2.1. *[15] Assume $h \mid (q^k+1)$. Then for any $\alpha \in \mathbb{F}_r^*$,*

$$S(\alpha) = \begin{cases} (-1)^l \sqrt{r} & \text{if } \alpha \notin C_{h_0}, \\ (-1)^{l-1}(h-1)\sqrt{r} & \text{if } \alpha \in C_{h_0}, \end{cases}$$

where

$$h_0 = \begin{cases} h/2 & \text{if } p > 2,\ l \text{ odd, and } (q^k+1)/h \text{ odd}, \\ 0 & \text{otherwise.} \end{cases}$$

Let $n = (r-1)/h(q-1)$. Note that $\mathbb{F}_q^* \subset C_0$. Define $d_i = g^{h(i-1)}$ for $i = 1, \ldots, n$. Then

$$\{d_1, d_2, \ldots, d_n\} \text{ is a complete set of coset representatives of } C_0/\mathbb{F}_q^*. \tag{3}$$

We are now ready to present the construction of the two-weight codes. For each $\alpha \in \mathbb{F}_r$, define the vector

$$\mathbf{c}(\alpha) = (\mathrm{Tr}_{r/q}(\alpha d_1), \mathrm{Tr}_{r/q}(\alpha d_2), \ldots, \mathrm{Tr}_{r/q}(\alpha d_n)).$$

We then define the code

$$\mathcal{C} = \{\mathbf{c}(\alpha) : \alpha \in \mathbb{F}_r\}.$$

Theorem 2.1. *The set $\mathcal{C}$ defined above is an $[n, m]$ two-weight linear code over $\mathbb{F}_q$ and has the following weight distribution:*

weight	*frequency*
$\frac{r+(-1)^l(h-1)\sqrt{r}}{qh}$	$\frac{r-1}{h}$
$\frac{r+(-1)^{l-1}\sqrt{r}}{qh}$	$\frac{(h-1)(r-1)}{h}$
0	1

Furthermore, the dual code $\mathcal{C}^{\perp}$ of $\mathcal{C}$ is an $[n, n-m, d^{\perp}]$ linear code with minimum distance $d^{\perp} \geq 3$.

Proof. Clearly, $\mathcal{C}$ is a linear code over $\mathbb{F}_q$ of length n. We now determine the dimension and weight distribution of $\mathcal{C}$. For any $\alpha \in \mathbb{F}_r^*$, the Hamming weight $w_H(\mathbf{c}(\alpha))$ of the codeword $\mathbf{c}(\alpha) \in \mathcal{C}$ is given by

$$\begin{aligned} w_H(\mathbf{c}(\alpha)) &= n - \left|\{i : 1 \leq i \leq n, \operatorname{Tr}_{r/q}(\alpha d_i) = 0\}\right| \\ &= n - \tfrac{1}{q} \sum_{i=1}^{n} \sum_{t \in \mathbb{F}_q} \zeta_p^{\operatorname{Tr}_{q/p}(t \operatorname{Tr}_{r/q}(\alpha d_i))} \\ &= n - \tfrac{n}{q} - \tfrac{1}{q} \sum_{t \in \mathbb{F}_q^*} \sum_{i=1}^{n} \chi(\alpha t d_i) \\ &= n - \tfrac{n}{q} - \tfrac{1}{q} \sum_{j=1}^{(r-1)/h} \chi(\alpha g^{jh}) \\ &= n - \tfrac{n}{q} - \tfrac{1}{qh} \left(S(\alpha) - 1\right), \end{aligned}$$

where the third equality holds by the transitivity of the trace (see [10, Theorem 2.26]) and the fourth equality holds by property (3).

It then follows from Lemma 2.1 that

$$w_H(\mathbf{c}(\alpha)) = n - \frac{n}{q} - \frac{1}{qh}(S(\alpha) - 1) = \begin{cases} \frac{r+(-1)^l(h-1)\sqrt{r}}{qh} & \text{if } \alpha \in C_{h_0}, \\ \frac{r+(-1)^{l-1}\sqrt{r}}{qh} & \text{otherwise.} \end{cases}$$

Since $h < \sqrt{r} + 1$, we have $w_H(\mathbf{c}(\alpha)) \neq 0$ for all nonzero $\alpha \in \mathbb{F}_r$. Hence the $\mathbb{F}_q$-linear map $\alpha \in \mathbb{F}_r \mapsto \mathbf{c}(\alpha)$ is injective. Then the conclusions about the dimension and weight distribution of $\mathcal{C}$ follow.

For the dual code $\mathcal{C}^{\perp}$, we only need to prove the lower bound for the minimum distance $d^{\perp}$. Suppose on the contrary that $d^{\perp} \leq 2$. By the well-known characterization of the minimum distance of a linear code in terms of a parity-check matrix (see [10, Lemma 9.14]), there exist integers $1 \leq i < j \leq n$ such that the ith column and the jth column of a parity-check matrix of $\mathcal{C}^{\perp}$ (which is the same as a generator matrix of $\mathcal{C}$) are linearly dependent over $\mathbb{F}_q$. If $\theta_1, \ldots, \theta_m$ form an $\mathbb{F}_q$-basis of $\mathbb{F}_r$, then $\mathbf{c}(\theta_1), \ldots, \mathbf{c}(\theta_m)$ form an $\mathbb{F}_q$-basis of $\mathcal{C}$. It follows that there exists $(c_1, c_2) \in \mathbb{F}_q^2 \setminus \{(0,0)\}$ such that

$$c_1 \operatorname{Tr}_{r/q}(\theta_u d_i) + c_2 \operatorname{Tr}_{r/q}(\theta_u d_j) = 0 \qquad \text{for } 1 \leq u \leq m.$$

This implies

$$\operatorname{Tr}_{r/q}(\theta_u(c_1 d_i + c_2 d_j)) = 0 \qquad \text{for } 1 \leq u \leq m.$$

It follows now from [10, Theorem 2.24] that

$$c_1 d_i + c_2 d_j = 0.$$

We may assume $c_1 \neq 0$. Hence $\frac{d_i}{d_j} = -\frac{c_2}{c_1} \in \mathbb{F}_q$, which is a contradiction to (3). This completes the proof. □

For the code $\mathcal{C}$ and its dual code $\mathcal{C}^\perp$, we have the following remarks.

(1) Theorem 7 in reference [5] is the special case of Theorem 2.1 with $h = 3$, $q \equiv 2 \bmod 3$, and $k = 1$.
(2) Numerical results show that the code $\mathcal{C}$ is not equivalent to any cyclic code in general (see reference [5] for details).
(3) If $l = k = 1$, $h < q+1$, and $h|(q+1)$, then $\mathcal{C}$ is a $[(q+1)/h, 2, (q+1)/h-1]$ MDS code over $\mathbb{F}_q$.
(4) If $l = 2$, $k = 1$, and $h = q+1$, then $\mathcal{C}$ is a $[q^2+1, 4, q(q-1)]$ optimal code over $\mathbb{F}_q$ with respect to the Griesmer bound.
(5) If $(p, s, l, k, h) = (3, 1, 1, 2, 2)$, then $\mathcal{C}$ is a $[20, 4, 12]$ optimal code over $\mathbb{F}_3$ according to [8]. Its dual is a $[20, 16, 3]$ optimal code over $\mathbb{F}_3$ according to [8].
(6) If $(p, s, l, k, h) = (5, 1, 1, 2, 2)$, then $\mathcal{C}$ is a $[78, 4, 60]$ optimal code over $\mathbb{F}_5$ according to [8]. Its dual is a $[78, 74, 3]$ optimal code over $\mathbb{F}_5$ according to [8].

3. Concluding remarks

The parameters of the code $\mathcal{C}$ in Theorem 2.1 are the same as those of a code in Example SU2 of reference [3]. It is open whether the two codes are equivalent. The reader is invited to attack the equivalence problem. At any rate, the construction of the code in Theorem 2.1 is much simpler than that in [3].

The weight distribution of the codes in Theorem 2.1 can be obtained from that of the irreducible cyclic codes in the semi-primitive case [1,16]. Here we prefer a direct proof using Lemma 2.1.

While the irreducible cyclic codes in the semi-primitive case are not good, the punctured codes in Theorem 2.1 are optimal or almost optimal compared with the best codes known in terms of error-correcting capability when the information rate is fixed. This shows a clear difference between the original codes and the punctured version presented in this paper. So it is worthwhile to report these codes and their duals.

References

1. L. D. Baumert and R. J. McEliece, Weights of irreducible cyclic codes, *Information and Control* 20(2) (1972), 158–175.
2. L. D. Baumert and J. Mykkeltveit, Weight distributions of some irreducible cyclic codes, *DSN Progress Report* 16 (1973), 128–131.
3. R. Calderbank and W. M. Kantor, The geometry of two-weight codes, *Bull. London Math. Soc.* 18 (1986), 97–122.
4. P. Delsarte and J. M. Goethals, Irreducible binary cyclic codes of even dimension, in: *Proc. Second Chapel Hill Conf. on Combinatorial Mathematics and Its Applications* (Univ. North Carolina, Chapel Hill, NC, 1970), 100–113.
5. C. Ding and H. Niederreiter, Cyclotomic linear codes of order 3, *IEEE Trans. Inf. Theory* 53(6) (2007), 2274–2277.
6. R. W. Fitzgerald and J. L. Yucas, Sums of Gauss sums and weights of irreducible codes, *Finite Fields and Their Appl.* 11 (2005), 89–110.
7. J. M. Goethals and H. C. A. van Tilborg, Uniformly packed codes, *Philips Res. Rep.* 30 (1975), 9–36.
8. M. Grassl, Bounds on the minimum distance of linear codes, available online at `http://www.codetables.de`.
9. P. Langevin, A new class of two weight codes, in: S. Cohen and H. Niederreiter, eds., *Finite Fields and Applications,* 181–187, Cambridge: Cambridge University Press, 1996.
10. R. Lidl and H. Niederreiter, *Finite Fields*, Cambridge: Cambridge University Press, 1997.
11. F. MacWilliams and J. Seery, The weight distributions of some minimal cyclic codes, *IEEE Trans. Inf. Theory* 27(6) (1981), 796–806.
12. R. J. McEliece, A class of two-weight codes, *Jet Propulsion Laboratory Space Program Summary* 37–41, vol. IV, 264–266.
13. R. J. McEliece, Irreducible cyclic codes and Gauss sums, in: *Combinatorics, Part 1: Theory of Designs, Finite Geometry and Coding Theory,* Math. Centre Tracts, no. 55, 179–196, Amsterdam: Math. Centrum, 1974.
14. R. J. McEliece and H. Rumsey Jr., Euler products, cyclotomy, and coding, *J. Number Theory* 4 (1972), 302–311.
15. M. J. Moisio, A note on evaluations of some exponential sums, *Acta Arith.* 93 (2000), 117–119.
16. B. Schmidt and C. White, All two-weight irreducible cyclic codes?, *Finite Fields and Their Appl.* 8 (2002), 1–17.
17. M. van der Vlugt, Hasse-Davenport curves, Gauss sums, and weight distributions of irreducible cyclic codes, *J. Number Theory* 55 (1995), 145–159.
18. G. Vega and J. Wolfmann, New classes of 2-weight cyclic codes, *Des. Codes Cryptogr.* 42 (2007), 327–334.
19. J. Wolfmann, Codes projectifs à deux poids, “caps” complets et ensembles de différences, *J. Comb. Theory Ser. A* 23 (1977), 208–222.
20. J. Wolfmann, Are 2-weight projective cyclic codes irreducible?, *IEEE Trans. Inf. Theory* 51 (2005), 733–737.
21. J. Wolfmann, Projective two-weight irreducible cyclic and constacyclic codes, *Finite Fields and Their Appl.*, accepted for publication.

On the Joint Linear Complexity of Linear Recurring Multisequences

Fang-Wei Fu
Chern Institute of Mathematics, Nankai University,
Tianjin 300071, China
E-mail: fwfu@nankai.edu.cn

Harald Niederreiter
Department of Mathematics, National University of Singapore,
2 Science Drive 2, Singapore 117543, Republic of Singapore
E-mail: nied@math.nus.edu.sg

Ferruh Özbudak
Division of Mathematical Sciences, School of Physical and Mathematical Sciences,
Nanyang Technological University, Singapore 637616, Republic of Singapore, and
Department of Mathematics and Institute of Applied Mathematics,
Middle East Technical University, Ankara, 06531, Ankara, Turkey
E-mail: ozbudak@metu.edu.tr

We study the joint linear complexity of linear recurring multisequences, i.e., of multisequences consisting of linear recurring sequences. The expectation and variance of the joint linear complexity of random linear recurring multisequences are determined. These results extend the corresponding results on the expectation and variance of the joint linear complexity of random periodic multisequences. Then we enumerate the linear recurring multisequences with fixed joint linear complexity and determine the generating polynomial for the distribution of joint linear complexities. The proofs use new methods that enable us to obtain results of great generality.

Keywords: Multisequence; linear recurring sequence; joint linear complexity; expectation; variance.

1. Introduction

The linear complexity of sequences is one of the important security measures for stream cipher systems (see [1,4,11,15,16]). The linear complexity of a finite or periodic sequence is the length of the shortest linear feedback shift register that can generate it. When a sequence is used in stream ciphers as a keystream, it must have high linear complexity to resist an attack by the Berlekamp-Massey algorithm, since otherwise one can use the Berlekamp-Massey algorithm to generate the whole sequence from a few initial terms. It is well known that a stream cipher system is completely secure if the keystream is a "truly random" (uniformly distributed) sequence.

A fundamental research problem in stream ciphers is to determine the expectation and variance of the linear complexity of random sequences that are uniformly distributed. Recently, motivated by the study of vectorized stream cipher systems (see [3,7]), the analogous problem for the joint linear complexity of multisequences has been investigated. Dai, Imamura, and Yang [2], Feng and Dai [5], Niederreiter [12], and Niederreiter and Wang [13,14,17] studied the expectation and variance and counting function of the joint linear complexity of finite multisequences over a finite field. Using the generalized discrete Fourier transform for multisequences, Meidl and Niederreiter [9] derived general formulas for the expectation of the joint linear complexity of random periodic multisequences. They also determined the counting function of the joint linear complexity of periodic multisequences for certain special values of the period. Using again the generalized discrete Fourier transform for multisequences, Fu, Niederreiter, and Su [6] derived a general formula for the variance of the joint linear complexity of random periodic multisequences.

Let q be an arbitrary prime power and $\mathbb{F}_q$ be a finite field with q elements. For a finite set M, we denote by $|M|$ its cardinality.

A sequence $\sigma = (s_n)_{n=0}^{\infty}$ of elements of $\mathbb{F}_q$ is called a *linear recurring sequence* over $\mathbb{F}_q$ with characteristic polynomial

$$\sum_{i=0}^{\ell} a_i x^i \in \mathbb{F}_q[x]$$

if $a_\ell = 1$ and

$$\sum_{i=0}^{\ell} a_i s_{n+i} = 0 \qquad \text{for } n = 0, 1, \ldots.$$

Here ℓ is an arbitrary nonnegative integer.

Let m be an arbitrary positive integer. For an m-fold multisequence $\boldsymbol{S} = (\sigma_1, \ldots, \sigma_m)$ consisting of linear recurring sequences $\sigma_1, \ldots, \sigma_m$ over $\mathbb{F}_q$ (that is, for a linear recurring multisequence $\boldsymbol{S}$ over $\mathbb{F}_q$), its *joint minimal polynomial* $P_{\boldsymbol{S}} \in \mathbb{F}_q[x]$ is defined to be the (uniquely determined) monic polynomial of the least degree such that $P_{\boldsymbol{S}}$ is a characteristic polynomial of σ_i for each $1 \leq i \leq m$. The *joint linear complexity* $L(\boldsymbol{S})$ of $\boldsymbol{S}$ is defined to be $L(\boldsymbol{S}) = \deg(P_{\boldsymbol{S}})$. For $1 \leq i \leq m$, let

$$\sigma_i = (s_{i,n})_{n=0}^{\infty},$$

and assume that σ_i is not the zero sequence for some $1 \leq i \leq m$. Then, equivalently, the joint linear complexity $L(\boldsymbol{S})$ is the smallest positive integer c for which there exist coefficients $a_1, a_2, \ldots, a_c \in \mathbb{F}_q$ such that for each $1 \leq i \leq m$, we have

$$s_{i,n} + a_1 s_{i,n-1} + \cdots + a_c s_{i,n-c} = 0 \qquad \text{for all } n \geq c.$$

For a monic polynomial $f \in \mathbb{F}_q[x]$, let $\mathcal{M}^{(m)}(f)$ be the set of m-fold multisequences $\boldsymbol{S} = (\sigma_1, \ldots, \sigma_m)$ such that for each $1 \leq i \leq m$, σ_i is a linear recurring sequence over $\mathbb{F}_q$ with characteristic polynomial f. It follows that

$$\left|\mathcal{M}^{(m)}(f)\right| = q^{m \deg(f)}.$$

In this paper we determine the expectation $\mathrm{E}^{(m)}(f)$ and the variance $\mathrm{Var}^{(m)}(f)$ of the joint linear complexity of random m-fold multisequences from $\mathcal{M}^{(m)}(f)$, which are uniformly distributed over $\mathcal{M}^{(m)}(f)$. For a monic polynomial $f \in \mathbb{F}_q[x]$ of positive degree and an arbitrary positive integer m, we derive a general formula for the counting function $\mathcal{N}^{(m)}(f;t)$ which is defined as the number of m-fold multisequences from $\mathcal{M}^{(m)}(f)$ with a given joint linear complexity t. We determine closed-form expressions for $\mathcal{N}^{(m)}(f;\deg(f))$, $\mathcal{N}^{(m)}(f;\deg(f)-1)$, and $\mathcal{N}^{(m)}(f;\deg(f)-2)$. We give concrete examples determining the counting functions in closed form in some special cases. Moreover, we determine the generating polynomial $\mathcal{G}^{(m)}(f;z)$ for the distribution of joint linear complexities of m-fold multisequences from $\mathcal{M}^{(m)}(f)$.

In the literature (cf. [6,9]) there are results for the special case $f(x) = x^N - 1$. In this paper we use new methods that enable us to obtain results of greater generality and also in a simpler way.

The rest of the paper is organized as follows. In Section 2 we give some definitions and we derive some technical results that we use later. The expectation $\mathrm{E}^{(m)}(f)$ and the variance $\mathrm{Var}^{(m)}(f)$ are determined in Section 3. The results on the counting function $\mathcal{N}^{(m)}(f;t)$ and the generating polynomial $\mathcal{G}^{(m)}(f;z)$ are given in Section 4.

2. Preliminaries

In this section we give some definitions and we derive some technical results that we use later in this paper. We assume that m is a positive integer.

For a monic polynomial $f \in \mathbb{F}_q[x]$ with $\deg(f) \geq 1$, let

$$\begin{aligned} C(f) &:= \{h \in \mathbb{F}_q[x] : \deg(h) < \deg(f)\}, \\ R^{(m)}(f) &:= \{(h_1, \ldots, h_m) \in C(f)^m : \gcd(h_1, \ldots, h_m, f) = 1\}, \\ \Phi_q^{(m)}(f) &:= \left| R^{(m)}(f) \right|, \text{ and } \Phi_q^{(m)}(1) := 1. \end{aligned} \tag{1}$$

We demonstrate some properties of $\Phi_q^{(m)}(f)$. We begin with the following lemma.

Lemma 2.1. *Let $f \in \mathbb{F}_q[x]$ be a monic polynomial. Then*

$$\sum_{d|f} \Phi_q^{(m)}(d) = q^{m \deg(f)},$$

where the summation is over all monic polynomials $d \in \mathbb{F}_q[x]$ dividing f.

Proof. For $f = 1$, the result is trivial. Assume that $\deg(f) \geq 1$. For a monic polynomial $d \in \mathbb{F}_q[x]$ dividing f, let

$$T^{(m)}(f, d) := \{(h_1, \ldots, h_m) \in C(f)^m : \gcd(h_1, \ldots, h_m, f) = d\}.$$

It is clear that we have the disjoint union

$$C(f)^m = \bigcup_{d|f} T^{(m)}(f, d),$$

where the union is taken over all monic polynomials $d \in \mathbb{F}_q[x]$ dividing f. Moreover, it follows from the definition that

$$T^{(m)}(f, d) = \left\{(dh_1, \ldots, dh_m) : (h_1, \ldots, h_m) \in R^{(m)}(f/d)\right\}.$$

Using (1) we complete the proof. □

We define the Möbius function μ_q on the set of monic polynomials from $\mathbb{F}_q[x]$ as follows. We put $\mu_q(1) := 1$. If $f \in \mathbb{F}_q[x]$ is a monic polynomial with $\deg(f) \geq 1$ and if $f = r_1^{e_1} r_2^{e_2} \cdots r_k^{e_k}$ is the canonical factorization of f into monic irreducible polynomials over $\mathbb{F}_q$, then we define

$$\mu_q(f) := \begin{cases} (-1)^k & \text{if } e_1 = \cdots = e_k = 1, \\ 0 & \text{if } e_i \geq 2 \text{ for some } 1 \leq i \leq k. \end{cases}$$

An argument similar to that for the classical Möbius function on the set of positive integers (cf. [8, Lemma 3.23]) implies that for a monic polynomial

$f \in \mathbb{F}_q[x]$ we have

$$\sum_{d|f} \mu_q(d) = \begin{cases} 1 & \text{if } f = 1, \\ 0 & \text{otherwise,} \end{cases} \tag{2}$$

where the summation is over all monic polynomials $d \in \mathbb{F}_q[x]$ dividing f.

In the next lemma we establish further properties of $\Phi_q^{(m)}(f)$.

Lemma 2.2. *Let $f, f_1, f_2 \in \mathbb{F}_q[x]$ be monic polynomials.*

(i) We have

$$\Phi_q^{(m)}(f) = \sum_{d|f} \mu_q(d) q^{m(\deg(f)-\deg(d))},$$

where the summation is over all monic polynomials $d \in \mathbb{F}_q[x]$ dividing f.

(ii) If $\gcd(f_1, f_2) = 1$, *then*

$$\Phi_q^{(m)}(f_1 f_2) = \Phi_q^{(m)}(f_1)\Phi_q^{(m)}(f_2).$$

(iii) If $f = r_1^{e_1} r_2^{e_2} \cdots r_k^{e_k}$ is the canonical factorization of f into monic irreducible polynomials over $\mathbb{F}_q$, then

$$\Phi_q^{(m)}(f) = q^{m \deg(f)} \prod_{i=1}^{k} \left(1 - q^{-m \deg(r_i)}\right).$$

Proof. Using Lemma 2.1 we get

$$\sum_{d|f} \mu_q(d) q^{m(\deg(f)-\deg(d))} = \sum_{d|f} \mu_q\left(\frac{f}{d}\right) q^{m \deg(d)} = \sum_{d|f} \mu_q\left(\frac{f}{d}\right) \sum_{d_1|d} \Phi_q^{(m)}(d_1)$$

which is equal to

$$\sum_{d_1|f} \Phi_q^{(m)}(d_1) \sum_{g|\frac{f}{d_1}} \mu_q(g). \tag{3}$$

By (2) we obtain that the inner sum in (3) is always 0, unless $d_1 = f$, in which case it is 1. Hence

$$\sum_{d|f} \mu_q(d) q^{m(\deg(f)-\deg(d))} = \Phi_q^{(m)}(f),$$

which proves (i). If $\gcd(f_1, f_2) = 1$, then it follows from the definition that

$$\mu_q(f_1 f_2) = \mu_q(f_1)\mu_q(f_2). \tag{4}$$

Hence (ii) follows from (i) and (4). If $r \in \mathbb{F}_q[x]$ is a monic irreducible polynomial and m and e are positive integers, then by the definition of the Möbius function μ_q and (i) we have

$$\Phi_q^{(m)}(r^e) = q^{me\deg(r)}\left(1 - q^{-m\deg(r)}\right). \tag{5}$$

Using (5) and (ii) with induction on the number of distinct monic irreducible factors in the canonical factorization of f over $\mathbb{F}_q$, we obtain (iii). □

3. Expectation and variance

Let $f \in \mathbb{F}_q[x]$ be a monic polynomial with $\deg(f) \geq 1$. In this section we determine the expectation $\mathrm{E}^{(m)}(f)$ and the variance $\mathrm{Var}^{(m)}(f)$ of the joint linear complexity of random m-fold multisequences from $\mathcal{M}^{(m)}(f)$.

First we define

$$S_1^{(m)}(f) := \sum_{d|f} \Phi_q^{(m)}(d)\deg(d) \tag{6}$$

and

$$S_2^{(m)}(f) := \sum_{d|f} \Phi_q^{(m)}(d)\left(\deg(d)\right)^2, \tag{7}$$

where the summations are over all monic polynomials $d \in \mathbb{F}_q[x]$ dividing f. These definitions are used in the following theorem.

Theorem 3.1. *Let $f \in \mathbb{F}_q[x]$ be a monic polynomial with $\deg(f) \geq 1$. The expectation $\mathrm{E}^{(m)}(f)$ and the variance $\mathrm{Var}^{(m)}(f)$ of the joint linear complexity of random m-fold multisequences from $\mathcal{M}^{(m)}(f)$ are given by*

$$\mathrm{E}^{(m)}(f) = \frac{S_1^{(m)}(f)}{q^{m\deg(f)}}$$

and

$$\mathrm{Var}^{(m)}(f) = \frac{S_2^{(m)}(f)}{q^{m\deg(f)}} - \left(\frac{S_1^{(m)}(f)}{q^{m\deg(f)}}\right)^2.$$

Proof. Let $\boldsymbol{S} = (\sigma_1, \ldots, \sigma_m)$ be an m-fold multisequences from $\mathcal{M}^{(m)}(f)$. Using [10, Lemma 1] we obtain a uniquely determined m-tuple $\left(\frac{g_1}{f}, \frac{g_2}{f}, \ldots, \frac{g_m}{f}\right)$ of rational functions with $g_i \in \mathbb{F}_q[x]$ and $\deg(g_i) < \deg(f)$ for $1 \leq i \leq m$, and different choices of $\boldsymbol{S}$ yield different m-tuples of such rational functions. The joint minimal polynomial of $\boldsymbol{S}$ is the monic polynomial $d \in \mathbb{F}_q[x]$ such that there exist $h_1, \ldots, h_m \in \mathbb{F}_q[x]$ with

$$\frac{g_i}{f} = \frac{h_i}{d} \quad \text{for } 1 \leq i \leq m \text{ and } \gcd(h_1, \ldots, h_m, d) = 1.$$

Hence for each monic polynomial $d \in \mathbb{F}_q[x]$ dividing f, it follows that $\Phi_q^{(m)}(d)$ is the number of m-fold multisequences in $\mathcal{M}^{(m)}(f)$ with joint minimal polynomial d. Therefore using (6) we obtain the formula for the expectation $\mathrm{E}^{(m)}(f)$.

The variance $\mathrm{Var}^{(m)}(f)$ is given by

$$\mathrm{Var}^{(m)}(f) = \frac{1}{q^{m \deg(f)}} \sum_{\boldsymbol{S} \in \mathcal{M}^{(m)}(f)} L(\boldsymbol{S})^2 - \mathrm{E}^{(m)}(f)^2.$$

Then using (7) we complete the proof similarly. □

We establish useful properties of $S_1^{(m)}(f)$ and $S_2^{(m)}(f)$ in the following proposition.

Proposition 3.1. *For a monic polynomial* $f \in \mathbb{F}_q[x]$ *with* $\deg(f) \geq 1$, *let*

$$f = r_1^{e_1} r_2^{e_2} \cdots r_k^{e_k}$$

be the canonical factorization of f *into monic irreducible polynomials over* $\mathbb{F}_q$. *For* $1 \leq i \leq k$, *let* $\alpha_i = q^{m \deg(r_i)}$. *Then*

$$S_1^{(m)}(f) = q^{m \deg(f)} \sum_{i=1}^{k} \frac{S_1^{(m)}(r_i^{e_i})}{\alpha_i^{e_i}} \tag{8}$$

and

$$S_2^{(m)}(f) = q^{m \deg(f)} \left(\sum_{i=1}^{k} \frac{S_2^{(m)}(r_i^{e_i})}{\alpha_i^{e_i}} + 2 \sum_{1 \leq i < j \leq k} \frac{S_1^{(m)}(r_i^{e_i})}{\alpha_i^{e_i}} \frac{S_1^{(m)}(r_j^{e_j})}{\alpha_j^{e_j}} \right).$$

Proof. Let $f_1, f_2 \in \mathbb{F}_q[x]$ be monic polynomials with $\gcd(f_1, f_2) = 1$. Using (1), (6), and Lemmas 2.1 and 2.2(ii), we obtain that

$$\begin{aligned} S_1^{(m)}(f_1 f_2) &= \sum_{d_1 | f_1} \sum_{d_2 | f_2} \Phi_q^{(m)}(d_1 d_2) \left(\deg(d_1) + \deg(d_2)\right) \\ &= S_1^{(m)}(f_1) \sum_{d_2 | f_2} \Phi_q^{(m)}(d_2) + S_1^{(m)}(f_2) \sum_{d_1 | f_1} \Phi_q^{(m)}(d_1) \qquad (9) \\ &= q^{m \deg(f_2)} S_1^{(m)}(f_1) + q^{m \deg(f_1)} S_1^{(m)}(f_2). \end{aligned}$$

Similarly using (1), (7), and Lemmas 2.1 and 2.2(ii), we also obtain that

$$S_2^{(m)}(f_1 f_2) = q^{m\deg(f_2)} S_2^{(m)}(f_1) + q^{m\deg(f_1)} S_2^{(m)}(f_2) \\ +2S_1^{(m)}(f_1)S_1^{(m)}(f_2). \tag{10}$$

We complete the proof using (9), (10), and induction on the number of distinct monic irreducible factors in the canonical factorization of f over $\mathbb{F}_q$. □

We obtain the following corollary.

Corollary 3.1. *For a monic polynomial $f \in \mathbb{F}_q[x]$ as in Proposition 3.1, we have*

$$\mathrm{E}^{(m)}(f) = \sum_{i=1}^{k} \mathrm{E}^{(m)}(r_i^{e_i}) \quad \textit{and} \quad \mathrm{Var}^{(m)}(f) = \sum_{i=1}^{k} \mathrm{Var}^{(m)}(r_i^{e_i}).$$

Proof. The formula for $\mathrm{E}^{(m)}(f)$ follows from Theorem 3.1 and (8). From Theorem 3.1 and Proposition 3.1 we get

$$\begin{aligned}\mathrm{Var}^{(m)}(f) &= \sum_{i=1}^{k} \frac{S_2^{(m)}(r_i^{e_i})}{\alpha_i^{e_i}} + 2 \sum_{1 \le i < j \le k} \frac{S_1^{(m)}(r_i^{e_i})}{\alpha_i^{e_i}} \frac{S_1^{(m)}(r_j^{e_j})}{\alpha_j^{e_j}} \\ &\quad - \sum_{i=1}^{k} \frac{S_1^{(m)}(r_i^{e_i})}{\alpha_i^{e_i}} \sum_{j=1}^{k} \frac{S_1^{(m)}(r_j^{e_j})}{\alpha_j^{e_j}} \\ &= \sum_{i=1}^{k} \left(\frac{S_2^{(m)}(r_i^{e_i})}{\alpha_i^{e_i}} - \left(\frac{S_1^{(m)}(r_i^{e_i})}{\alpha_i^{e_i}} \right)^2 \right).\end{aligned}$$

We complete the proof using Theorem 3.1. □

Now we determine the expectation and the variance of the joint linear complexity of random m-fold multisequences from $\mathcal{M}^{(m)}(f)$.

Theorem 3.2. *For a monic polynomial $f \in \mathbb{F}_q[x]$ with* $\deg(f) \ge 1$, *let*

$$f = r_1^{e_1} r_2^{e_2} \cdots r_k^{e_k}$$

be the canonical factorization of f into monic irreducible polynomials over $\mathbb{F}_q$. For $1 \le i \le k$, let $\alpha_i = q^{m\deg(r_i)}$. For an arbitrary positive integer

m, the expectation $\mathrm{E}^{(m)}(f)$ and the variance $\mathrm{Var}^{(m)}(f)$ of the joint linear complexity of random m-fold multisequences from $\mathcal{M}^{(m)}(f)$ are given by

$$\mathrm{E}^{(m)}(f) = \deg(f) - \sum_{i=1}^{k} \frac{1-\alpha_i^{-e_i}}{\alpha_i - 1} \deg(r_i),$$

$$\mathrm{Var}^{(m)}(f) = \sum_{i=1}^{k} \left(\frac{\deg(r_i)}{1-\alpha_i^{-1}}\right)^2 \left((2e_i+1)\left(\alpha_i^{-e_i-2} - \alpha_i^{-e_i-1}\right) - \alpha_i^{-2e_i-2} + \alpha_i^{-1}\right).$$

Proof. For $1 \le i \le k$, it follows from (6) and Lemma 2.2(iii) that

$$S_1^{(m)}(r_i^{e_i}) = \sum_{j=0}^{e_i} \alpha_i^j \left(1 - \frac{1}{\alpha_i}\right) j \deg(r_i). \tag{11}$$

For any positive integer e and any real number $z \neq 1$, we have the identity

$$\sum_{j=0}^{e} j z^j = \frac{e z^{e+2} - (e+1) z^{e+1} + z}{(z-1)^2}. \tag{12}$$

For $1 \le i \le k$, using (11) and (12) we obtain that

$$S_1^{(m)}(r_i^{e_i}) = \deg(r_i) \left(\frac{\alpha_i - 1}{\alpha_i}\right) \frac{e_i \alpha_i^{e_i+2} - (e_i+1)\alpha_i^{e_i+1} + \alpha_i}{(\alpha_i - 1)^2}$$

and hence

$$\frac{S_1^{(m)}(r_i^{e_i})}{\alpha_i^{e_i}} = \deg(r_i) \left(e_i - \frac{1-\alpha_i^{-e_i}}{\alpha_i - 1}\right). \tag{13}$$

Then the formula for $\mathrm{E}^{(m)}(f)$ follows from Theorem 3.1, Corollary 3.1, and (13).

Similarly, for $1 \le i \le k$, it follows from (7) and Lemma 2.2(iii) that

$$S_2^{(m)}(r_i^{e_i}) = \sum_{j=0}^{e_i} \alpha_i^j \left(1 - \frac{1}{\alpha_i}\right) j^2 (\deg(r_i))^2. \tag{14}$$

For any positive integer e and any real number $z \neq 1$, we have the identity

$$\sum_{j=0}^{e} j^2 z^j = \frac{e^2 z^{e+3} - (2e^2 + 2e - 1) z^{e+2} + (e+1)^2 z^{e+1} - z^2 - z}{(z-1)^3}. \tag{15}$$

For $1 \le i \le k$, from (14) and (15) we obtain that

$$\begin{aligned} \frac{S_2^{(m)}(r_i^{e_i})}{\alpha_i^{e_i}} &= \left(\frac{\deg(r_i)}{\alpha_i - 1}\right)^2 \\ &\quad \times \left(e_i^2 \alpha_i^2 - (2e_i^2 + 2e_i - 1)\alpha_i + (e_i+1)^2 - \alpha_i^{1-e_i} - \alpha_i^{-e_i}\right). \end{aligned} \tag{16}$$

Using Theorem 3.1, (13), and (16), for $1 \le i \le k$ we get that

$$\mathrm{Var}^{(m)}\left(r_i^{e_i}\right) = \left(\frac{\deg(r_i)}{\alpha_i - 1}\right)^2 \left(2e_i\left(\alpha_i^{-e_i} - \alpha_i^{1-e_i}\right) + \alpha_i - \alpha_i^{1-e_i} - \alpha_i^{-2e_i} + \alpha_i^{-e_i}\right)$$

$$= \left(\frac{\deg(r_i)}{1 - \alpha_i^{-1}}\right)^2 \left((2e_i + 1)\left(\alpha_i^{-e_i-2} - \alpha_i^{-e_i-1}\right) - \alpha_i^{-2e_i-2} + \alpha_i^{-1}\right).$$

Then we obtain the formula for $\mathrm{Var}^{(m)}(f)$ using Corollary 3.1. □

Remark 3.1. When $f(x) = x^N - 1 \in \mathbb{F}_q[x]$ and N is an arbitrary positive integer, Theorem 3.2 yields the corresponding results of [9] and [6] by a simpler method. These corresponding results are the general formulas for the expectation and the variance of the joint linear complexity of random m-fold N-periodic multisequences over $\mathbb{F}_q$.

4. Counting function

Let $f \in \mathbb{F}_q[x]$ be a monic polynomial with $\deg(f) \ge 1$ and t be a nonnegative integer. Recall that the counting function $\mathcal{N}^{(m)}(f;t)$ gives the number of m-fold multisequences from $\mathcal{M}^{(m)}(f)$ with joint linear complexity t. In this section we derive a general formula for the counting function $\mathcal{N}^{(m)}(f;t)$. We determine closed-form expressions for $\mathcal{N}^{(m)}(f;t)$ if t is $\deg(f)$, $\deg(f) - 1$, or $\deg(f) - 2$. In general, it is quite hard to obtain simple expressions for the exact values of $\mathcal{N}^{(m)}(f;t)$. We compute $\mathcal{N}^{(m)}(f;t)$ in closed form in some cases. Moreover, we determine the generating polynomial $\mathcal{G}^{(m)}(f;z)$ for the distribution of joint linear complexities of m-fold multisequences from $\mathcal{M}^{(m)}(f)$.

Theorem 4.1. *Let $f \in \mathbb{F}_q[x]$ be a monic polynomial with $\deg(f) \ge 1$. Let t be a nonnegative integer with $t \le \deg(f)$. Then for the counting function $\mathcal{N}^{(m)}(f;t)$ we have*

$$\mathcal{N}^{(m)}(f;t) = \sum_{\substack{d \mid f \\ \deg(d)=t}} \Phi_q^{(m)}(d),$$

where the summation is over all monic polynomials $d \in \mathbb{F}_q[x]$ of degree t and dividing f.

Proof. Using the arguments in the proof of Theorem 3.1, we obtain that for a monic polynomial $d \in \mathbb{F}_q[x]$ dividing f, the number of m-fold multisequences in $\mathcal{M}^{(m)}(f)$ with joint minimal polynomial d is $\Phi_q^{(m)}(d)$. Note that for an m-fold multisequence $\boldsymbol{S} \in \mathcal{M}^{(m)}(f)$, the joint linear complexity of $\boldsymbol{S}$ is the degree of the joint minimal polynomial of $\boldsymbol{S}$. This completes the proof. □

The following lemma will be used in the proof of the next theorem.

Lemma 4.1. *Let $u \leq e$ be positive integers and $r \in \mathbb{F}_q[x]$ be a monic irreducible polynomial. Then for any positive integer m we have*

$$\frac{\Phi_q^{(m)}(r^{e-u})}{\Phi_q^{(m)}(r^e)} = \begin{cases} \dfrac{1}{q^{um\deg(r)}\left(1 - q^{-m\deg(r)}\right)} & \text{if } u = e, \\ \dfrac{1}{q^{um\deg(r)}} & \text{if } u < e. \end{cases}$$

Proof. By (5) we have

$$\Phi_q^{(m)}\left(r^e\right) = q^{em\deg(r)}\left(1 - q^{-m\deg(r)}\right).$$

Moreover, if $u = e$, then $\Phi_q^{(m)}\left(r^{e-u}\right) = 1$, and if $u < e$, then

$$\Phi_q^{(m)}\left(r^{e-u}\right) = q^{(e-u)m\deg(r)}\left(1 - q^{-m\deg(r)}\right).$$

This completes the proof. □

Assume that $f \in \mathbb{F}_q[x]$ is a monic polynomial of positive degree with the canonical factorization

$$f = r_1^{e_1} r_2^{e_2} \cdots r_k^{e_k}$$

into monic irreducibles over $\mathbb{F}_q$. We introduce further notation which will be used in the next theorem. Let $\Lambda(1)$ and $\Lambda(2)$ be the (possibly empty) sets consisting of the monic polynomials over $\mathbb{F}_q$ of degree 1 and 2, respectively, and dividing f. Let

$$\Lambda(1;1) := \{r \in \Lambda(1) : r^2 \nmid f\}$$

and $\Lambda(1;0) := \Lambda(1) \setminus \Lambda(1;1)$. Similarly, let

$$\begin{aligned} \Lambda(2;1) &:= \{r^2 \in \Lambda(2) : r \text{ monic}, \ r^3 \nmid f\}, \\ \Lambda(2;2) &:= \{r_1 r_2 \in \Lambda(2) : r_1^2 \nmid f, \ r_2^2 \mid f, \ r_2 \text{ monic}, \ \deg(r_2) = 1\}, \\ \Lambda(2;3) &:= \{r_1 r_2 \in \Lambda(2) : r_1^2 \nmid f, \ r_2^2 \nmid f, \ r_2 \text{ monic}, \ \deg(r_2) = 1\}, \\ \Lambda(2;4) &:= \{r \in \Lambda(2) : r \text{ is irreducible}, \ r^2 \nmid f\}, \end{aligned}$$

and $\Lambda(2;0) := \Lambda(2) \setminus \bigcup_{i=1}^{4} \Lambda(2;i)$.

Theorem 4.2. *Let $f \in \mathbb{F}_q[x]$ be a monic polynomial of positive degree and let m be an arbitrary positive integer. With the notation above, we have the following results for the counting function:*

(i) $\mathcal{N}^{(m)}(f;\deg(f)) = \Phi_q^{(m)}(f) = q^{m\deg(f)}\prod_{i=1}^{k}\left(1-q^{-m\deg(r_i)}\right).$

(ii) $\dfrac{\mathcal{N}^{(m)}(f;\deg(f)-1)}{\Phi_q^{(m)}(f)} = \dfrac{|\Lambda(1;0)|}{q^m} + \dfrac{|\Lambda(1;1)|}{q^m-1}.$

(iii) If $\deg(f)\geq 2$*, then we have*

$$\frac{\mathcal{N}^{(m)}(f;\deg(f)-2)}{\Phi_q^{(m)}(f)} = \frac{|\Lambda(2;0)|}{q^{2m}} + \frac{|\Lambda(2;1)|+|\Lambda(2;2)|}{q^{2m}-q^m} + \frac{|\Lambda(2;3)|}{(q^m-1)^2} + \frac{|\Lambda(2;4)|}{q^{2m}-1}.$$

Proof. Using Theorem 4.1 and Lemma 2.2(iii), we obtain that

$$\mathcal{N}^{(m)}(f;\deg(f)) = \Phi_q^{(m)}(f) = q^{m\deg(f)}\prod_{i=1}^{k}\left(1-q^{-m\deg(r_i)}\right),$$

which completes the proof of (i).

By Theorem 4.1, we also have

$$\mathcal{N}^{(m)}(f;\deg(f)-1) = \sum_{\substack{d\mid f\\ \deg(d)=\deg(f)-1}} \Phi_q^{(m)}(d) = \sum_{\substack{d\mid f\\ \deg(d)=1}} \Phi_q^{(m)}\left(\frac{f}{d}\right).$$

For a monic irreducible polynomial $r\in\mathbb{F}_q[x]$, let e_r be the nonnegative integer such that $r^{e_r}\mid f$ and $r^{e_r+1}\nmid f$. Using Lemma 2.2(ii), we obtain that

$$\frac{\mathcal{N}^{(m)}(f;\deg(f)-1)}{\Phi_q^{(m)}(f)} = \sum_{r\in\Lambda(1)} \frac{\Phi_q^{(m)}\left(r^{e_r-1}\right)}{\Phi_q^{(m)}\left(r^{e_r}\right)}.$$

Then by Lemma 4.1 we get

$$\frac{\mathcal{N}^{(m)}(f;\deg(f)-1)}{\Phi_q^{(m)}(f)} = \sum_{r\in\Lambda(1;0)}\frac{1}{q^m} + \sum_{r\in\Lambda(1;1)}\frac{1}{q^m-1},$$

which completes the proof of (ii).

Now we consider the proof of (iii). If $r\in\mathbb{F}_q[x]$ is a monic polynomial with $r^2\in\Lambda(2;1)$, then using Lemma 4.1 we get

$$\frac{\Phi_q^{(m)}\left(r^{e_r-2}\right)}{\Phi_q^{(m)}(r^{e_r})} = \frac{1}{q^{2m}\left(1-q^{-m}\right)}.$$

If $r_1, r_2 \in \mathbb{F}_q[x]$ are monic irreducible polynomials with $r_1 r_2 \in \Lambda(2;2)$, then using Lemma 4.1 we get

$$\frac{\Phi_q^{(m)}\left(r_1^{e_{r_1}-1} r_2^{e_{r_2}-1}\right)}{\Phi_q^{(m)}(r_1^{e_{r_1}} r_2^{e_{r_2}})} = \frac{1}{q^{2m}\,(1-q^{-m})}.$$

If $r_1, r_2 \in \mathbb{F}_q[x]$ are monic irreducible polynomials with $r_1 r_2 \in \Lambda(2;3)$, then using Lemma 4.1 we get

$$\frac{\Phi_q^{(m)}\left(r_1^{e_{r_1}-1} r_2^{e_{r_2}-1}\right)}{\Phi_q^{(m)}(r_1^{e_{r_1}} r_2^{e_{r_2}})} = \left(\frac{1}{q^{m}\,(1-q^{-m})}\right)^2.$$

Also if $r \in \Lambda(2;4)$, again using Lemma 4.1 we get

$$\frac{\Phi_q^{(m)}\left(r^{e_r-1}\right)}{\Phi_q^{(m)}(r^{e_r})} = \frac{1}{q^{2m}\,(1-q^{-2m})}.$$

Finally, for $d \in \Lambda(2;0)$ we have

$$\frac{\Phi_q^{(m)}(f/d)}{\Phi_q^{(m)}(f)} = \frac{1}{q^{2m}}.$$

We complete the proof of (iii) using similar arguments as in the proof of (ii). □

We note that for any monic polynomial $f \in \mathbb{F}_q[x]$ with $\deg(f) \geq 1$, it follows from Theorem 4.1 that $\mathcal{N}^{(m)}(f;0) = 1$. For $t \geq 1$, we compute $\mathcal{N}^{(m)}(f;t)$ in closed form in some special cases in the following examples.

Example 4.1. Let $f \in \mathbb{F}_q[x]$ be a monic polynomial with $\deg(f) \geq 1$ and let m be a positive integer. Assume that

$$f = r^e$$

is the canonical factorization of f into monic irreducibles over $\mathbb{F}_q$. For the counting function $\mathcal{N}^{(m)}(f;t)$ and $t \geq 1$, we have

$$\mathcal{N}^{(m)}(f;t) = \begin{cases} q^{mt}(1-q^{-m\deg(r)}) & \text{if } \deg(r) \mid t \text{ and } 1 \leq t/\deg(r) \leq e, \\ 0 & \text{otherwise.} \end{cases}$$

Example 4.2. Let $f \in \mathbb{F}_q[x]$ be a monic polynomial with $\deg(f) \geq 1$ and let m be a positive integer. Assume that the canonical factorization of f into monic irreducibles over $\mathbb{F}_q$ is given by

$$f = r_1^{e_1} r_2^{e_2} \quad \text{with } 1 \leq e_1 \leq e_2 \text{ and } \deg(r_1) = \deg(r_2) = \bar{n}.$$

If $\bar{n} \nmid t$ or $t > \bar{n}(e_1 + e_2)$, then $\mathcal{N}^{(m)}(f;t) = 0$. If $\bar{n} \mid t$ and $1 \leq t/\bar{n} \leq e_1$, then

$$\frac{\mathcal{N}^{(m)}(f;t)}{q^{mt}} = 2\left(1 - q^{-m\bar{n}}\right) + \left(\frac{t}{\bar{n}} - 1\right)\left(1 - q^{-m\bar{n}}\right)^2.$$

If $\bar{n} \mid t$ and $e_1 < t/\bar{n} \leq e_2$, then

$$\frac{\mathcal{N}^{(m)}(f;t)}{q^{mt}} = 1 - q^{-m\bar{n}} + e_1\left(1 - q^{-m\bar{n}}\right)^2.$$

If $\bar{n} \mid t$ and $e_2 < t/\bar{n} \leq e_1 + e_2$, then

$$\frac{\mathcal{N}^{(m)}(f;t)}{q^{mt}} = \left(e_1 + e_2 + 1 - \frac{t}{\bar{n}}\right)\left(1 - q^{-m\bar{n}}\right)^2.$$

Now we prove the formulas for $\mathcal{N}^{(m)}(f;t)$ in the current example. If $d \in \mathbb{F}_q[x]$ is a monic polynomial with $\deg(d) = t$ and $d \mid f$, then $d = r_1^{i_1} r_2^{i_2}$ for integers $0 \leq i_1 \leq e_1$, $0 \leq i_2 \leq e_2$, and hence $\bar{n} \mid t = \bar{n}(i_1 + i_2)$. It follows that $\mathcal{N}^{(m)}(f;t) = 0$ if $\bar{n} \nmid t$ or $t > \bar{n}(e_1 + e_2)$. If $\bar{n} \mid t$ and $1 \leq t/\bar{n} \leq e_1 + e_2$, using Theorem 4.1 we obtain that

$$\mathcal{N}^{(m)}(f;t) = \sum_{i=\max(0,t/\bar{n}-e_2)}^{\min(e_1,t/\bar{n})} \Phi_q^{(m)}\left(r_1^i r_2^{t/\bar{n}-i}\right). \tag{17}$$

Recall that $1 \leq e_1 \leq e_2$. Furthermore, we have

$$\max(0, t/\bar{n} - e_2) = \begin{cases} 0 & \text{if } 1 \leq t/\bar{n} \leq e_2, \\ t/\bar{n} - e_2 & \text{if } e_2 < t/\bar{n} \leq e_1 + e_2, \end{cases} \tag{18}$$

and

$$\min(e_1, t/\bar{n}) = \begin{cases} t/\bar{n} & \text{if } 1 \leq t/\bar{n} \leq e_1, \\ e_1 & \text{if } e_1 < t/\bar{n} \leq e_1 + e_2. \end{cases} \tag{19}$$

The formulas for $\mathcal{N}^{(m)}(f;t)$ above follow from (17), (18), (19), and Lemma 2.2(iii).

We define the generating polynomial $\mathcal{G}^{(m)}(f;z)$ for the distribution of joint linear complexities of m-fold multisequences from $\mathcal{M}^{(m)}(f)$ as the univariate polynomial in the variable z with nonnegative integer coefficients given by

$$\mathcal{G}^{(m)}(f;z) := \sum_{t \geq 0} \mathcal{N}^{(m)}(f;t) z^t. \tag{20}$$

Using Theorem 4.1 and Example 4.1, we now determine $\mathcal{G}^{(m)}(f;z)$ as a product of certain polynomials in z depending on the canonical factorization of f into monic irreducibles over $\mathbb{F}_q$.

Theorem 4.3. *Let $f \in \mathbb{F}_q[x]$ be a monic polynomial with* $\deg(f) \geq 1$. *Let m be a positive integer.*

(i) If $f = f_1 f_2$, where $f_1, f_2 \in \mathbb{F}_q[x]$ are monic polynomials with $\deg(f_1), \deg(f_2) \geq 1$, *and* $\gcd(f_1, f_2) = 1$, *then*

$$\mathcal{G}^{(m)}(f; z) = \mathcal{G}^{(m)}(f_1; z)\mathcal{G}^{(m)}(f_2; z).$$

(ii) If $f = r_1^{e_1} r_2^{e_2} \cdots r_k^{e_k}$ is the canonical factorization of f into monic irreducibles over $\mathbb{F}_q$, then

$$\mathcal{G}^{(m)}(f; z) = \prod_{j=1}^{k} \left(1 + (1 - \alpha_j^{-1}) \frac{\left(\alpha_j z^{\deg(r_j)}\right)^{e_j+1} - \alpha_j z^{\deg(r_j)}}{\alpha_j z^{\deg(r_j)} - 1} \right),$$

where $\alpha_j = q^{m \deg(r_j)}$ for $1 \leq j \leq k$.

Proof. By the definition (20), the statement (i) is equivalent to the statement that for each integer $t \geq 0$ we have

$$\mathcal{N}^{(m)}(f; t) = \sum_{0 \leq t_1 \leq t} \mathcal{N}^{(m)}(f_1; t_1)\mathcal{N}^{(m)}(f_2; t - t_1).$$

If $t > \deg(f)$, then this is trivial. Assume that $0 \leq t \leq \deg(f)$. By Theorem 4.1 we have

$$\mathcal{N}^{(m)}(f; t) = \sum_{\substack{d | f \\ \deg(d) = t}} \Phi_q^{(m)}(d).$$

Using Lemma 2.2(ii), we obtain that

$$\sum_{\substack{d | f \\ \deg(d) = t}} \Phi_q^{(m)}(d) = \sum_{0 \leq t_1 \leq t} \sum_{\substack{d_1 | f_1 \\ \deg(d_1) = t_1}} \Phi_q^{(m)}(d_1) \sum_{\substack{d_2 | f_2 \\ \deg(d_2) = t - t_1}} \Phi_q^{(m)}(d_2).$$

Hence (i) follows from Theorem 4.1.

Now we prove (ii). For $1 \leq j \leq k$, using Example 4.1 we get

$$\mathcal{G}^{(m)}(r_j^{e_j}; z) = 1 + (1 - \alpha_j^{-1}) \sum_{\ell=1}^{e_j} \left(\alpha_j z^{\deg(r_j)}\right)^{\ell}$$

$$= 1 + (1 - \alpha_j^{-1}) \alpha_j z^{\deg(r_j)} \frac{\left(\alpha_j z^{\deg(r_j)}\right)^{e_j} - 1}{\alpha_j z^{\deg(r_j)} - 1}.$$

Then we complete the proof using (i) and induction on k. □

In the following remark, we recall a well-known and useful fact (cf. [6, Remark 1] and [8, Theorems 2.45 and 2.47]).

Remark 4.1. Assume that $n \geq 1$ is an integer with $\gcd(n,q)=1$. Let $\phi(\ell)$ be the Euler totient function on the positive integers, i.e., $\phi(\ell)$ is the number of nonnegative integers less than ℓ and coprime to ℓ. For each positive integer d dividing n, let $H_q(d)$ be the multiplicative order of q modulo d, i.e., the least positive integer h such that $q^h \equiv 1 \mod d$. There is a one-to-one correspondence between the set of cyclotomic cosets modulo n relative to the powers of q and the set of monic irreducible polynomials dividing $x^n - 1$ over $\mathbb{F}_q$. For each such cyclotomic coset, there exists a uniquely determined positive integer d dividing n. This is the positive integer d such that the monic irreducible polynomial corresponding to the given cyclotomic coset is a factor of the dth cyclotomic polynomial over $\mathbb{F}_q$ (cf. [8, Definition 2.44]). Conversely, for each positive integer d dividing n, there are exactly $\phi(d)/H_q(d)$ distinct cyclotomic cosets modulo n corresponding to the given d, and each of these cosets has the same size $H_q(d)$.

For $N \geq 1$, recall that the set of m-fold N-periodic multisequences over $\mathbb{F}_q$ is the same as the set $\mathcal{M}^{(m)}(f)$, where $f(x) = x^N - 1 \in \mathbb{F}_q[x]$. Using Theorem 4.3 and Remark 4.1, in the next corollary we determine the generating polynomial of random periodic multisequences over $\mathbb{F}_q$.

Corollary 4.1. *Let $m, N \geq 1$ be integers and p be the characteristic of $\mathbb{F}_q$. Let $n \geq 1$ and $\nu \geq 0$ be the integers such that $N = p^\nu n$ and $\gcd(n,p)=1$. Assume that $f(x) = x^N - 1 \in \mathbb{F}_q[x]$. For each positive integer d dividing n, let $\phi(d)$ be the Euler totient function and $H_q(d)$ be the multiplicative order of q modulo d (cf. Remark 4.1). Then we have*

$$\mathcal{G}^{(m)}(f;z)$$
$$= \prod_{d|n} \left(1 + \left(1 - q^{-mH_q(d)}\right) \frac{\left(q^{mH_q(d)} z^{H_q(d)}\right)^{p^\nu+1} - q^{mH_q(d)} z^{H_q(d)}}{q^{mH_q(d)} z^{H_q(d)} - 1} \right)^{\phi(d)/H_q(d)} .$$

Proof. Using Remark 4.1, the canonical factorization of $x^N - 1$ into monic irreducibles over $\mathbb{F}_q$ is obtained as

$$x^N - 1 = \prod_{d|n} \left(r_{d,1}^{p^\nu} r_{d,2}^{p^\nu} \cdots r_{d,\phi(d)/H_q(d)}^{p^\nu} \right),$$

where $\deg(r_{d,1}) = \cdots = \deg\left(r_{d,\phi(d)/H_q(d)}\right) = H_q(d)$. Then the result follows from Theorem 4.3(ii). □

Remark 4.2. Note that the counting function $\mathcal{N}^{(m)}(f;t)$ is the coefficient of the term z^t of the generating polynomial $\mathcal{G}^{(m)}(f;z)$. Theorem 4.3 and

Corollary 4.1 determine $\mathcal{G}^{(m)}(f;z)$ as a product of certain polynomials in z. However, even for $f(x) = x^N - 1$, i.e., the periodic case, it is difficult in general to obtain the coefficient of the term z^t from the product in Corollary 4.1.

Acknowledgments

A part of this paper was written while Fang-Wei Fu and Ferruh Özbudak were with Temasek Laboratories, National University of Singapore. They would like to express their thanks to Temasek Laboratories and the Department of Mathematics at the National University of Singapore for the hospitality.

This research was supported by the DSTA grant R-394-000-025-422 with Temasek Laboratories in Singapore.

References

1. T.W. Cusick, C. Ding, A. Renvall, Stream Ciphers and Number Theory, Elsevier, Amsterdam, 1998.
2. Z. Dai, K. Imamura, J. Yang, Asymptotic behavior of normalized linear complexity of multi-sequences, in: T. Helleseth et al. (Eds.), Sequences and Their Applications — SETA 2004, Lecture Notes in Computer Science, Vol. 3486, Springer, Berlin, 2005, pp. 129–142.
3. E. Dawson, L. Simpson, Analysis and design issues for synchronous stream ciphers, in: H. Niederreiter (Ed.), Coding Theory and Cryptology, World Scientific, Singapore, 2002, pp. 49–90.
4. C. Ding, G. Xiao, W. Shan, The Stability Theory of Stream Ciphers, Lecture Notes in Computer Science, Vol. 561, Springer, Berlin, 1991.
5. X. Feng, Z. Dai, Expected value of the linear complexity of two-dimensional binary sequences, in: T. Helleseth et al. (Eds.), Sequences and Their Applications — SETA 2004, Lecture Notes in Computer Science, Vol. 3486, Springer, Berlin, 2005, pp. 113–128.
6. F.-W. Fu, H. Niederreiter, M. Su, The expectation and variance of the joint linear complexity of random periodic multisequences, J. Complexity 21 (2005) 804–822.
7. P. Hawkes, G.G. Rose, Exploiting multiples of the connection polynomial in word-oriented stream ciphers, in: T. Okamoto (Ed.), Advances in Cryptology — ASIACRYPT 2000, Lecture Notes in Computer Science, Vol. 1976, Springer, Berlin, 2000, pp. 303–316.
8. R. Lidl, H. Niederreiter, Finite Fields, Cambridge University Press, Cambridge, 1997.
9. W. Meidl, H. Niederreiter, The expected value of the joint linear complexity of periodic multisequences, J. Complexity 19 (2003) 61–72.
10. H. Niederreiter, Sequences with almost perfect linear complexity profile, in: D. Chaum, W.L. Price (Eds.), Advances in Cryptology — EUROCRYPT '87,

Lecture Notes in Computer Science, Vol. 304, Springer, Berlin, 1988, pp. 37–51.
11. H. Niederreiter, Linear complexity and related complexity measures for sequences, in: T. Johansson, S. Maitra (Eds.), Progress in Cryptology — INDOCRYPT 2003, Lecture Notes in Computer Science, Vol. 2904, Springer, Berlin, 2003, pp. 1–17.
12. H. Niederreiter, The probabilistic theory of the joint linear complexity of multisequences, in: G. Gong et al. (Eds.), Sequences and Their Applications — SETA 2006, Lecture Notes in Computer Science, Vol. 4086, Springer, Berlin, 2006, pp. 5–16.
13. H. Niederreiter, L.P. Wang, Proof of a conjecture on the joint linear complexity profile of multisequences, in: S. Maitra et al. (Eds.), Progress in Cryptology — INDOCRYPT 2005, Lecture Notes in Computer Science, Vol. 3797, Springer, Berlin, 2005, pp. 13–22.
14. H. Niederreiter, L.P. Wang, The asymptotic behavior of the joint linear complexity profile of multisequences, Monatsh. Math. 150 (2007) 141–155.
15. R.A. Rueppel, Analysis and Design of Stream Ciphers, Springer, Berlin, 1986.
16. R.A. Rueppel, Stream ciphers, in: G.J. Simmons (Ed.), Contemporary Cryptology: The Science of Information Integrity, IEEE Press, New York, 1992, pp. 65–134.
17. L.P. Wang, H. Niederreiter, Enumeration results on the joint linear complexity of multisequences, Finite Fields Appl. 12 (2006) 613–637.

Research on P2P Worn Detection Based on Information Correlation-PWDIC*

Huaping Hu
Key Lab of Network Security and Cryptology,
Fujian Normal University Fuzhou, Fujian, China

Jing Zhang, Fengtao Xiao, and Bo Liu
School of Computer Science,
National University of Defense Technology Changsha, Hunan, China

Along with the fast increase of P2P users, P2P worm has become a severe threat to the P2P network and Internet. The P2P worm detection based on Information Correlation-PWDIC is presented in this paper. According to the information correlation, this paper establishes a series of filter rules to realize the detection and containment for P2P worm. Finally, a simulation experiment is given. The result shows that this P2P worm detection method has a good effect on P2P worm detection and also shows the distribution of resources has a great influence on the effect of containment for P2P worm spread.

Keywords: Peer-to-Peer, P2P worm, Spread, Detection, Containment

1. Introduction

Peer-to-peer network, also called P2P network, is a network model, in which every node is a peer. Nodes in the P2P network have identical responsibility and ability. They accomplish tasks in coordination, make direct interconnection logically, and share the information resources, processor resources, store resources and even cache resources. In P2P network, the applied software lacks in uniform specification and can not guarantee threat-tolerance in program and management, while convenience share and fast routing mechanism in P2P network environment offer more opportunity for network attackers in which P2P worm is a severe threat.

P2P worm has two meanings [1] :

*This work was supported by the open funds of Key Lab of Fujian Province University Network Security and Cryptology (No. 07A004).

(1) As hosts in P2P systems maintains a certain number of neighbors for routing , worm fired into the P2P network can easily propagate it to its neighbors without scanning stage. So infection rate will be increased, and then raise the spread speed of worm.
(2) The worm body which constructed by using the P2P principle. A propagating approach of active worm which could achieve fast and self propagation performances by division strategy is proposed. This worm also makes the existing detection method reduce effectiveness.

P2P worm, no matter which infecting way it uses, has brought severe security threat to network society.

In this paper, the research of worm detection method mainly aims at the first kind of P2P worm. The paper is organized as follows: the brief introduction of P2P and P2P worm is given in Section I. In Section 2, a simple introduction of the present research related with P2P worm is given. A worm detection method of PWDIC is proposed in Section 3. In Section 4, the experiment for the detection method and analysis is given. Conclusion of this paper and future work are discussed in Section 5.

2. P2P Worm

Now, the effective scanning mechanism which Internet worm adopts is the routing scanning combined with random scanning. The Worm gets large IP address information by attaching BGP, and then constructs hit-list according this information. After the worm infects all hosts in hit-list, random scanning is adopted. But P2P worm adopts a new way, which will be shown later. Stanford [2] pointed out that P2P system has the property that suits for worm spreading. Jayanthkumar [3] has proposed a worm which constructs hit-list through excavating P2P system application vulnerability and collecting the P2P neighbor. Zhou L etc. [4] puts forward a worm using P2P logical topology, which makes the propagation more conceal and makes the attack more effective, also makes the most defense mechanisms based on the worm's scanning course invalid. Luo Xingrui etc. [1] present the worm constructed by P2P principle, can achieve fast and self propagation performances by division strategy. The worm let the detection based on signature through packet regrouping become less effective. The scale of P2P network is so large and P2P worm can directly get valid IP list from node. This worm has three main characteristics [2]: rapidly propagating speed, highly successful spreading rate and being difficultly checked. So P2P worm can infect many computers in very short time. And with the

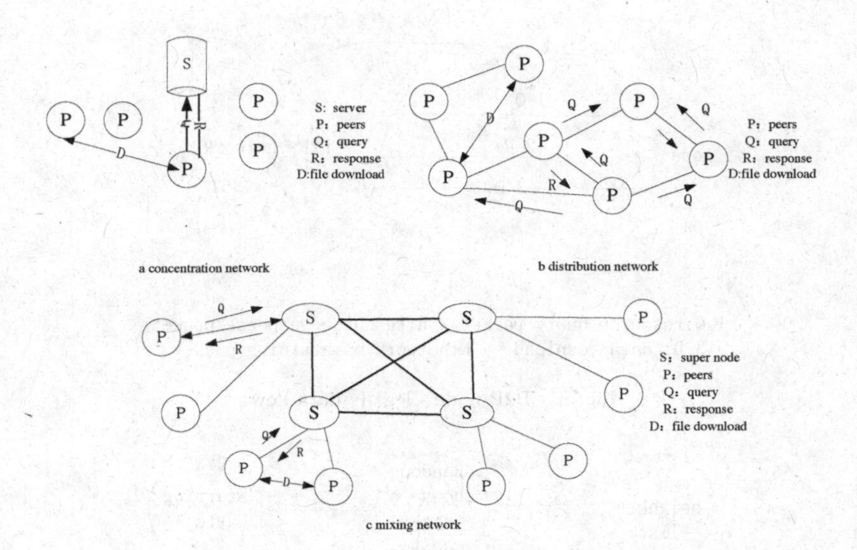

Fig. 1. Three kinds of p2p network in [5]

large scale of P2P network, it becomes a potential great threat to the Internet. Now, P2P network has offered technology and network environment support for the spread of worm, which makes worm improve in against presented detection method and get high spreading speed. So, designing an efficiency P2P worm detection method is imperative.

3. P2P Worm Detection Based On Information Correlation-PWDIC

3.1. *The Behaviors of P2P Network*

P2P network is formed as a self organized overlay network. In this network, each node is peers. Now P2P network has three types [5]: the concentration of directory services network, completely distribution network and mixing network, which are shown as Figure. 1. From the structure and composition of the three kinds P2P network, we can find out that the packet in P2P network mainly includes resource query, resource responses, data transmission, and certain network information is needed to maintain the network.

3.2. *P2P Information Correlation Analysis*

From the analysis in section A, we conclude that the packet in P2P network between nodes are resource query, response information, data transfer and

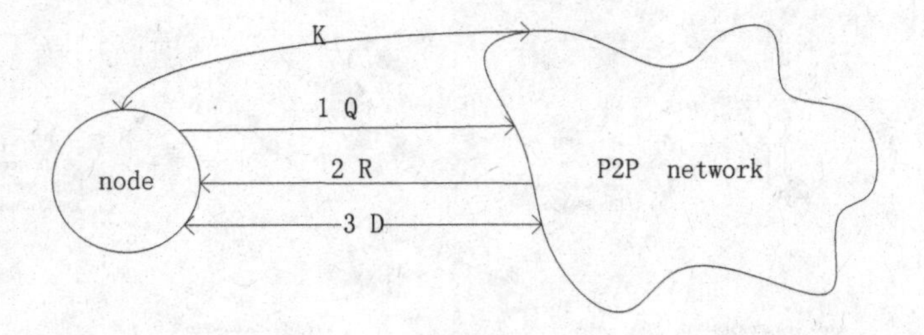

Fig. 2. P2P node's legally data flow

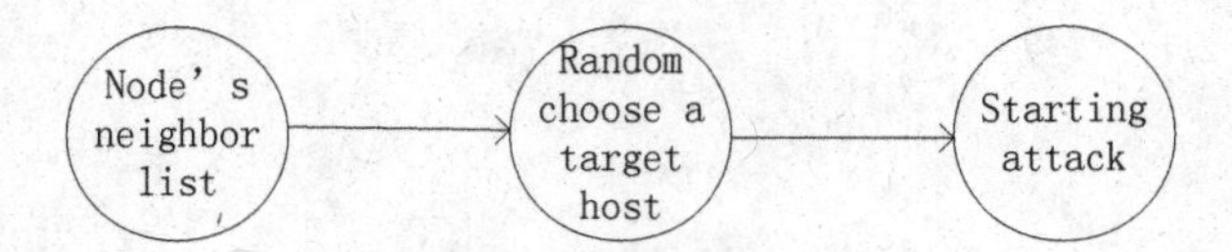

Fig. 3. One propagation of P2P worm

P2P network maintain information. The network maintaining information has no relations with other three kinds of information. But the correlation in resource query, response information and data transmission is very close. Now excluding P2P data flow, other data flow classification and identification has some effective methods, therefore what this paper considers is the detection of worm in pure P2P network. After a node in P2P network carries out resource query, it constructs a list of P2P node that possesses the resource from response information. Then the node chooses nodes in this list according to certain mechanism, which will arouse data transmission among those nodes, including sending and receiving information. From this, we can find out that the transmission of data is aroused by resource query and response information. Basis filters the legal data [6]. We conclude the PWDIC detection method in this paper. Data flow between node and P2P network is shown in Figure 2, which is set according the packet's presenting order. The propagation of P2P worm is shown in Figure 3.

3.3. *The Design of PWDIC Detection Model*

The PWDIC detection model which based on the analysis of correlation among network packets, shown as Figure 4, includes three layers: message classification layer, rule generation layer and the warning layer, addition-

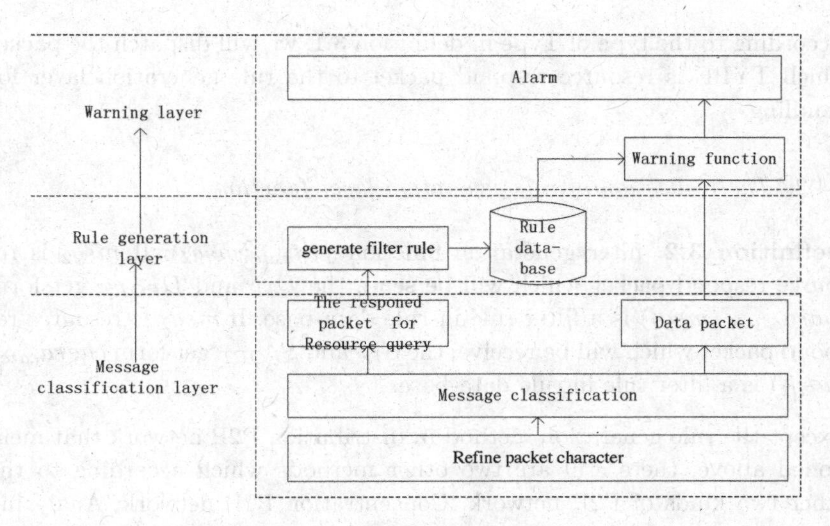

Fig. 4. The PWDIC Model

ally, still including a filter rule data-base. The relation between layers has reflected sequential movement. Message classification layer offers data for rule generation. The common action of message classification layer, ruler data-base and the function of warning reflects the specifically detection of this method. This detection method is arranged on the node but not on the network.

3.4. *The Realization of PWDIC*

3.4.1. *The realization of message classification layer function*

The network maintains packet, resource query packet and respond packet of each kind of P2P application software has its own characteristic form. Therefore according to its packet character to filtrate network maintain information, resource query and response packet, and analyze the resource respond packet for resource query to establish filter rule. Here the respond message includes send and receive.

Definition 3.1. $chara_{catch}(msg_1) = (Type, D_{IP}, D_{PORT}, S_{IP}, S_{PORT})$: msg_1 represents a packet, Type expresses the type of this packet (network maintains packet, resource query packet, respond packet and data packet), D_{IP}, D_{PORT} expresses the destination address and port of this packet. S_{IP}, S_{PORT} the source address and port.

According to the type of Type in definition 3.1, we will dispatch the packet which TYPE is resource respond packet to the rule generation layer for handling.

3.4.2. *The realization of rule generation layer function*

Definition 3.2. filter generation function $filt_{rule}(msg_2)$: If msg_2 is resource respond packet which will be send, the D_{IP} and D_{PORT} get form $chara_{catch}$ (msg_2) is a filter rule in rule data-base. If msg_2 is resource respond packet which will be receive, the S_{IP} and S_{PORT} get form $chara_{catch}$ (msg_2) is a filter rule in rule data-base.

Except the rule generation method in distribution P2P network that mentioned above, there still are two other methods which according to the other two kinds of P2P network. Concentration P2P network: Analyzing the packet received from the server of directory services which respond for

---

Algorithm 3.1 Warning Function

Parameter: msg: packet;
Catch(msg): the function get (D_{IP}, D_{PORT}) from $chara_{catch}(msg)$;
$delay_{queue}$: the queue of illegal packet;
$throt_{length}$: the threshold values of the $delay_{queue}$.

1: $(D_{ip}, D_{port}) = Catch(msg)$
2: **if** (D_{ip}, D_{port}) in $filt_{rule}$ **then**
3:     send msg directly
4: **else**
5:     push msg in $delay_{queue}$ and delay send
6: **end if**
7: **if** $delay_{queue} \rightarrow length > throt_{length}$ **then**
8:     alarm
9:     block the communication
10: **else**
11:     **for all** msg $(delay_{queue})$ **do**
12:         **if** msg$\rightarrow$ $delay_{time}$==0 **then**
13:             send msg
14:             $delay_{queue} \rightarrow length$ decrease
15:         **end if**
16:     **end for**
17: **end if**

---

resource query, generating and establishing filter rule. Mixing P2P network: Analyzing the packet received from super node which respond to resource query, generating and establishing filter rule. And the filter rule updates in certain time which according to the P2P application software.

3.4.3. *The realization of warning layer function*

Let the packet which gets from message classification layer and Type is data packet match the rule data-base, the packet which matches is legal data flows, otherwise apply the warning function on the packet. The warning function consists of warning and delay. Postpone the sending of the illegal packet, if no alarm happen, and then transmit the packet, that realizes the containment of worm's propagation. The warning function was shown in Algorithm 3.1.

3.4.4. *The flow of detection*

The handling of detection method process is divided into three steps, which is shown in Figure 5.

(1) Monitor the packet which responded for the resource query no matter sending or receiving, then establishment filter rule, all node are same.
(2) Draw the destination IP and Port from the data packet ,then match the rule data-base.
(3) Apply the warning function on the packet.

4. Simulated Experiment and the Result

4.1. *Experiment environment*

Under the experiment environment of PERL combined with NWS, a distributing P2P network with 5000 nodes is build. According the above analysis, in this network we only simulate simple three type of packet: resource query, respond packet and data packet. The P2P network in experiment is stabilized, therefore do not have network maintain packet. The P2P worm is random to select neighbor node for infecting. In the experiment, the time which infects one host costs is called a STEP. In a STEP, all active worms will send infected packet. And filter rule will update periodically in certain STEPs.

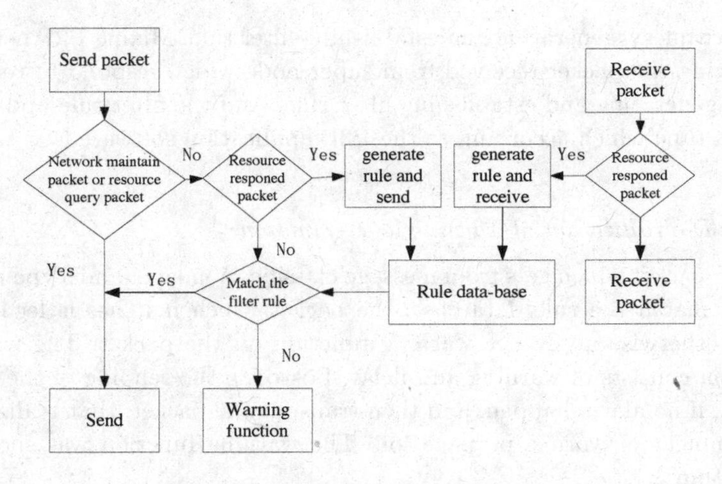

Fig. 5. Handling flow chart

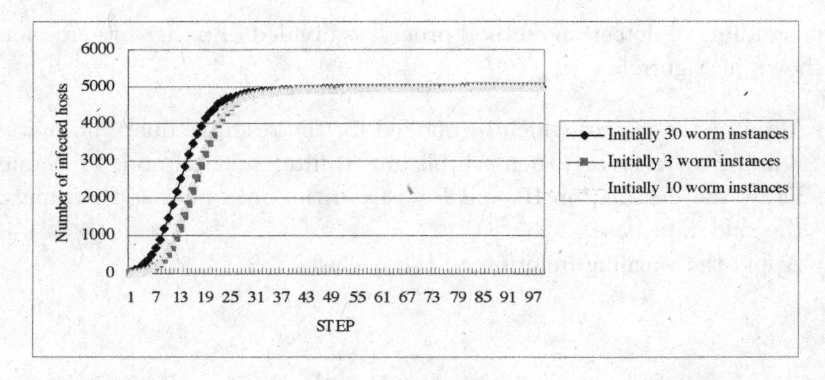

Fig. 6. P2P worm spread with different number of initial worm instances

4.2. *Experimental Simulation*

4.2.1. *Effect of the number of initial worm instances*

From Figure 6, we can find out that the number of infected host P2P network which infected by different initial worm instances is basically identical after certain time. The initial worm number only influent the initial speed of worm spread. The bigger number of initial worms, the less time it costs to infect identical number of hosts.

4.2.2. *The effect of detection and containment*

Arrange this detection method on each P2P node, then evaluates the effect of the method. The way of resource's distribution is random. The number of initial worm instances is 10. We must consider all occasions which may appear in real world, including two aspects: one is the way of distributing resource is random, the other is certain proportion resource distributed on neighbor nodes. So different resource distribution can make the PWDIC have different detection and containment effect. The detection and containment effect under different distribution resource are listed in Table 1.

The proportion of resource distributed on neighbor node	Number of Alarm Infected	Number of detecting host	Number of infected host	Fail to detect probability
90%	2903	2903	2931	0.009
80%	2439	2439	2459	0.008
60%	1226	1226	1237	0.008
50%	104	104	104	0.009
40%	74	74	74	0
30%	12	12	12	0
Random distribution	11	11	11	0

Table 1: the effect of detection and containment

The results show that different resource distribution makes this detection method for P2P worm have different effect in detection and containment. Under the condition of below 50For example, M represents the number of resource, the neighbor node number of certain node is N, and the proportion resource distributed on neighbor node is Q. The worm spread successfully once needs satisfy two conditions:

(1) The node which worm selects possess this resource, the probability is $Q * M/N$.
(2) The node which worm selects also carry out data exchange, know to count study the law of summing up calculation this probability is $(1 + M)/2M$.

Since worm propagates is random selecting target node, therefore calculate

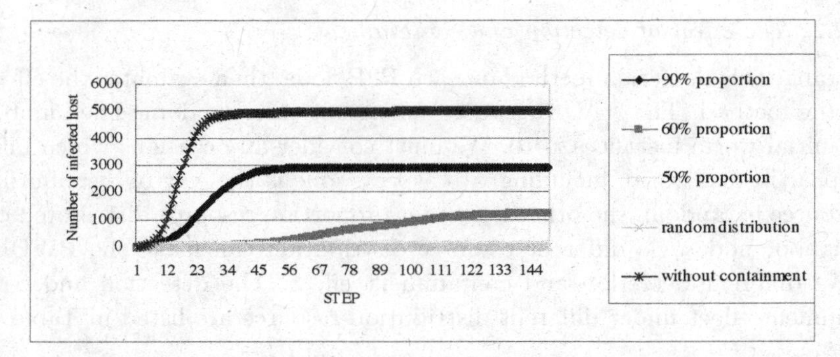

Fig. 7. The effect of detection and containment under different proportion distributed on the neighbor node

the probability of its spread success is $Q * (M + 1)/2N^2$, this has explained why the less resource possessed by the neighbor, the better effect this detection method has. The detection method that this paper has put forward does not only dispatch the data which the node will send, but also dispatch the data which the node will receive. The data of Tab.1 implies that in simulated experiment, the effect of this method is very good, has no false alarm, and only failing to alarm and the proportion is also very low. In actual network environment, the distribution of resource, selecting way of peers which posses this resource and the neighbor number etc. may make the proportion of failing to alarm big. Because of the complex network environment of P2P networks, for instance, the rubbish data can make this method have false alarm. The effect of detection and containment is showed in Figure 7.

4.3. *Analyzed the Experiment Results*

From the experiment result we can find out that the P2P worm detection method based on information correlation presented in this paper has good effect. In the aspect of containing P2P worm's spread, the distribution of resource has a great influence on the result. If the neighbor node of P2P node contains higher proportion resource, the PWDIC does not have good containment effect but also have better detection effect.

5. Conclusions

Above all, comparing to other existing methods, the presented method constructs the filter rule according to the node's network behavior not the information of all nodes, it is simple and effective. The presented detection method is based on the information correlation in P2P network. From the result of experiment, we can find out that the PWDIC has good detection effect. In the containment in P2P worm's spread, the distribution of resource has a great effect on the result. The future work includes: let this detection method combine with automatic signature generation technology, and establish network issued mechanism, make the detection method has good containment effect in any distribution condition of resource.

References

1. Luo Xingrui, Yao Yu and Gao Fuxiang, Journal on Communications11A, 53-58(2006).
2. Staniford S, Paxson V and Weaver N, Proceedings of the 1lth VSENZX Security Symposium, 149-167(2002).
3. Kannan J, California: CS294-4 Project, 2003(2003).
4. Zhou L, Zhang L and MeSherry F, Proceedings of the Peer-to-Peer Systems 4th International Workshop Ithaca, 24-25(2005).
5. Wei Yu, Proceedings of the Eighth IEEE International Symposium on High Assurance Systems Engineering, 308-309 (2004)
6. David Whyte, Evangelos Kranakis and P.C. van Oorschot, the Proceedings of the 12th Annual Network and Distributed System Security Symposium, 181-195(2005).

On the Relation among Various Security Models for Certificateless Cryptography

Qiong Huang and Duncan S. Wong

Department of Computer Science
City University of Hong Kong
Hong Kong, China
E-mail: {*csqhuang,duncan*}*@cityu.edu.hk*

Certificateless Cryptography is a promising technology for solving the key escrow problem in identity-based cryptography. However, the lack of unified set of definitions and security models currently hinders its progression as much effort has been put on refining the definitions and looking for an appropriate and practical security models. In this paper, we target to contribute on unifying the definitions and security models for certificateless encryption (CLE) schemes. First, we show that the original seven-algorithm definition is equivalent to a simplified five-algorithm definition. We believe that this simplified definition may lead to more compact and efficient implementations in practice and also help in the effort of standardization of CLE. Second, we show that a CLE scheme cannot be both *Malicious KGC Secure* and *Strongly Secure* in the standard model. Due to the practicality and attacking significance of Malicious KGC Security, and the uncertainty of how practical the Strong Security is, we therefore suggest constructing practical CLE schemes to be Malicious KGC Secure. Third, we propose to formalize a new adversarial capability called Partial Key Replacement attack into the security models and show that our generic scheme proposed recently is secure against this new type of attacks.

Keywords: Certificateless Cryptography, Encryption, Security Models

1. Introduction

In traditional public key cryptography, a user selects a public/secret key pair (pk, sk) and publishes pk. A *certificate*, which essentially is a signature on the user's identity and pk issued by a *certification authority* (CA), will then be employed for vouching the relationship between the user and pk. This method works under the public key infrastructure (PKI) involves a lot of additional work for verifying certificates and managing certificates that include issuance, revocation, storage and distribution.

In 1984, Shamir[1] introduced the notion of identity-based cryptography, aiming for alleviating the usability problems in PKI by getting rid of certificates. The first identity-based encryption algorithm was proposed by Boneh and Franklin [2] in 2001. In an identity-based cryptosystem, a user can use an email address, an IP address or any other information which can uniquely identify the user, as his public key. There is a trusted party, called Key Generation Center (KGC), which is in charge of the user secret key generation. The advantage of an identity-based cryptosystem is that anyone can simply use the user's identity to encrypt messages. This can be done even before the user gets its secret key from the KGC. However, the KGC should not impersonate the user or decrypt any of the user's ciphertexts, as the *key escrow* feature is inherent in identity-based cryptography.

In 2003, Al-Riyami and Paterson [3] introduced *certificateless cryptography*, which is intended to solve the key escrow problem of the identity-based cryptography without using certificates. In a certificateless cryptosystem, the KGC is involved in issuing a user *partial* key psk_{ID} for a user with identity ID. The user independently generates a user public/secret key pair $(upk_{\mathsf{ID}}, usk_{\mathsf{ID}})$, and publishes upk_{ID}. A message will then be encrypted under *both* upk_{ID} and ID. To decrypt a ciphertext, the user must have the knowledge of both psk_{ID} and usk_{ID}. Knowing only one of them does not allow the recovery of the plaintext. One feature that is lost when comparing with identity-based encryption is that a sender is not able to encrypt messages to a receiver if the receiver has not generated a user public key. Interestingly, the sender *can* encrypt messages to the receiver once the receiver's user public key is known. This can be done before the receiver gets a user partial key from the KGC.

1.1. *A Summary of Results in This Paper*

1. Simplification of CLE Definition. The original definition for CLE (Certificateless Encryption) [3] consists of *seven* algorithms. We propose a simplified *five*-algorithm definition and show that this simplified definition is equivalent to the original one. The significance of this simplification is the elimination of unnecessary complication of the original definition. This may lead to more compact and efficient actual implementations in practice and help in the effort of standardization of CLE in the near future.

2. Malicious KGC Security vs. Strong Security. Since the introduction of CLE, the security models for CLE schemes have never been settled and there have been many versions proposed [3–8]. The two prominent and strongest ones ever proposed are *Strong Security* [3] and *Malicious KGC*

Security [6–8]. For Strong Security, the games (Type-I and Type-II) require the simulators to provide correct decryption to adversaries even after the user public key is replaced by the adversaries but the corresponding user secret key is not given to the simulators. The model has been considered as the strongest one and has usually been considered as the ultimate security goal of constructing a CLE scheme [3,4,9]. However, we are uncertain on how realistic such a strong decryption oracle is. In [6] and followed in [7,8], security against *malicious-but-passive KGC* attacks was introduced. It considers the adversary in Type-II game to be malicious-but-passive at the very beginning of a game, rather than making an assumption that the adversary is benign when generating the master public key and turns malicious once after. This type of attacks is significant because the attacks are natural and very practical. Many schemes have been shown insecure against malicious-but-passive KGC attacks [6,8].

Recently, schemes which are Strong Type-I Secure and Strong Type-II Secure [9] and schemes which are Type-I Secure and Malicious KGC Type-II Secure [4,7,8] in the standard model have been proposed. On the other hand, there is no scheme which has been shown secure against a *mixture* of Strong Security and Malicious KGC Security. For example, a problem that is still open is whether it is possible to construct a scheme which is both *Strong Type-I Secure* and *Malicious KGC Type-II Secure.* In [4,6], researchers considered this as an open problem and even suspected its feasibility. In [8], Huang and Wong conjectured that the *Strong Type-I Security* and *Malicious KGC Type-II Security* are self-contradictory. However, until now, there is no affirmative answer to this problem.

In this paper, we show that it is **impossible** to construct a CLE which is both *Strong Type-I Secure* and *Malicious KGC Type-II Secure* in the standard model. We also show that *Strong Security* and *Malicious KGC Security* cannot co-exist in the standard model. Since Malicious KGC Security is very significant in practice while Strong Security is still uncertain on its practicality, to strive for a balance between the strongest adversarial model and its extent of reality, we therefore suggest to put effort on constructing CLE schemes which satisfy *Malicious KGC Security.*

3. User Partial Key Replacement Attacks. In all the current models for CLE schemes, adversaries can only replace user public keys but not user partial keys. In practice, the user partial key of a user may need to be regenerated, for example, after the user loses his previous user partial key. Also, the malicious-but-passive KGC may try to change the user partial key with respect to the user public key for the purpose of compromising a particular user's secrecy. It is therefore necessary and tends to make the models stronger by introducing a new adversarial capability called *User Partial Key Replacement.* We introduce this new adversarial capability and show that our generic CLE scheme proposed in [8] is secure against it.

Paper Organization. In the next section, we describe a five-algorithm definition for CLE and show that it is equivalent to the original seven-algorithm definition. In Sec. 3, we review the two strongest security notions ever proposed, namely Strong Security and Malicious KGC Security, and show that they cannot co-exist on a secure CLE scheme in the standard model. For promoting practical significance, we formalize *Type-I Security* and *Malicious KGC Type-II Security* and consider them as the strongest *practical* security requirements for a secure CLE scheme. In Sec. 4, we introduce and formalize user partial key replacement attacks. This is followed by Sec. 5, in which we show that our generic CLE scheme proposed in [8] is secure against this new type of attacks. The paper is concluded in Sec. 6.

2. Definitions

A certificateless encryption (CLE) scheme consists of five probabilistic polynomial-time (PPT) algorithms:

- MasterKeyGen: On input 1^k where $k \in \mathbb{N}$ is a security parameter, it generates a master public/secret key pair (mpk, msk).
- PartialKeyGen: On input msk and an identity ID, it generates a user partial key psk_{ID}.
- UserKeyGen: On input mpk and ID, it generates a user public/secret key pair $(upk_{\mathsf{ID}}, usk_{\mathsf{ID}})$.
- Enc: On input mpk, ID, upk_{ID} and a message m, it returns a ciphertext c or a symbol $\perp$ indicating the failure of encryption. Message space $\mathcal{M}(mpk, upk_{\mathsf{ID}})$ is defined under mpk and upk_{ID}.
- Dec: On input psk_{ID}, usk_{ID}, and a ciphertext c, it returns a plaintext $m \in \mathcal{M}(mpk, upk_{\mathsf{ID}})$ or a symbol $\perp$ indicating the failure of decryption.

In practice, the KGC (Key Generation Center) performs the first two algorithms: MasterKeyGen and PartialKeyGen. The master public key mpk is assumed to be publicly known and all the user partial keys are issued to the corresponding users securely so that no one except the intended individual users can get them. Every user in the system also performs UserKeyGen for generating its own public/secret key pair and publishes the public key.

For correctness, we require that for any $k \in \mathbb{N}$, $(mpk, msk) \leftarrow \mathsf{MasterKeyGen}(1^k)$, $\mathsf{ID} \in \{0,1\}^*$, $psk_{\mathsf{ID}} \leftarrow \mathsf{PartialKeyGen}(msk, \mathsf{ID})$, $(upk_{\mathsf{ID}}, usk_{\mathsf{ID}}) \leftarrow \mathsf{UserKeyGen}(mpk, \mathsf{ID})$, and $m \in \mathcal{M}(mpk, upk_{\mathsf{ID}})$, we have

$$m \leftarrow \mathsf{Dec}(psk_{\mathsf{ID}}, usk_{\mathsf{ID}}, \mathsf{Enc}(mpk, \mathsf{ID}, upk_{\mathsf{ID}}, m)).$$

2.1. *Equivalence to the Original Seven-Algorithm Definition*

The definition above is based on the simplification approach first proposed by Hu, Wong, Zhang and Deng in [10,11] and later adopted to defining CLE and related schemes in [6–8]. In the original definition of CLE proposed by Al-Riyami and Paterson [3], there are seven algorithms. In the following, we show that our simplified five-algorithm definition is equivalent to the original one. To ease comparison, in the following, we first briefly review the original seven-algorithm definition given in [3] and use the notations denoted above rather than the notations used originally in [3]. A certificateless encryption ($\mathsf{CLE}^{\mathsf{old}}$) scheme consists of seven PPT algorithms:

- $\mathsf{MasterKeyGen}^{\mathsf{old}}$: On input 1^k, it generates (mpk, msk).
- $\mathsf{PartialKeyGen}^{\mathsf{old}}$: On input msk and ID, it generates psk_{ID}.
- $\mathsf{UserSecretKeyGen}^{\mathsf{old}}$: On input mpk and ID, it generates usk_{ID}.
- $\mathsf{UserPublicKeyGen}^{\mathsf{old}}$: On input usk_{ID}, it generates upk_{ID}.
- $\mathsf{UserPrivateKeyGen}^{\mathsf{old}}$: On input psk_{ID} and usk_{ID}, it generates a *full* user private key usk_{ID}^{full}.
- $\mathsf{Enc}^{\mathsf{old}}$: On input mpk, ID, upk_{ID} and a message m, it returns a ciphertext c or a symbol $\perp$ indicating the failure of encryption.
- $\mathsf{Dec}^{\mathsf{old}}$: On input usk_{ID}^{full}, and a ciphertext c, it returns a plaintext $m \in \mathcal{M}(mpk, upk_{\mathsf{ID}})$ or a symbol $\perp$ indicating the failure of decryption.

The correctness requirement is defined in a straightforward way similar to above and therefore is omitted here.

Theorem 2.1. *For any* $\mathsf{CLE}^{\mathsf{old}}$ *scheme, it is a* CLE *scheme. The reverse is also true, namely, for any* CLE *scheme, it is also a* $\mathsf{CLE}^{\mathsf{old}}$ *scheme.*

Proof. First, we show that any $\mathsf{CLE}^{\mathsf{old}}$ scheme can be converted to a scheme satisfying the definition of CLE.

Set MasterKeyGen, PartialKeyGen and Enc to $\mathsf{MasterKeyGen}^{\mathsf{old}}$, $\mathsf{PartialKeyGen}^{\mathsf{old}}$ and $\mathsf{Enc}^{\mathsf{old}}$, respectively. UserKeyGen and Dec are constructed as follows.

$\mathsf{UserKeyGen}$: On input mpk, ID,

(1) run $usk_{\mathsf{ID}} \leftarrow \mathsf{UserSecretKeyGen}^{\mathsf{old}}(mpk, \mathsf{ID})$;
(2) run $upk_{\mathsf{ID}} \leftarrow \mathsf{UserPublicKeyGen}^{\mathsf{old}}(usk_{\mathsf{ID}})$; and
(3) output $(upk_{\mathsf{ID}}, usk_{\mathsf{ID}})$.

Dec: On input psk_{ID}, usk_{ID}, and a ciphertext c,

(1) run $usk_{\mathsf{ID}}^{full} \leftarrow \mathsf{UserPrivateKeyGen}^{\mathsf{old}}(psk_{\mathsf{ID}}, usk_{\mathsf{ID}})$; and
(2) output $\mathsf{Dec}^{\mathsf{old}}(usk_{\mathsf{ID}}^{full}, c)$.

It can be seen that the correctness requirement is also followed. We now show that any CLE scheme can be converted to a scheme satisfying the definition of $\mathsf{CLE}^{\mathsf{old}}$.

Set $\mathsf{MasterKeyGen}^{\mathsf{old}}$, $\mathsf{PartialKeyGen}^{\mathsf{old}}$ and $\mathsf{Enc}^{\mathsf{old}}$ to MasterKeyGen, PartialKeyGen and Enc, respectively. $\mathsf{UserSecretKeyGen}^{\mathsf{old}}$, $\mathsf{UserPublicKeyGen}^{\mathsf{old}}$ and $\mathsf{UserPrivateKeyGen}^{\mathsf{old}}$ are constructed as follows.

$\mathsf{UserSecretKeyGen}^{\mathsf{old}}$: On input mpk and ID,

(1) run $(upk_{\mathsf{ID}}, usk_{\mathsf{ID}}) \leftarrow \mathsf{UserKeyGen}(mpk, \mathsf{ID}; \omega)$, where ω denotes the random coin tosses;
(2) output a user secret key $usk'_{\mathsf{ID}} := (usk_{\mathsf{ID}}, \omega)$.

$\mathsf{UserPublicKeyGen}^{\mathsf{old}}$: On input $usk'_{\mathsf{ID}} := (usk_{\mathsf{ID}}, \omega)$,

(1) run $(upk_{\mathsf{ID}}, usk_{\mathsf{ID}}) \leftarrow \mathsf{UserKeyGen}(mpk, \mathsf{ID}; \omega)$; and
(2) output upk_{ID}.

$\mathsf{UserPrivateKeyGen}^{\mathsf{old}}$: On input psk_{ID} and $usk'_{\mathsf{ID}} := (usk_{\mathsf{ID}}, \omega)$, set $usk_{\mathsf{ID}}^{full} := (psk_{\mathsf{ID}}, usk_{\mathsf{ID}})$.

Finally, $\mathsf{Dec}^{\mathsf{old}}$ can be constructed as follows.

$\mathsf{Dec}^{\mathsf{old}}$: On input $usk_{\mathsf{ID}}^{full} := (psk_{\mathsf{ID}}, usk_{\mathsf{ID}})$ and a ciphertext c, output $\mathsf{Dec}(psk_{\mathsf{ID}}, usk_{\mathsf{ID}}, c)$.

Similarly, the correctness requirement is also satisfied. □

The first significance of this simplification is the elimination of unnecessary complication on functionalities of the original definition. In the original seven-algorithm definition, the user public key and user secret key are generated separately using two algorithms, $\mathsf{UserSecretKeyGen}^{\mathsf{old}}$ and $\mathsf{UserPublicKeyGen}^{\mathsf{old}}$. However, it is hardly convincing to require the generation of multiple user public keys with respect to the same user secret key by running $\mathsf{UserPublicKeyGen}^{\mathsf{old}}$ for multiple times. Instead, it is more natural and consistent with almost all other cryptographic settings (such as CA-based PKI and identity-based cryptography) to use one single function, that is UserKeyGen, to generate a user public/secret key pair. Furthermore, UserPrivateKeyGen^{old} may only serve as a conceptual purpose, by explicitly indicating that both user partial key and user secret key should be used for decryption. It barely has any practical significance. Therefore, we suggest to remove it as well for yielding the new five-algorithm definition.

This simplified five-algorithm definition may also lead to more compact and efficient actual implementations in practice and help in the effort of CLE standardization in the near future.

3. Security Models

The security of a CLE scheme is typically formalized using two games, **Game-I** and **Game-II**, which are played against adversaries $\mathcal{A}_1$ and $\mathcal{A}_2$, respectively. In both games, the adversaries are trying to break the IND-CCA security. Adversary $\mathcal{A}_1$ models a malicious adversary which can compromise the user secret key usk_{ID} or replace the user public key upk_{ID}, however, cannot compromise the master secret key msk nor get access to the user partial key psk_{ID}. Adversary $\mathcal{A}_2$ models a *malicious-but-passive* KGC [6,8] which controls the generation of the master public/secret key pair, and that of any user partial key psk_{ID}. Below are the oracles which can be accessed by the adversaries.

- CreateUser: On input an identity ID, if ID has not been created, the oracle runs $psk_{\mathsf{ID}} \leftarrow \mathsf{PartialKeyGen}(msk, \mathsf{ID})$ and $(upk_{\mathsf{ID}}, usk_{\mathsf{ID}}) \leftarrow \mathsf{UserKeyGen}(mpk, \mathsf{ID})$. It then stores $(\mathsf{ID}, psk_{\mathsf{ID}}, upk_{\mathsf{ID}}, usk_{\mathsf{ID}})$ into a list List[a]. In both cases, upk_{ID} is returned. ID is said to have been *created* if it is in List.
- RevealPartialKey: On input an identity ID, the oracle searches List for an entry corresponding to ID. If it is not found, $\perp$ is returned; otherwise, the corresponding psk_{ID} is returned.
- RevealSecretKey: On input an identity ID, the oracle searches List for the entry of ID. If it is not found, $\perp$ is returned; otherwise, the corresponding usk_{ID} is returned.
- ReplacePublicKey: On input an identity ID along with a user public/secret key pair (upk', usk'), the oracle searches List for the entry

[a] List is shared among all oracles and is assumed to be initialized to an empty list at the beginning of a game by the game simulator.

of ID. If it is not found, nothing will be carried out. If $usk' = \perp$, the oracle sets $usk' = usk_{\mathsf{ID}}$. Then, it updates $(\mathsf{ID}, psk_{\mathsf{ID}}, upk_{\mathsf{ID}}, usk_{\mathsf{ID}})$ in List to $(\mathsf{ID}, psk_{\mathsf{ID}}, upk', usk')$.

- Decryption: On input an identity ID and a ciphertext c, the oracle searches List for the entry of ID. If it is not found, $\perp$ is returned. Otherwise, it runs $m \leftarrow \mathsf{Dec}(psk_{\mathsf{ID}}, usk_{\mathsf{ID}}, c)$ and returns m. Note that the original upk_{ID} (which is returned by CreateUser oracle) may have been replaced by the adversary.

Let $\mathcal{C}_1$ and $\mathcal{C}_2$ be the game simulators of **Game-I** and **Game-II**, respectively. Let $k \in \mathbb{N}$ be a security parameter. The two games run as follows.

Game-I :

(1) $\mathcal{C}_1$ generates a master public key mpk, then invokes $\mathcal{A}_1$ on input 1^k and mpk. Note that $\mathcal{C}_1$ may simply run $(mpk, msk) \leftarrow \mathsf{MasterKeyGen}(1^k)$ to generate mpk. However, this is not mandatory. The only requirement is that mpk generated by $\mathcal{C}_1$ should be computationally indistinguishable from the output of $\mathsf{MasterKeyGen}(1^k)$.
(2) Upon receiving mpk, $\mathcal{A}_1$ can start issuing queries to CreateUser, RevealPartialKey, RevealSecretKey, ReplacePublicKey and Decryption in an adaptive manner.
(3) $\mathcal{A}_1$ submits two equal-length messages (m_0, m_1) along with a target identity ID^*.
(4) $\mathcal{C}_1$ selects a random bit $b \in \{0, 1\}$, computes a challenge ciphertext c^* by running $c^* \leftarrow \mathsf{Enc}(mpk, \mathsf{ID}^*, upk_{\mathsf{ID}^*}, m_b)$, and returns c^* to $\mathcal{A}_1$, where upk_{ID^*} is the user public key currently in List for ID^*.
(5) $\mathcal{A}_1$ continues to issue queries as above. Finally it outputs a bit b'.

$\mathcal{A}_1$ wins the game if $b' = b$, and (1) $\mathcal{A}_1$ did not query RevealPartialKey on ID^*, (2) $\mathcal{A}_1$ did not query Decryption on (ID^*, c^*). We denote by $\Pr[\mathcal{A}_1 \text{ Succ}]$ the probability that $\mathcal{A}_1$ wins the game, and define the *advantage* of $\mathcal{A}_1$ in **Game-I** to be $\mathsf{Adv}_{\mathcal{A}_1} = \left|\Pr[\mathcal{A}_1 \text{ Succ}] - \frac{1}{2}\right|$.

Game-II : $\mathcal{C}_2$ invokes $\mathcal{A}_2$ on input 1^k directly.

(1) $\mathcal{A}_2$ returns a master public key mpk to $\mathcal{C}_2$. It is required that mpk generated by $\mathcal{A}_2$ is computationally indistinguishable from the output of $\mathsf{MasterKeyGen}(1^k)$. At this stage, $\mathcal{A}_2$ is not allowed to make any oracle query[b].

[b]One exception is that if a scheme is analyzed under the random oracle model, $\mathcal{A}_2$ can

(2) $\mathcal{A}_2$ then starts querying oracles CreateUser, RevealSecretKey, ReplacePublicKey and Decryption. Oracle RevealPartialKey is no longer needed as $\mathcal{A}_2$ is the one who generates the user partial key for a user. One thing to notice is that when $\mathcal{A}_2$ issues a query to CreateUser oracle, it should additionally provide the user partial key psk_{ID}. In other words, oracle CreateUser for this game is revised as follows.

> CreateUser: On input an identity ID and a user partial key psk_{ID}, if there is no entry in List starting with ID (or, $\mathsf{ID} \notin \mathsf{List}$), the oracle runs $(upk_{\mathsf{ID}}, usk_{\mathsf{ID}}) \leftarrow \mathsf{UserKeyGen}(mpk, \mathsf{ID})$ and adds (ID, psk_{ID}, upk_{ID}, usk_{ID}) into List. upk_{ID} is returned.

(3) $\mathcal{A}_2$ submits two equal-length messages (m_0, m_1) along with a target identity ID^*.

(4) $\mathcal{C}_2$ randomly selects a bit b, and computes the challenge ciphertext c^* by running $c^* \leftarrow \mathsf{Enc}(mpk, \mathsf{ID}^*, upk_{\mathsf{ID}^*}, m_b)$. It returns c^* to $\mathcal{A}_2$.

(5) $\mathcal{A}_2$ continues making queries as in step 1. Finally, it outputs a bit b'.

$\mathcal{A}_2$ wins the game if $b' = b$, and (1) $\mathcal{A}_2$ did not query RevealSecretKey on ID^*, (2) $\mathcal{A}_2$ did not query ReplacePublicKey on $(\mathsf{ID}^*, \cdot, \cdot)$ to replace upk_{ID^*}, (3) $\mathcal{A}_2$ did not query Decryption on (ID^*, c^*). Similarly, we denote by $\Pr[\mathcal{A}_2 \ \mathsf{Succ}]$ the probability that $\mathcal{A}_2$ wins the game, and define the *advantage* of $\mathcal{A}_2$ in **Game-II** to be $\mathsf{Adv}_{\mathcal{A}_2} = \left|\Pr[\mathcal{A}_2 \ \mathsf{Succ}] - \frac{1}{2}\right|$.

Definition 3.1. A certificateless encryption scheme CLE is *Type-I secure* (resp. *Malicious KGC Type-II secure*) if there is no PPT adversary $\mathcal{A}_1$ (resp. $\mathcal{A}_2$) which wins ***Game-I*** (resp. ***Game-II***) with non-negligible advantage. CLE is said to be *IND-CCA secure* if it is both *Type-I secure* and *Malicious KGC Type-II secure.*

3.1. *Malicious KGC Security vs. Strong Security*

In the games above, we target to model attacks against a *realistic* user. Therefore, the Decryption oracle is simulated such that the decryption is performed using the user's *current* knowledge of the user secret key regardless whether the user public key has been replaced or not. Through ReplacePublicKey, the user secret key can also be replaced accordingly.

---

query the random oracle. In this paper, we do not consider this and focus ourselves to the standard model only.

In the original security models for CLE proposed by Al-Riyami and Paterson [3], on the other hand, the Decryption oracle is defined such that it should (always) provide correct decryption of a well-formed ciphertext as long as the current user public key is well-formed. In particular, this definition requires Decryption oracle to return corresponding plaintext even if the user public key is chosen and replaced by the adversary while the corresponding user secret key is not given. As first termed by Dent [4], if the Decryption oracle is defined in this way, the corresponding types of security are called *Strong Type-I Security* and *Strong Type-II Security.* Note that in [3], malicious-but-passive KGC attacks are not captured. If the **Game-II** above has the Decryption oracle be replaced by this *strong* version, then the corresponding security is called *Strong Malicious KGC Type-II Security.*

By considering the security of a CLE scheme against *malicious-but-passive KGC attacks* or not; and *strong* chosen ciphertext attack or not, we have altogether two Type-I Security levels (*Type-I Security, Strong Type-I Security*) and four Type-II Security levels (which are specified in Table 1).

Table 1. Levels of Type-II Security

	Realistic CCA	Strong CCA
Honest-but-curious KGC	*Type-II Security*	*sType-II Security*
Malicious-but-passive KGC	*mType-II Security*	*smType-II Security*

Legends:
Realistic CCA — Decryption oracle defined above
Strong CCA — Decryption oracle defined in [3]
sType-II — Strong Type-II
mType-II — Malicious KGC Type-II
smType-II — Strong Malicious KGC Type-II

We now study the relationship among these security levels. First of all, the following relations are easy to see.

- *Strong Type-I Security* $\Rightarrow$ *Type-I Security*
- *Malicious KGC Type-II Security* $\Rightarrow$ *Type-II Security*
- *Strong Type-II Security* $\Rightarrow$ *Type-II Security*
- *Strong Malicious KGC Type-II Security* $\Rightarrow$ *Malicious KGC Type-II Security* $\wedge$ *Strong Type-II Security*

In the above, *A Security* $\Rightarrow$ *B Security* $\wedge$ *C Security* means that if a CLE scheme which is *A Secure*, then the scheme is also *B Secure* and *C Secure.*

In Def. 3.1, we require an IND-CCA secure CLE to be *Type-I Secure* and *Malicious KGC Type-II Secure.* In [4,7,8], CLE schemes have been shown to satisfy this security definition in the standard model. In [9], a CLE scheme proven to be *Strong Type-I Secure* and *Strong Type-II Secure* in the standard model was proposed.

A problem that still remains open is whether it is possible to construct a scheme which is secure against a *mixture* of malicious-but-passive KGC attacks and strong chosen ciphertext attacks. For example, a scheme which is both *Strong Type-I Secure* and *Malicious KGC Type-II Secure.* The following theorem states that it is **impossible** to build such a scheme in the standard model. We will also show that *Strong Security* and *Malicious KGC Security* **cannot co-exist** on any CLE scheme in the standard model.

Theorem 3.1. *In the standard model, if a* CLE *scheme is* Strong Type-I Secure*, then it must not be* Malicious KGC Type-II Secure. *In other words,* Strong Type-I Security *and* Malicious KGC Type-II Security ***cannot co-exist*** *on any* CLE *scheme in the standard model.*

Intuitively, it seems to be somewhat *self-contradictory* if we require a CLE scheme to be both *Strong Type-I Secure* and *Malicious KGC Type-II Secure.* Suppose that it is secure against Type-I adversaries with access to a *strong* Decryption oracle. The game simulator, which also plays the role of the KGC, has to simulate the *strong* Decryption oracle. In order to provide correct decryption without the user secret key, the simulator usually needs to play tricks when setting *mpk*. However, in the same way, the Type-II adversary which behaves as a *malicious-but-passive* KGC should also be able to play the same tricks with *mpk* so that it can decrypt without knowing the user secret key.

In the proof below, let **Strong Game-I** be **Game-I** with the Decryption oracle being replaced by StrongDecryption oracle which corresponds to the decryption oracle defined in [3].

Proof. Suppose there exists a CLE scheme which is *Strong Type-I Secure*. This implies that there is an interactive PPT game simulator $\mathcal{C}_1^{Strong}$ for **Strong Game-I** such that for any PPT $\mathcal{A}_1^{Strong}$, the advantage of $\mathcal{A}_1^{Strong}$ (that is defined similarly to $\mathsf{Adv}_{\mathcal{A}_1}$) is negligible, and $\mathcal{C}_1^{Strong}$ successfully simulate **Strong Game-I** with some non-negligible probability ε. This particularly implies that $\mathcal{C}_1^{Strong}$ can simulate StrongDecryption successfully with non-negligible probability for requests made by any PPT $\mathcal{A}_1^{Strong}$. Below are the major functionalities provided by $\mathcal{C}_1^{Strong}$.

F1. On input 1^k, $\mathcal{C}_1^{Strong}$ generates a master public key *mpk* which is computationally indistinguishable from the output of $\mathsf{MasterKeyGen}(1^k)$.
F2. Simulate CreateUser: $upk_{\mathsf{ID}} \leftarrow \mathsf{CreateUser}(\mathsf{ID})$

F3. Simulate $\mathsf{RevealPartialKey}$: $psk_{\mathsf{ID}} \leftarrow \mathsf{RevealPartialKey}(\mathsf{ID})$
F4. Simulate $\mathsf{RevealSecretKey}$: $usk_{\mathsf{ID}} \leftarrow \mathsf{RevealSecretKey}(\mathsf{ID})$
F5. Simulate $\mathsf{ReplacePublicKey}$: $\mathsf{ReplacePublicKey}(\mathsf{ID}, upk', usk')$
F6. Simulate challenge ciphertext generation: $b \leftarrow 0/1$ and $c^* \leftarrow \mathsf{Enc}(mpk, \mathsf{ID}^*, upk_{\mathsf{ID}^*}, m_b)$
F7. Simulate $\mathsf{StrongDecryption}$: $m/\perp \leftarrow \mathsf{StrongDecryption}(\mathsf{ID}, c)$

Given $\mathcal{C}_1^{Strong}$, we construct a Type-II adversary $\mathcal{A}_2$ which targets to win **Game-II** (page 161) with non-negligible advantage. Note that $\mathcal{C}_2$ is the game simulator of **Game-II**.

At the beginning of **Game-II**, $\mathcal{C}_2$ invokes $\mathcal{A}_2(1^k)$. Below is the execution procedure of $\mathcal{A}_2$.

(1) $\mathcal{A}_2$ invokes $\mathcal{C}_1^{Strong}$ and gets mpk. This part corresponds to function **F1** of $\mathcal{C}_1^{Strong}$.
(2) $\mathcal{A}_2$ randomly picks an identity $\mathsf{ID}^* \in_R \{0,1\}^k$.
(3) $\mathcal{A}_2$ (acts as the adversary $\mathcal{A}_1^{Strong}$ of **Strong Game-I** simulated by $\mathcal{C}_1^{Strong}$ and) interacts with $\mathcal{C}_1^{Strong}$ for function **F2** on ID^*.
(4) $\mathcal{A}_2$ interacts with $\mathcal{C}_1^{Strong}$ for function **F3** on ID^*. Let the user partial key returned by $\mathcal{C}_1^{Strong}$ be psk_{ID^*}.
(5) $\mathcal{A}_2$ queries $\mathsf{CreateUser}$ (simulated by $\mathcal{C}_2$) on ID^* and receives user public key upk'_{ID^*}.
(6) $\mathcal{A}_2$ interacts with $\mathcal{C}_1^{Strong}$ for function **F5** on $(\mathsf{ID}^*, upk'_{\mathsf{ID}^*}, \perp)$.
(7) $\mathcal{A}_2$ picks two equal-length messages $m_0, m_1 \in_R \mathcal{M}(mpk, upk'_{\mathsf{ID}^*})$ randomly and submits them along with ID^* to $\mathcal{C}_2$. Suppose the received challenge ciphertext is c^* which is computed as $\mathsf{Enc}(mpk, \mathsf{ID}^*, upk'_{\mathsf{ID}^*}, m_b)$ for a random bit $b \leftarrow 1/0$.
(8) $\mathcal{A}_2$ interacts with $\mathcal{C}_1^{Strong}$ for function **F7** on (ID^*, c^*). Suppose the return of $\mathcal{C}_1^{Strong}$ is m^*.
(9) If $m^* = m_0$, $\mathcal{A}_2$ returns 0. If $m^* = m_1$, $\mathcal{A}_2$ returns 1. Otherwise, $\mathcal{A}_2$ halts with outputting a random bit.

(*Analysis*). First, we can see that $\mathcal{A}_2$ runs in polynomial time. This is because $\mathcal{C}_1^{Strong}$ is a PPT algorithm and all other operations in the execution procedure of $\mathcal{A}_2$ above can be completed in polynomial time.

Second, we show that $\mathcal{C}_1^{Strong}$ will provide the correct decryption of c^* to $\mathcal{A}_2$ with non-negligible probability. Since all operations in the execution procedure of $\mathcal{A}_2$ are in polynomial time, this satisfies the condition that $\mathcal{C}_1^{Strong}$ can simulate **Strong Game-I** successfully with non-negligible probability. Therefore, with non-negligible probability, $\mathcal{C}_1^{Strong}$ generates mpk (via **F1**), creates a user with identity ID^* (via **F2**), reveals psk_{ID^*} to

$\mathcal{A}_2$ (via **F3**), gets user public key of ID^* replaced by upk'_{ID^*} (via **F5**), and finally simulates StrongDecryption for decrypting the challenge ciphertext c^* with respect to ID^* and upk'_{ID^*} (via **F7**). Hence with some non-negligible probability ε' (which is obviously no less than ε), $\mathfrak{C}_1^{Strong}$ will recover the message encrypted in c^* successfully.

Finally, the above result implies that $\mathcal{A}_2$ wins **Game-II** with the probability at least

$$\varepsilon' + \frac{1}{2}(1 - \varepsilon') = \frac{1}{2} + \frac{\varepsilon'}{2} \geq \frac{1}{2} + \frac{\varepsilon}{2}$$

which is non-negligibly greater than 1/2. This concludes the proof. □

Theorem 3.2. *In the standard model, if a* CLE *scheme is* Strong Type-II Secure*, then it must not be* Malicious KGC Type-II Secure*. In other words,* Strong Type-II Security *and* Malicious KGC Type-II Security ***cannot co-exist*** *on any* CLE *scheme in the standard model.*

The proof is similar to that for Theorem 3.1 and is skipped.

Corollary 3.1. *In the standard model, regardless of Type-I Security or Type-II Security, there is* ***no*** CLE *scheme which can be* Strong Type-I/II Secure *and* Malicious KGC Type-II Secure*. In other words,* Strong Security ***cannot co-exist*** *with* Malicious KGC Security *on any* CLE *scheme.*

Since Malicious KGC Security and Strong Security cannot co-exist, we have to make a choice on the highest security level that a CLE scheme should achieve. Malicious-but-passive KGC attacks are important and practical while strong decryption oracle is hardly realistic. We therefore suggest to put effort on constructing CLE schemes which are *Type-I Secure* and *Malicious KGC Type-II Secure*, that is, fulfilling Def. 3.1. In fact, there are already some good results appeared [4,7,8]. Interestingly, all the current constructions are generic ones.

Remark 3.1. It is also interesting to see that in [12], Libert and Quisquater proposed a CLE scheme that is proven both Strong Type-I secure and Strong Type-II secure in the *random oracle model* [13], and in [6], Au *et al.* showed that this scheme is also Malicious KGC Type-II secure in the *random oracle model.* This result does not contradict our conclusion above, since Libert-Quisquater and Au *et al.*'s results are in the random oracle model, in which the game simulator is always assumed to have the full control of the output of random oracles, while in the standard model, no such power is given to the simulator.

4. User Partial Key Replacement Attacks

We introduce a new oracle called ReplacePartialKey to the games. Through this oracle, adversaries can *replace* user partial keys (i.e. psk_{ID}). In all the comparable models [3–6,10,11], adversaries can only replace user public keys (i.e. upk_{ID}). Replacing upk_{ID} simulates the scenario that a self-generated user public key published on some public domain is being replaced. This is feasible because upk_{ID} is not authenticated. A more powerful attacking scenario related to user public key replacement is when the random source of a user has been compromised so that the corresponding user secret key (i.e. usk_{ID}) is also known or being controlled by the adversary (e.g. via side-channel attacks).

Regarding the user partial key, psk_{ID}, of a user, it may also be *regenerated* in a real case, for example, after the user has lost his previous user partial key. In another case, the malicious-but-passive KGC may also want to try out different values of user partial key after learning the user public key of a particular user, with the purpose of compromising the user's secrecy. In this case, the KGC might send the user a specific user partial key, which when combined with the user secret key, might leak information that increases the KGC's chance of decrypting a ciphertext of the user. Note that the difference between replacing a user partial key (i.e. ReplacePartialKey) and revealing a user partial key (i.e. RevealPartialKey on page 160) is that the former one is adaptive while the latter one is static.

We note that in [4], Dent proposed another variant of decryption oracle called Weak PPK Decrypt for the access of Type-II adversary. The oracle takes not only ID and ciphertext c, it also takes a user partial key psk'_{ID}. The decryption of c is done using the current user secret key usk_{ID} and this temporary user partial key psk'_{ID}, rather than the current user partial key psk_{ID}. In fact, new attacks enabled by Weak PPK Decrypt are equivalent to that by ReplacePartialKey. To see that Weak PPK Decrypt oracle can be captured by ReplacePartialKey, we can first query ReplacePartialKey to change the user partial key to the temporary one psk'_{ID} and then query Decryption to perform decryption of the ciphertext c. After this is done, we query ReplacePartialKey again for resuming the original user partial key psk_{ID}. For the other direction, namely, ReplacePartialKey introduces new attacks that can also be captured by Weak PPK Decrypt, we notice that ReplacePartialKey does not affect the generation of challenge ciphertext c^* from one of the messages m_0 and m_1. Also, ReplacePartialKey has to work jointly with Decryption for giving any information back to the adversary. These two queries are carried out in one query of Weak PPK Decrypt. Below is the complete description of the ReplacePartialKey oracle.

ReplacePartialKey: On input an identity ID and a user partial key psk', the oracle searches List for the entry of ID. If it is not found, nothing will be carried out. If found, it updates $(\mathsf{ID}, psk_{\mathsf{ID}}, upk_{\mathsf{ID}}, usk_{\mathsf{ID}})$ in List with $(\mathsf{ID}, psk', upk_{\mathsf{ID}}, usk_{\mathsf{ID}})$.

For **Game-I**, one additional condition needs to be added for $\mathcal{A}_1$ winning the game: $\mathcal{A}_1$ should not have queried ReplacePartialKey on ID^*. There is no additional condition to be added for $\mathcal{A}_2$ to win **Game-II**.

In [4], an attack is given to show that a Malicious KGC Type-II secure CLE scheme may not be secure if Weak PPK Decrypt oracle is allowed to access. As Weak PPK Decrypt gives the same effect as having ReplacePartialKey in the game, the example also illustrates that a Malicious KGC Type-II secure CLE scheme may not be secure if access to ReplacePartialKey oracle is granted.

5. Our Scheme

In [8], we proposed a generic CLE construction and showed that it is IND-CCA secure (Def. 3.1). In this section, we show that the scheme is also secure if **Game-I** and **Game-II** in Sec. 3 have the user partial key replacement attacks described in Sec. 4 above included.

Let $\mathsf{IBE} = (\mathsf{KG}, \mathsf{Extract}, \mathsf{Enc}, \mathsf{Dec})$ be an IND-ID-CCA secure identity-based encryption scheme, $\mathsf{PKE} = (\mathsf{KG}, \mathsf{Enc}, \mathsf{Dec})$ an IND-CCA secure public key encryption scheme, and $\mathsf{S} = (\mathsf{KG}, \mathsf{Sign}, \mathsf{Vrfy})$ a strong one-time signature scheme. In the following, we review our generic CLE scheme[8].

- MasterKeyGen: The KGC runs $(mpk, msk) \leftarrow \mathsf{IBE.KG}(1^k)$, publishes mpk and keeps msk secret.
- PartialKeyGen: On input an identity ID, the KGC runs $psk_{\mathsf{ID}} \leftarrow \mathsf{IBE.Extract}(msk, \mathsf{ID})$ and returns psk_{ID}.
- UserKeyGen: The user (with identity ID) runs $(upk_{\mathsf{ID}}, usk_{\mathsf{ID}}) \leftarrow \mathsf{PKE.KG}(1^k)$, publishes $(\mathsf{ID}, upk_{\mathsf{ID}})$ and stores usk_{ID}.
- Enc: To encrypt a message m for user ID, the encryptor computes the following and returns $c \stackrel{\text{def}}{=} (c_2, \sigma, vk)$:

$$\begin{aligned} (vk, sk) &\leftarrow \mathsf{S.KG}(1^k) \\ c_1 &\leftarrow \mathsf{IBE.Enc}(mpk, \mathsf{ID}, m \| vk) \\ c_2 &\leftarrow \mathsf{PKE.Enc}(upk_{\mathsf{ID}}, c_1) \\ \sigma &\leftarrow \mathsf{S.Sign}(sk, c_2) \end{aligned}$$

- Dec: On input an identity ID and a ciphertext $c = (c_2, \sigma, vk)$, if $0 \leftarrow \mathsf{S.Vrfy}(vk, \sigma, c_2)$, $\perp$ is returned. Otherwise, the decryptor computes the following:

$$c_1 \leftarrow \mathsf{PK.Dec}(usk_{\mathsf{ID}}, c_2)$$
$$m \| vk' \leftarrow \mathsf{IBE.Dec}(psk_{\mathsf{ID}}, \mathsf{ID}, c_1)$$

If $vk' \neq vk$, the decryptor outputs $\perp$; otherwise, it outputs m.

Theorem 5.1. *The certificateless encryption scheme* CLE *is* Type-I secure, *provided that the underlying identity-based encryption scheme* IBE *is* IND-ID-CCA *secure, and the one-time signature scheme* S *is* strongly unforgeable.

Theorem 5.2. *The certificateless encryption scheme* CLE *is* Type-II secure *if the underlying public key encryption scheme* PKE *is* IND-CCA *secure and the one-time signature scheme* S *is* strongly unforgeable.

Due to page limitation, we skip all the proofs here. The corollary below is obtained directly from Theorem 5.1 and Theorem 5.2.

Corollary 5.1. *The certificateless encryption scheme* CLE *described above is* IND-CCA secure.

Please refer to [8] for detailed discussions on how to implement this generic scheme efficiently.

6. Conclusion

Our target in this paper is to contribute on unifying the definitions and security models for certificateless encryption (CLE) schemes. We showed that the original seven-algorithm definition is equivalent to a simplified five-algorithm definition, which may lead to building more compact and efficient implementations in practice and help in the effort of standardizing CLE. We showed that a CLE scheme cannot be both *Malicious KGC Secure* and Strongly Secure in the standard model. Due to the practicality and attacking significance of Malicious KGC Security, we suggest constructing CLE schemes to be Malicious KGC Secure. We also proposed a new adversarial capability called Partial Key Replacement attacks which have practical implications and showed that our generic scheme proposed recently is also secure against this new type of attacks.

Acknowledgement

The work was supported by CityU grant (Project No. 7002001).

References

1. A. Shamir, Identity-based cryptosystems and signature schemes, in *Proc. CRYPTO 84*, (Springer-Verlag, 1984). LNCS 196.
2. D. Boneh and M. Franklin, Identity-based encryption from the Weil pairing, in *Proc. CRYPTO 2001*, (Springer-Verlag, 2001). LNCS 2139.
3. S. S. Al-Riyami and K. G. Paterson, Certificateless public key cryptography, in *Proc. ASIACRYPT 2003*, (Springer-Verlag, 2003). LNCS 2894.
4. A. W. Dent, A survey of certificateless encryption schemes and security models Cryptology ePrint Archive, Report 2006/211, (2007), `http://eprint.iacr.org/2006/211` (Last revised 14 December 2007).
5. J. Baek, R. Safavi-Naini and W. Susilo, Certificateless public key encryption without pairing, in *8th International Conference on Information Security (ISC 2005)*, (Springer-Verlag, 2005). LNCS 3650.
6. M. H. Au, J. Chen, J. K. Liu, Y. Mu, D. S. Wong and G. Yang, Malicious KGC attacks in certificateless cryptography, in *ACM ASIACCS'07*, (ACM, 2007). Also at `http://eprint.iacr.org/2006/255`.
7. Q. Huang and D. S. Wong, Generic certificateless key encapsulation mechanism, in *Information Security and Privacy: 12th Australian Conference, ACISP 2007*, (Springer-Verlag, 2007). LNCS 4586.
8. Q. Huang and D. S. Wong, Generic certificateless encryption in the standard model, in *2nd International Workshop on Security (IWSEC 2007)*, (Springer-Verlag, 2007). LNCS 4752. Also at `http://eprint.iacr.org/2007/095`.
9. A. W. Dent, B. Libert and K. G. Paterson, Certificateless encryption schemes strongly secure in the standard model Cryptology ePrint Archive, Report 2007/121, (2007), `http://eprint.iacr.org/2007/121`.
10. B. C. Hu, D. S. Wong, Z. Zhang and X. Deng, Key replacement attack against a generic construction of certificateless signature, in *Information Security and Privacy: 11th Australian Conference, ACISP 2006*, (Springer-Verlag, 2006). LNCS 4058.
11. B. C. Hu, D. S. Wong, Z. Zhang and X. Deng, *Designs, Codes, and Cryptography* **42**, 109 (2007).
12. B. Libert and J.-J. Quisquater, On constructing certificateless cryptosystems from identity based encryption, in *9th International Conference on Theory and Practice in Public Key Cryptography, PKC 2006*, (Springer-Verlag, 2006). LNCS 3958.
13. M. Bellare and P. Rogaway, Random oracles are practical: A paradigm for designing efficient protocols, in *First ACM Conference on Computer and Communications Security*, (ACM, Fairfax, 1993).

Distance-Preserving Mappings

Torleiv Kløve

Department of Informatics
University of Bergen, Norway
E-mail: Torleiv.Klove@ii.uib.no

This article gives a survey of the known distance-preserving mappings.

Keywords: Distance-preserving mapping (DPM), distance-increasing mapping (DIM), permutation array (PA), powerline communication

1. Introduction

Permutation arrays as combinatorial objects have been studied for many years. However, a few years ago, Ferreira and Vinck [6] found applications in powerline communication: permutations arrays can be used as error correcting codes. For a given length m, a permutation array (of length m) is a set of permutations of the set $\{1, 2, \ldots, m\}$. The minimum distance d of the permutation array is, as usual, the smallest Hamming distance between the permutations. For the application, there is the usual trade off between minimum distance and size of the code (permutation array).

Because of the application in powerline communication, there has been a renewed interest in permutation arrays, and a substantial number of papers with new and better constructions have appeared during the last 6-7 years.

One way to construct permutation arrays, introduced by Ferreira and Vinck [6] is to use the image of codes under a distance-preserving mapping from binary vectors to permutations. A mapping from the set of all binary vectors of length n to the set of all permutations of $\{1, 2, \ldots, m\}$ is called a distance-preserving mapping (DPM) if every two distinct vectors are mapped to permutations with the same or larger Hamming mutual distance than that of the mapped vectors. Since the mapping is distance-preserving, the minimum distance of the image (which is a permutation array) is lower bounded by the minimum distance of the code. A number of papers[1–3,5–13,19]. have studied various constructions of DPMs, with variations. The permutation arrays constructed by this method are *the best known* for many values of the parameters m and d.

Recently, DPM from ternary mapping have been found. The PAs constructed using these are in many cases better than those obtained from mappings from binary vectors.

No proofs are given in this paper, but we give references for all results.

2. Basic notations and applications

Let S_m denote the set of all $m!$ permutations of $F_m = \{1, 2, \ldots, m\}$. A permutation array (PA) of length m is subset of S_m.

A permutation $\pi : F_m \to F_m$ is represented by an m-tuple $\pi = (\pi_1, \pi_2, \ldots, \pi_m)$ where $\pi_i = \pi(i)$.

For a permutation $\rho = (\rho_1, \rho_2, \ldots, \rho_m) \in S_m$ and an integer a, let $\rho + a = (\rho_1 + a, \rho_2 + a, \ldots, \rho_m + a)$. This is a permutation of $\{a+1, a+2, \ldots, a+m\}$.

Let Z_q^n denote the set of all q-ary vectors of length n, where

$$Z_q = \{0, 1, \ldots, q-1\}.$$

The Hamming distance between two n-tuples $\mathbf{a} = (a_1, a_2, \ldots, a_n)$ and $\mathbf{b} = (b_1, b_2, \ldots, b_n)$ is denoted by $d_H(\mathbf{a}, \mathbf{b})$ and is defined as

$$d_H(\mathbf{a}, \mathbf{b}) = |\{j \in F_n : a_j \neq b_j\}|.$$

One construction idea for PAs is as follows: map some known good code to permutations. We will make this more precise below.

Let n and m be natural numbers and k a non-negative integer. An $(n, m, k; q)$-distance-preserving mapping (for short: an $(n, m, k; q)$-DPM) is a mapping $f : Z_q^n \to S_m$ such that

$$d_H(f(\mathbf{x}), f(\mathbf{y})) \geq \min\{m, d_H(\mathbf{x}, \mathbf{y}) + k\} \text{ for all distinct } \mathbf{x}, \mathbf{y} \in Z_q^n.$$

If $k \geq 1$, the mapping is also known as a distance increasing (DIM).

The set of $(n, m, k; q)$-DPM is denoted by $\mathcal{F}(q, n, m, k)$.

For $f \in \mathcal{F}(q, n, m, k)$, $D_{i,j} = D_{i,j}(f)$ denotes the number of pairs $(\mathbf{x}, \mathbf{y}) \in Z_q^n$ such that

$$d_H(\mathbf{x}, \mathbf{y}) = i \text{ and } d_H(f(\mathbf{x}), f(\mathbf{y})) = j.$$

Let $f \in \mathcal{F}(q, n, m, k)$. Some trivial observations are:

- $D_{0,0}(f) = q^n$, $D_{0,j}(f) = 0$ for $j > 0$.
- if $0 < i \leq n$, and $0 \leq j < \min\{m, i + k\}$, then $D_{i,j}(f) = 0$.

Obviously, if $q^n > m!$, then $\mathcal{F}(q,n,m,k)$ is empty. For example, $|\mathcal{F}(2,3,3,0)| = 0$.

Very little is known about how large $\mathcal{F}(q,n,m,k)$ is in other cases. Two known result [5] are:

$$|\mathcal{F}(2,4,4,0)| = 146\,964\,039\,552.$$

$$|\mathcal{F}(2,n+1,n+1,0)| \geq n^{2^n} |\mathcal{F}(2,n,n,0)| \quad \text{for} \quad n \geq 4.$$

2.1. *Construction of permutation arrays by DPMs*

The maximal size of a PA of length m and minimum distance d is denoted by $P(m,d)$. The maximal size of a code of length n and minimum distance d over Z_q is denoted by $A_q(n,d)$.

Let f be a mapping from Z_q^n to S_m such that

$$D_{i,j} = 0 \text{ for } d \leq i \leq n \text{ and } d' \leq j \leq m.$$

If C is an $(n,d;q)$ code of size $A_q(n,d)$, then $f(C)$ is an (m,d') permutation array of the same size. Hence $P(m,d') \geq A_q(n,d)$. This lower bound on $P(m,d')$ is the best known for many combinations of n and d', and this has been a main motivation for the study of DPMs.

2.2. *Total distance increase*

Let $f \in \mathcal{F}(q,n,m,0)$. Consider

$$\Delta(f) = \sum_{\mathbf{x},\mathbf{y} \in Z_q^n} \Big\{ d_H(f(\mathbf{x}), f(\mathbf{y})) - d_H(\mathbf{x},\mathbf{y}) \Big\} = \Delta_1(f) - \Delta_0,$$

where

$$\Delta_0 = \sum_{\mathbf{x},\mathbf{y} \in Z_q^n} d_H(\mathbf{x},\mathbf{y}) = q^n \sum_{i=1}^{n} i \binom{n}{i} (q-1)^i = n(q-1)q^{2n-1},$$

and

$$\Delta_1(f) = \sum_{\mathbf{x},\mathbf{y} \in Z_q^n} d_H(f(\mathbf{x}), f(\mathbf{y})) = \sum_{i=1}^{n} \sum_{j=i+1}^{m} (j-i) D_{i,j}.$$

Swart, de Beer, and Ferreira[17,18] gave the following upper bound on $\Delta_1(f)$. Let $\alpha = \lfloor q^n/m \rfloor$ and $\beta = q^n \mod m$. Then

$$\Delta_1(f) \leq m\Big(q^{2n} - (2\alpha\beta + \beta + \alpha^2 m)\Big).$$

In particular, if $m = q^r$, where $r \leq n$, then

$$\Delta_1(f) \leq q^{2n}(q^r - 1).$$

3. Particular constructions of DPMs from binary vectors

A number of DPMs from binary vectors have been constructed. We will give references and, in some cases describe them, in the order they have been published. The mappings are given in one of three forms:

- The values of the mapping are explicitly listed in a *table.*
- An *algorithm* is given that on any argument $\mathbf{x}$ as input computes $f(\mathbf{x})$.
- A *recursive* procedure is given for constructing a mapping from mappings of vectors of shorter lengths. The construction requires some (one or more) mappings to start the recursion.

The first examples of DPMs were given by Ferreira and Vinck [6]. They gave explicit examples with $n \leq 8$ in the following cases:

- $k = 0$, $m = n$, $4 \leq n \leq 8$
- $k = 1$, $m = n + 1$, $2 \leq n \leq 4$
- $k = 2$, $m = n + 2$, $4 \leq n \leq 6$

The mappings were found by a combination of search and recursion, and they were listed as tables.

3.1. *Mappings with $k = 0$*

Chang, Chen, Kløve, and Tsai[5] gave both recursive and algorithmic constructions for $m = n \geq 4$ and $k = 0$. We give the main constructions.

First recursive construction

Let $g \in \mathcal{F}(2, n-1, n-1, 0)$ and $p : Z_2^{n-1} \to F_m$. For $\mathbf{x} \in Z_2^{n-1}$, let $(\pi_1, \pi_2, \ldots, \pi_{n-1}) = g(\mathbf{x})$ and $i = p(\mathbf{x})$.

Define $f : Z_2^n \to S_n$ by

$$f(\mathbf{x}, 0) = (\pi_1, \pi_2, \ldots, \pi_{n-1}, n)$$
$$f(\mathbf{x}, 1) = (\pi_1, \ldots, \pi_{i-1}, n, \pi_{i+1}, \ldots, \pi_{n-1}, \pi_i)$$

Then $f \in \mathcal{F}(2, n, n, 0)$.

For the construction, one needs some initial mapping for $n = 4$ to start the recursion. Such mappings were found by computer search (one can alternatively use the algorithmic mapping given below).

Second recursive construction

Let $f \in \mathcal{F}(2, n, n, 0)$ and $g \in \mathcal{F}(2, \nu, \nu, 0)$ and define $f \diamond g$ by

$$(f \diamond g)(\mathbf{x}, \mathbf{x}') = (f(\mathbf{x}), g(\mathbf{x}') + n).$$

Then $f \diamond g \in \mathcal{F}(2, n+\nu, n+\nu, 0)$.

For the construction, one again needs some initial mappings (for $4 \le n \le 7$) to start the recursion. Such mappings can for example be constructed by the first construction.

An algorithmic mapping

Let a be an odd positive integer such that $2a \le n$. The algorithm to compute the mapping, which depends on a and n, is as follows:

Input: $(x_1, x_2, \dots, x_n) \in Z_2^n$.
Output: $(\pi_1, \pi_2, \dots, \pi_n) = f(x_1, x_2, \dots, x_n)$.
{
  $(\pi_1, \pi_2, \dots, \pi_n) \leftarrow (1, 2, \dots, n)$;
  for i **from** 1 to a **do if** $x_i = 1$ **then** swap(π_{2i-1},π_{2i});
  for i **from** $a+1$ to $2a$ **do if** $x_i = 1$ **then** swap(π_{i-a},π_i);
  for i **from** $2a+1$ **to** n **do if** $x_i = 1$ **then** swap(π_1,π_i);
}

It was shown that

$$D_{i,i} = 0 \text{ for } 1 \le i < 2a.$$

In particular, if $n = 4$ or $n \equiv 2 \pmod 4$, then this implies that $k = 1$ for this mapping, that is, it is distance increasing. Also, in general for this class of mappings, $D_{i,j} = 0$ for $j \ge 2i + 2$. For $n = 5$, $D_{i,j}$ are given by:

	$j=1$	$j=2$	$j=3$	$j=4$	$j=5$
$i=1$	0	80	0	0	0
$i=2$	0	0	96	64	0
$i=3$	0	0	0	112	48
$i=4$	0	0	0	16	64
$i=5$	0	0	0	0	16

Lee[10] gave another algorithmic construction with $k = 0$ and $m = n \ge 5$ for n odd. The mapping has $D_{i,i} = 0$ for $i \le n - 4$. The algorithm is as follows.

```
Input:   (x_1, x_2, ..., x_n) ∈ Z_2^n.
Output: (π_1, π_2, ..., π_n) = f(x_1, x_2, ..., x_n).
{
  ν ← (n − 1)/2;
  if x_n = 0 then
  {
    (π_1, π_2, ..., π_n) ← (1, 2, ..., n);
    for i from 1 to n − 1 do if x_i = 1 then swap(π_i, π_{i+1});
  }
  else
  {
    (π_1, π_2, ..., π_n) ← (ν + 1, ν + 2, ..., n, 1, 2, ..., ν);
    for i from 1 to n − 1 do if x_i = 0 then swap(π_i, π_{i+1});
  }
}
```

For $n = 5$, $D_{i,j}$ are given by:

	$j=1$	$j=2$	$j=3$	$j=4$	$j=5$
$i=1$	0	64	6	2	8
$i=2$	0	4	68	64	24
$i=3$	0	0	14	76	70
$i=4$	0	0	0	22	58
$i=5$	0	0	0	0	16

Swart and Ferreira[19] gave a class of "multilevel" constructions for $n = m \geq 4$. The structure of the mapping is as follows. For convenience, permutations of Z_m rather than F_m are considered. Let $L = \lceil \log_2 m \rceil$. For $a, b \in Z_m$, let $a = \sum_{i=1}^{L} a_i 2^{i-1}$ and $b = \sum_{i=1}^{L} b_i 2^{i-1}$ be the binary expansions of a and b. The swap of a and b is said to be on level l if

$$a_l \neq b_l \text{ and } a_i = b_i \text{ for } l+1 \leq i \leq L.$$

The input is divided into L blocks, with n_1 bits in the first block, followed by n_2 bits in the second block, etc, where $n_1 + n_2 + \cdots + n_L = n$. For each bit in the l'th block, a sequence of one or more swaps on level l is determined, and they must satisfy some conditions. The mapping is algorithmic. Initially, $(\pi_1, \pi_2, \ldots, \pi_m) = (0, 1, \ldots, m-1)$. Running through the input bits, if the input bit is 1, the corresponding sequence of swaps is performed.

It was shown that by suitable choices of the sequences of swaps corresponding to each input bit, for the resulting mapping the values of $\Delta_1(f)$ are higher than for other constructions (for lengths up to 16). In particular, for $n = 2^r$, the upper bound $2^{2n}(2^r - 1)$ on $\Delta_1(f)$ is reached for these constructions.

The construction method can also be used to construct mappings with $k > 0$.

3.2. *Mappings with* k = 1 *(DIM)*

As noted above, Chang et al.[5] found mappings with $k = 1$ and $n = m$ for $n = 4$ and for $n \equiv 2 \pmod 4$. The first constructions with $k = 1$ for all $m = n \geq 4$ were given by Chang[1]. The constructions are recursive. We give one of them as an example. We denote it by f_n. To start the recursion we need mappings for $4 \leq n \leq 6$. For $n = 4$ and $n = 6$ one can use the mappings found by Chang et al.[5]. For $n = 5$, a mapping was found by search and listed as a table. For $n \geq 7$, the mapping is given recursively by the following algorithm.

Input: $(x_1, x_2, \ldots, x_n) \in Z_2^n$.
Output: $(\pi_1, \pi_2, \ldots, \pi_n) = f_n(x_1, x_2, \ldots, x_n)$.
{
  $\nu \leftarrow ((n+2) \bmod 3) + 4$;
  $(\pi_1, \pi_2, \ldots, \pi_\nu) \leftarrow f_\nu(1, 2, \ldots, \nu)$;
  $(\pi_{\nu+1}, \pi_{\nu+2}, \ldots, \pi_n) \leftarrow (\nu+1, \nu+2, \ldots, n)$;
  if $(\nu < n)$ **then**
  {
    if $(x_\nu = 1)$ **then** swap(π_ν,$\pi_{\nu+1}$);
    for i **from** $\nu + 1$ to n **do if** $x_i = 1$ **then**
    {
      if $((i - \nu) \bmod 3) = 0)$ **then** {swap(π_{i-2},π_i); swap(π_i,π_{i+1});}
      if $((i - \nu) \bmod 3) = 1)$ **then** swap(π_{i+1},π_{i+2});
      if $((i - \nu) \bmod 3) = 2)$ **then** swap(π_{i-2},π_i);
    }
  }
  if $(x_n = 1)$ **then** swap(π_{n-2},π_n);
}

Further constructions with $k = 1$ for $n = m \geq 4$, both recursive and algorithmic without table look up have been found by Lee[10,11] (in two papers, one for even n and one for odd n), Chang[2], and Lin, Chang, and Chen[13].

3.3. *Mappings with k = 2*

Huang, Tsai, and Wu[9] gave a recursive construction with $k = 2$ and $m = n+1$ for $n \geq 6$. For $n = 6, 7, 8, 9$, the mappings were given by ad hoc algorithms. For $n \geq 10$, the functions $f_n \in \mathfrak{F}(2, n, n+1, 2)$ are given recursively by the following algorithm where $g : Z_2^4 \to S_6$ is an auxiliary function given by a table.

Input: $(x_1, x_2, \ldots, x_n) \in Z_2^n$.
Output: $(\pi_1, \pi_2, \ldots, \pi_n, \pi_{n+1}) = f_n(x_1, x_2, \ldots, x_n)$.
{
$(\pi_1, \pi_2, \ldots, \pi_{n-5}, a, b) \leftarrow f_{n-4}(x_1, x_2, \ldots, x_{n-4})$;
$(\tau_1, \tau_2, \ldots, \tau_6) \leftarrow g(x_1, x_2, x_3, x_4) + n - 5$;
$\tau_{\tau^{-1}(n-4)} \leftarrow a$; $\tau_{\tau^{-1}(n-3)} \leftarrow b$;
$(\pi_{n-4}, \pi_{n-3}, \ldots, \pi_{n+1}) \leftarrow (\tau_1, \tau_2, \ldots, \tau_6)$;
if $(x_1 = 1)$ **then** swap(π_1,π_n);
if $(x_2 = 1)$ **then** swap(π_2,π_{n+1});
}

3.4. *Mapping with arbitrary k*

Chang[3] gave two recursive constructions of mappings with $m = n$ for increasing values of k.

First construction. Let $f \in \mathfrak{F}(2, \nu, \nu, k)$ where $\nu > k \geq 1$ and let $g \in \mathfrak{F}(2, n, n, 1)$. Define $f \otimes g : Z_2^{\nu n} \to S_{\nu n}$ as follows: let

$$\mathbf{x} = (x_{1,1}, x_{1,2}, \ldots, x_{1,n}, x_{2,1}, x_{2,2}, \ldots, x_{\nu,n}) \in Z_2^{\nu n}.$$

Let

$$(u_{1,j}, u_{2,j}, \ldots, u_{\nu,j}) = f(x_{1,j}, x_{2,j}, \ldots, x_{\nu,j}) \text{ for } 1 \leq j \leq n$$
$$(v_{i,1}, v_{i,2}, \ldots, v_{i,n}) = g(x_{i,1}, x_{i,2}, \ldots, x_{i,n}) \text{ for } 1 \leq i \leq \nu$$

and let

$$\pi_{i,j} = (u_{i,j} - 1)n + v_{u_{i,j},j}.$$

Then

$$(f \otimes g)(\mathbf{x}) = (\pi_{1,1}, \ldots, \pi_{1,n}, \pi_{2,1}, \ldots, \pi_{\nu,n}).$$

Chang showed that $f \otimes g \in \mathfrak{F}(2, \nu n, \nu n, k+1)$.

Second construction. This construction is also recursive. The recursion is given as an algorithm.

Let $f \in \mathcal{F}(2, \nu, \nu, k)$ and $g \in \mathcal{F}(2, n, n, k)$, where $\min(\nu, n) \geq 2k \geq 4$. The construction gives a mapping $Z_2^{\nu+n} \rightarrow S_{\nu+n}$ that is proved to belong to $\mathcal{F}(2, \nu + n, \nu + n, k)$.

Input: $(x_1, x_2, \ldots, x_{\nu+n}) \in Z_2^{\nu+n}$.
Output: $(\pi_1, \pi_2, \ldots, \pi_{\nu+n}) \in S_{\nu+n}$.
{
  $(\pi_1, \pi_2, \ldots, \pi_\nu) \leftarrow f(x_1, x_2, \ldots, x_\nu)$;
  $(\pi_{\nu+1}, \pi_{\nu+2}, \ldots, \pi_{\nu+n}) \leftarrow g(x_{\nu+1}, x_{\nu+2}, \ldots, x_{\nu+n}) + \nu$;
  for i **from** 1 to k **do**
  {
    if $(x_i = 1)$ **then** swap(π_1,$\pi_{\nu+n+1-i}$);
    if $(x_{\nu+i} = 1)$ **then** swap($\pi_{\nu+i}$,$\pi_{\nu+1-i}$);
  }
}

Combining these two constructions, mappings in $\mathcal{F}(2, n, n, k)$ can be found for all $n \geq 20^{k-1} + 3 \cdot 5^{k-1} + 3 \cdot 4^{k-1} + 1$ when $k \geq 1$.

4. Mappings from ternary codes

4.1. *Construction by Lin, Chang, Chen, and Kløve*

Lin, Chang, Chen, Kløve[14,15] gave the first construction method for DPM from ternary codes. The construction is recursive. To describe it, we introduce one new notation: for any array $\mathbf{u} = (u_1, u_2, \ldots, u_n)$, we use the notation $\mathbf{u}_i$ to denote the element u_i in position i.

The recursive definition of functions from Z_3^n to S_m, where $m \geq 5$, is as follows. Let $f \in \mathcal{F}(3, n, m, k)$.

For $\mathbf{x} = (x_1, x_2, \ldots, x_n) \in Z_3^n$, let $f(\mathbf{x}) = (\varphi_1, \varphi_2, \ldots, \varphi_m)$. Suppose that the element $m - 4$ occurs in position r, that is $\varphi_r = m - 4$. Define g from Z_3^{n+1} to S_{m+1} as follows.

$$g(\mathbf{x}|0)_i = \begin{cases} m+1 \text{ for } i = m+1 \\ \varphi_i, \quad\ \ \text{for } i \neq m+1 \end{cases}$$

$$g(\mathbf{x}|1)_i = \begin{cases} m-4 \text{ for } i = m+1 \\ m+1 \text{ for } i = r \\ \varphi_i, \quad\ \ \text{for } i \notin \{r, m+1\} \end{cases}$$

If n is even and $x_n = 2$, then

$$g(\mathbf{x}|2)_i = \begin{cases} m+1 & \text{for } i = m-1 \\ \varphi_{m-1} & \text{for } i = m+1 \\ \varphi_i, & \text{for } i \notin \{m-1, m+1\} \end{cases}$$

otherwise (n is odd or $x_n < 2$), then

$$g(\mathbf{x}|2)_i = \begin{cases} m+1 & \text{for } i = m \\ \varphi_m & \text{for } i = m+1 \\ \varphi_i, & \text{for } i \notin \{m, m+1\} \end{cases}$$

It was shown that $g \in \mathcal{F}(3, n+1, m+1, k)$. Mappings to start the recursion were found, by a combination of computer search and ad-hoc constructions, in the following cases:

- $m = n + 2$, $k = 1$, $n \geq 3$
- $m = n + 1$, $k = 0$, $n \geq 9$
- $m = n$, $k = 0$, $n \geq 13$

They also observed that the recursion can be modified by changing $m-4$ to $m-t$ for some fixed t, where $3 \leq t < m$.

4.2. *Construction by Lin, Tsai, and Wu*

Lin, Tsai, and Wu[16] gave an algorithmic construction for DPMs from ternary vectors for $n = m \geq 16$ and $k = 0$. We write $n = 8\nu + \kappa$ where $\nu \geq 1$ and $0 \leq \kappa \leq 7$.

Input: $(x_1, x_2, \ldots, x_n) \in Z_3^n$.
Output: $(\pi_1, \pi_2, \ldots, \pi_n) = f(x_1, x_2, \ldots, x_n)$.
{ $(\pi_1, \pi_2, \ldots, \pi_n) \leftarrow (1, 2, \ldots, n)$;
  for i **from** 0 to $4\nu - 1$ **do if** $x_{2i+1} = 1$ **then** swap(π_{2i+1},π_{2i+2});
  for i **from** 0 to $4\nu - 1$ **do if** $x_{2i+2} = 1$ **then** swap(π_{2i+2},π_{2i+3});
  for i **from** 0 to $\nu - 1$ **do**
    for j **from** 1 to 4 **do if** $x_{8i+j} = 2$ **then** swap(π_{8i+j},π_{8i+4+j});
  for i **from** 0 to $\nu - 1$ **do**
    for j **from** 5 to 8 **do if** $x_{8i+j} = 1$ **then** swap(π_{8i+j},π_{8i+4+j});
  for i **from** 1 to κ **do**
  { **if** $x_{8\nu+i} = 1$ **then** swap($\pi_{8\nu+i}$,$\pi_{\pi^{-1}(i-3)}$);
    if $x_{8\nu+i} = 2$ **then** swap($\pi_{8\nu+i}$,π_i);
  }
}

5. Open problems

Some problems that are open in general (and in fact, in most cases) are:

- Given q, n, m, k, decide if $\mathcal{F}(q,n,m,k)$ empty or not.
- Given q, n, m, k, determine the size of $\mathcal{F}(q,n,m,k)$.
- Clearly, $\mathcal{F}(q,n,m,k+1) \subset \mathcal{F}(q,n,m,k)$. For given q, n, m, determine the maximal k such that $\mathcal{F}(q,n,m,k)$ is non-empty.
- If $\mathcal{F}(q,n,m,k)$ is non-empty, find a mapping in $\mathcal{F}(q,n,m,k)$.
- If $\mathcal{F}(q,n,m,k)$ is non-empty, find an $f \in \mathcal{F}(q,n,m,k)$ with maximal value of $\Delta_1(f)$.

References

1. J.-C. Chang, "Distance-increasing mappings from binary vectors to permutations", *IEEE Trans. on Inform. Theory*, vol. 51, no. 1, pp. 359–363, Jan. 2005.
2. J.-C. Chang, "New algorithms of distance-increasing mappings from binary vectors to permutations by swaps", *Designs, Codes and Cryptography*, vol. 39, pp. 335–345, Jan. 2006.
3. J.-C. Chang, "Distance-increasing mappings from binary vectors to permutations that increase Hamming distances by at least two", *IEEE Trans. on Inform. Theory*, vol. 52, no. 4, pp. 1683–1689, April 2006.
4. J.-C. Chang and S.-F. Chang, "Constructions of Distance-Almost-Increasing Mappings from Binary Vectors to Permutations", Report ICS, 2004.
5. J.-C. Chang, R.-J. Chen, T. Kløve, and S.-C. Tsai, "Distance-preserving mappings from binary vectors to permutations", *IEEE Trans. on Inform. Theory*, vol. 49, pp. 1054–1059, Apr. 2003.
6. H. C. Ferreira and A. J. H. Vinck, "Inference cancellation with permutation trellis arrays", *Proc. IEEE Vehicular Technology Conf.*, pp. 2401–2407, 2000.
7. H. C. Ferreira, A. J. H. Vinck, T. G. Swart, and I. de Beer, "Permutation trellis codes", *Proc. IEEE Trans. on Communications*, vol. 53, no. 11, pp. 1782–1789, Nov. 2005.
8. H. C. Ferreira, D. Wright, and A. L. Nel, "Hamming distance-preserving mappings and trellis codes with constrained binary symbols", *IEEE Trans. on Inform. Theory*, vol. 35, no. 5, pp. 1098–1103, Sept. 1989.
9. Y.-Y. Huang, S.-C. Tsai, H.-L. Wu, "On the construction of permutation arrays via mappings from binary vectors to permutations", *Designs, Codes and Cryptography*, vol. 40, pp. 139–155, 2006.
10. K. Lee, "New distance-preserving maps of odd length", *IEEE Trans. on Inform. Theory*, vol. 50, no. 10, pp. 2539–2543, Oct. 2004.
11. K. Lee, "Cyclic constructions of distance-preserving maps", *IEEE Trans. on Inform. Theory*, vol. 51, no. 12, pp. 4292–4396, Dec. 2005.
12. K. Lee, "Distance-increasing maps of all length by simple mapping algorithms", *IEEE Trans. on Inform. Theory*, vol. 52, no. 7, pp. 3344-3348, July 2006.

13. J.-S. Lin, J.-C. Chang, and R.-J. Chen, "New simple constructions of distance-increasing mappings from binary vectors to permutations", *Information Processing Letters*, vol. 100, iss. 2, pp. 83-89, Oct. 2006.
14. J.-S. Lin, J.-C. Chang, R.-J. Chen, T. Kløve, "Distance-preserving mappings from ternary vectors to permutations", *arXiv:* 0704.1358v1 [cs.DM]
15. J.-S. Lin, J.-C. Chang, R.-J. Chen, T. Kløve, "Distance-preserving and distance-increasing mappings from ternary vectors to permutations", *IEEE Trans. on Inform. Theory*, vol. 53, 2008, to appear.
16. T.-T. Lin, S.-C. Tsai, H.-L. Wu, "Distance-preserving mappings from ternary vectors to permutations", Manuscript 2007.
17. T. G. Swart, I. de Beer, and H. C. Ferreira, "On the optimality of permutation mappings", *Proc. IEEE Int. Symp. Information Theory*, Adelaide, Australia, Sept. 2005, pp. 1068-1072.
18. T. G. Swart and H. C. Ferreira, "A multilevel construction for mappings from binary sequences to permutation sequences", *Proc. IEEE Int. Symp. Information Theory*, Seattle, USA, July 2006, pp. 1895–1899.
19. T. G. Swart and H. C. Ferreira, "A generalized upper bound and a multilevel construction for distance-preserving mappings", *IEEE Trans. on Inform. Theory*, vol. 52, no. 8, pp. 3685–3695, August 2006.

Single Cycle Invertible Function and its Cryptographic Applications

Chao Li*, Bing Sun† and Qingping Dai

Department of Mathematics and System Sciences National University of Defence Technology Changsha 410073, China
University Key Laboratory of Network Security and Cryptography Technology of Fujian
** E-mail: Lichao_nudt@sina.com*
† E-mail: happy_come@163.com

Single cycle invertible functions(SCIF) are studied in this paper. We show that if $\beta + \gamma = k2^{n-a}$, $2^a\beta + \alpha \equiv 1 \pmod 4$, and $\alpha + k \equiv C \equiv 1 \pmod 2$, $g(x) = \alpha x + \beta(x >>> b) + \gamma(x >> b) + C$ is a single cycle invertible function and some cryptographic properties are studied; new ways based on SCIF and linear feedback shift registers are introduced to generate longer periods state which can be used as the state transition function in stream ciphers. At last, a new stream cipher proposal which is based on SCIF and LFSR is introduced and its cryptographic properties are studied.

Keywords: single cycle invertible function, T-function, stream cipher.

1. Introduction

Invertible transformations over n-bit words are essential ingredients in many cryptographic constructions such as block ciphers and hash functions. When n is small, we can compactly represent this transformation as an S-box, but when n is large, it is impossible or inconvenient to use such S-box. We usually represent it as a composition of several simpler operations. For example, the quarterround function of Salsa20[1] stream cipher is a composition of Xor, modular addition and left rotation. The usual ways to construct invertible transformations are: S-P networks, *Feistel* structures and permutation polynomials etc. In [2], T-function was introduced as a new component to build cryptographic primitives. In short, $f(x)$ is a T-function, if higher bits of x don't influence lower bits of $f(x)$[2].

The cycle structure of an invertible function is very important when the function is used as a state transition function in a stream cipher, since in this application we iterate the function many times while at the same time we don't want the sequence of generated states to be trapped in a short cycle, so the single cycle property of an invertible function is another necessity in many situations just as m-sequence in LFSR. If $f(x)$ is a T-function, there are a lot of results about its cycle property[2,3]. In[2], the author proved that $f(x) = x + (x^2 \vee C)$ is invertible if and only if $C \equiv 1 \pmod 2$ and $f(x)$ is an SCIF if and only if $C \equiv 5$ *or* $7 \pmod 8$. In[3], it is proved that a T-function is invertible if and only if it can be represented in the form $c + x + 2v(x)$ where c is a constant and $v(x)$ is a T-function; and that an invertible T-function defines a single cycle if and only if it can be represented in the form $1 + x + 2(v(x+1) - v(x))$ where $v(x)$ is a T-function. In[4], the author presented several functions some of which are not T-functions and claimed that they are all SCIFs except the last one which is an SCIF over odd numbers.

In general, there is no standard way to show that whether a complex function based on simple operations is invertible or not, and the single cycle property is even harder. In this paper, we use number theory[5] to prove the single cycle property of a new class of functions which are wide T-functions.

On the other hand, many stream ciphers based on T-functions are broken[6–9]. In this paper, a new method to generate keystreams which is different from prevois ones is given.

2. New Class of Single Cycle Invertible Function

The main work of this paper is based on[4], and some generalized results were proved.

In the following, we use $\lfloor x \rfloor$ to denote the greatest integer that doesn't exceed x; >>> means right rotation, >> means right shift and | means concatenation. Besides, we always assume $n \geq 4$ and use $f(x)$ instead of $f(x) \pmod{2^n}$.

The equation $a \equiv b \pmod c$ means $c|a-b$ and $a = b \pmod c$ means a equals to the value b modular c where $0 \leq b \pmod c \leq c-1$.

Definition 2.1. A function $f(x)$: $Z_{2^n} \to Z_{2^n}$ is called single cycle invertible function (SCIF) iff for $\forall\, x_0 \in Z_{2^n}$, $\{x_i | x_i = f(x_{i-1}), 1 \leq i \leq 2^n\} = Z_{2^n}$.

Lemma 2.1. *$f(x) = ax + C$ is a single cycle invertible function iff* $a \equiv 1 \pmod 4$ *and* $C \equiv 1 \pmod 2$.

Note that in[4], the author claimed that when a is of the form 5^t and C is an odd number, $f(x)$ is a SCIF, which is a special case of this lemma.

Proof. First, we can check that $f(x)$ is invertible $\Leftrightarrow a \equiv 1 \pmod 2$.

Let $x_{t+1} = f(x_t)$, so $x_t = a^t x_0 + C\sum_{i=0}^{t-1} a^i \pmod{2^n}$. Since $f(x)$ is a T-function, its cycle length must be a power of 2[2]. Besides, since a is odd, $(a, 2^n) = 1$, from *Euler* theorem, we have $a^{2^{n-1}} \equiv 1 \pmod{2^n}$, so $x_{2^{n-1}} = a^{2^{n-1}} x_0 + C \sum_{i=0}^{2^{n-1}-1} a^i \equiv x_0 + C \sum_{i=0}^{2^{n-1}-1} a^i \pmod{2^n}$.

$f(x)$ is a single cycle invertible function $\Leftrightarrow x_{2^{n-1}} \equiv x_0 + C \sum_{i=0}^{2^{n-1}-1} a^i \not\equiv x_0 \pmod{2^n} \Leftrightarrow C \sum_{i=0}^{2^{n-1}-1} a^i \not\equiv 0 \pmod{2^n}$.

Now, let's compute $C \sum_{i=0}^{2^{n-1}-1} a^i \pmod{2^n}$ as following:

$$C \sum_{i=0}^{2^{n-1}-1} a^i = C\frac{a^{2^{n-1}} - 1}{a-1} = C\prod_{i=0}^{n-2}(a^{2^i} + 1) \quad if\ a \neq 1$$

$$C \sum_{i=0}^{2^{n-1}-1} a^i = 2^{n-1}C \quad if\ a = 1$$

For the case of $a = 1$, it is easy to find that

$$C \sum_{i=0}^{2^{n-1}-1} a^i = 2^{n-1}C \not\equiv 0 \pmod{2^n} \Leftrightarrow C \equiv 1 \pmod{2}.$$

For the case of $a \neq 1$, since a is odd, $2|(a^{2^i} + 1)$, $C \sum_{i=0}^{2^{n-1}-1} a^i \not\equiv 0 \pmod{2^n} \Leftrightarrow C$ and $\frac{a^{2^i}+1}{2}$ must be odd $\Leftrightarrow a \equiv 1 \pmod{4}$ and $C \equiv 1 \pmod{2}$. □

Corollary 2.1. *If* $a \equiv 3 \pmod{4}$ *and* $C \equiv 1 \pmod{2}$, $f(x) = ax + C$ *is an invertible function which has two cycle of the same length* 2^{n-1}.

Lemma 2.2. *If* $A \equiv 1 \pmod{4}$ *and* $B \equiv 1 \pmod{2}$, $\sum_{x=0}^{2^n-1} \lfloor \frac{Ax+B}{2^n} \rfloor \equiv 1 \pmod{2}$

Proof. Let $f(x) = \frac{Ax+B}{2^n}$. Since A is odd, $Ax + B \equiv 0 \pmod{2^n}$ has a unique solution modular 2^n: $x = x_0 \pmod{2^n}$. Moreover, because B is an odd integer, x_0 must be an odd integer too. In other words, if $x \not\equiv x_0 \pmod{2^n}$, $f(x)$ can not be an integer.

We have $f(x_0+i)+f(x_0-i)=2f(x_0)\equiv 0 \pmod 2$. If $x_0 \pm i \not\equiv x_0 \pmod{2^n}$, neither $f(x_0+i)$ nor $f(x_0-i)$ is an integer, so

$$\lfloor f(x_0+i)\rfloor + \lfloor f(x_0-i)\rfloor \equiv 1 \pmod 2 \tag{1}$$

For any integer x, $f(x+2^n)=f(x)+A$, which indicates that $\lfloor f(x+2^n)\rfloor = \lfloor f(x)\rfloor + A$. Thus

$$\lfloor f(x+2^n)\rfloor + \lfloor f(x)\rfloor = 2\lfloor f(x)\rfloor + A \equiv 1 \pmod 2 \tag{2}$$

We compute $\sum\limits_{x=0}^{2^n-1} \lfloor f(x)\rfloor$ under the assumption that $x_0 < 2^{n-1}$:

$$\begin{aligned}\sum_{x=0}^{2^n-1} \lfloor f(x)\rfloor &= \sum_{i=1}^{x_0}(\lfloor f(x_0+i)\rfloor + \lfloor f(x_0-i)\rfloor)\\ &\quad + \sum_{i=x_0+1}^{2^{n-1}-1}(\lfloor f(x_0-i+2^n)\rfloor + \lfloor f(x_0+i)\rfloor)\\ &\quad + (\lfloor f(2^{n-1}+x_0)\rfloor + \lfloor f(x_0)\rfloor)\\ &\triangleq S_1+S_2+S_3\end{aligned}$$

According to (1),

$$S_1 = \sum_{i=1}^{x_0}(\lfloor f(x_0+i)\rfloor + \lfloor f(x_0-i)\rfloor) \equiv \sum_{i=1}^{x_0} 1 \equiv x_0 \equiv 1 \pmod 2 \tag{3}$$

According to (1) and (2):

$$\begin{aligned}S_2 &= \sum_{i=x_0+1}^{2^{n-1}-1}(\lfloor f(x_0-i+2^n)\rfloor + \lfloor f(x_0+i)\rfloor)\\ &\equiv \sum_{i=x_0+1}^{2^{n-1}-1}(\lfloor f(x_0-i+2^n)\rfloor\\ &\quad + \lfloor f(x_0-i)\rfloor) + (\lfloor f(x_0-i)\rfloor + \lfloor f(x_0+i)\rfloor)\\ &\equiv \sum_{i=x_0+1}^{2^{n-1}-1}(1+1) \equiv 0 \pmod 2\end{aligned} \tag{4}$$

Since $A \equiv 1 \pmod 4$, let $A = 4\alpha+1$, then:

$$\begin{aligned}S_3 &= \lfloor f(2^{n-1}+x_0)\rfloor + \lfloor f(x_0)\rfloor = 2f(x_0) + [2^{-1}(4\alpha+1)]\\ &= 2f(x_0)+2\alpha\\ &\equiv 0 \pmod 2\end{aligned} \tag{5}$$

From (3), (4) and (5) we can conclude that $\sum_{x=0}^{2^n-1} \lfloor f(x) \rfloor \equiv 1 \pmod 2$ when $x_0 < 2^{n-1}$.

If $x_0 > 2^{n-1}$, we can get the same conclusion. This ends our proof. □

Corollary 2.2. *If* $A \equiv 3 \pmod 4$ *and* $B \equiv 1 \pmod 2$, $\sum_{x=0}^{2^n-1} \lfloor \frac{Ax+B}{2^n} \rfloor \equiv 0 \pmod 2$.

Now, we will give the new class of single cycle invertible functions which are wide T-functions.

Theorem 2.1. $g(x) = x + (x >>> b) - (x >> b) + C$ *is a single cycle invertible function, if* $1 \leq b \leq n-2$ *and* $C \equiv 1 \pmod 2$.

Proof. We use $x^{(t)}$ to denote a t-bit integer, and $a = n - b$.

Case 1: $a < b$

Denote $x = x^{(a)}|x^{(b)} = 2^b x^{(a)} + x^{(b)}$, then $x >>> b = 2^a x^{(b)} + x^{(a)}$ and $x >> b = x^{(a)}$, $g(x) = (2^a+1)x^{(b)} + 2^b x^{(a)} + C - \beta 2^n$ ($\beta = 0$ *or* 1). Next, let $g(x) = y = y^{(a)}|y^{(b)} = 2^b y^{(a)} + y^{(b)}$, then:

$$2^b y^{(a)} + y^{(b)} = (2^a+1)x^{(b)} + 2^b x^{(a)} + C - \beta 2^n \tag{6}$$

Modular 2^b on both sides:

$$2^b y^{(a)} + y^{(b)} \equiv (2^a+1)x^{(b)} + 2^b x^{(a)} + C - \beta 2^n \pmod{2^b}$$

Since $a < b$, we can conclude that

$$y^{(b)} = (2^a+1)x^{(b)} + C \pmod{2^b} \tag{7}$$

From lemma 1, (7) is a single cycle invertible function if and only if $a \geq 2$. Let T_b denote the cycle length of $y^{(b)}$, therefore $T_b = 2^b$.

According to (7), $y^{(b)} = (2^a+1)x^{(b)} + C - \alpha 2^b$ ($\alpha = \lfloor \frac{(2^a+1)x^{(b)}+C}{2^b} \rfloor$), combined with (6): $2^b y^{(a)} + (2^a+1)x^{(b)} + C - \alpha 2^b = (2^a+1)x^{(b)} + 2^b x^{(a)} + C - \beta 2^n$, which means to $2^b y^{(a)} - \alpha 2^b = 2^b x^{(a)} - \beta 2^n$, in other words:

$$y^{(a)} = x^{(a)} + \alpha \pmod{2^a} \tag{8}$$

If the initial value is $x_0 = x_0^{(a)}|x_0^{(b)}$, let $\alpha_i = \lfloor \frac{(2^a+1)x_i^{(b)}+C}{2^b} \rfloor$, we can compute $x_t = x_t^{(a)}|x_t^{(b)}$ by (7) and (8):

$$\begin{cases} x_t^{(b)} = (2^a+1)x_{t-1}^{(b)} + C \pmod{2^b} \\ x_t^{(a)} = x_0^{(a)} + \sum_{i=0}^{t-1} \alpha_i \pmod{2^a} \end{cases} \tag{9}$$

Substitute t by T_b in the above equation: $x^{(a)}_{T_b} = x^{(a)}_0 + \sum_{i=0}^{T_b-1} \alpha_i \pmod{2^a}$. Since the single cycle property of (7) when $a \geq 2$, and combined with the result of lemma 2, we can conclude:

$$\sum_{i=0}^{T_b-1} \alpha_i = \sum_{i=0}^{2^b-1} \lfloor \frac{(2^a+1)x_i^{(b)} + C}{2^b} \rfloor = \sum_{i=0}^{2^b-1} \lfloor \frac{(2^a+1)i + C}{2^b} \rfloor \equiv 1 \pmod 2$$

So, if $x^{(a)}_{T_aT_b} = x^{(a)}_0 + T_a \sum_{i=0}^{T_b-1} \alpha_i \equiv x^{(a)}_0 \pmod{2^a}$, T_a must be 2^a and the cycle length of $g(x)$ is $T_aT_b = 2^n$.

Case 2: $a > b$

Modular 2^a on both sides of (6): $2^b y^{(a)} + y^{(b)} \equiv x^{(b)} + 2^b x^{(a)} + C \pmod{2^a}$, it is equivalent to $2^b(y^{(a)} - x^{(a)}) \equiv x^{(b)} + C - y^{(b)} \pmod{2^a}$ which indicates that

$$y^{(b)} = x^{(b)} + C \pmod{2^b} \tag{10}$$

Let $y^{(b)} = x^{(b)} + C - \omega 2^b$ $(\omega = \lfloor \frac{x^{(b)}+C}{2^b} \rfloor)$, use this to simplify (6):

$$y^{(a)} = x^{(a)} + 2^{a-b}x^{(b)} + \omega \pmod{2^a} \tag{11}$$

Let $\omega_i = \lfloor \frac{x_i^{(b)}+C}{2^b} \rfloor$, similar with the case $a < b$:

$$\begin{cases} x_t^{(b)} = x_{t-1}^{(b)} + C \pmod{2^b} \\ x_t^{(a)} = x_0^{(a)} + 2^{a-b} \sum_{i=0}^{t-1} x_i^{(b)} + \sum_{i=0}^{t-1} \omega_i \pmod{2^a} \end{cases} \tag{12}$$

(10) is a single cycle invertible function, so $T_b = 2^b$. Since C is an odd number, we can compute $\sum_{i=0}^{t-1} \omega_i \equiv 1 \pmod 2$, thus $T_a = 2^a$.

Case 3: $a = b$

The method is similar with case 1 and 2, so we omit the details. □

By using corollary 1 and 2, we can get:

Corollary 2.3. *If* $C \equiv 1 \pmod 2$, $f(x) = x + (x >>> (n-1)) - (x >> (n-1)) + C$ *is an invertible function which has two cycles of the same length* 2^{n-1}.

Note: In [4], the author claimed that when $a = 5^t$ and $C \equiv 1 \pmod 2$, $f(x) = ax + C$ is a single cycle invertible function which is a special case of lemma 1; the third function $h(x) = x + (x >>> b) + (\overline{x} >> b)$ is a special case of theorem 1 when $C = 2^a - 1$.

In fact, we can use a little more techniques to prove a more generalized proposition:

Theorem 2.2. *Let* $a = n - b$, *if* $1 \le a \le n-1$, $\beta + \gamma = k2^b$, $2^a\beta + \alpha \equiv 1 \pmod 4$, *and* $\alpha + k \equiv C \equiv 1 \pmod 2$, $g(x) = \alpha x + \beta(x >>> b) + \gamma(x >> b) + C$ *is a single cycle invertible function.*

The assumption in the theorem is equal to $2^{b+1} | \beta + \gamma$ and $2^a\beta + \alpha \equiv 1 \pmod 4$. And if $a \ge 2$, it can be replaced by $2^{b+1} | \beta + \gamma$ and $\alpha \equiv 1 \pmod 4$.

Proof. We only give a brief proof under the assumption that $a < b$.

As the technique we used in theorem 1, we can get

$$y^{(b)} = (2^a\beta + \alpha)x^{(b)} + C \pmod{2^b}$$

Since $2^a\beta + \alpha \equiv 1 \pmod 4$ and $C \equiv 1 \pmod 2$, according to what lemma 1 says, $y^{(b)}$ is a single cycle invertible function which means $T_b = 2^b$. Assume $y^{(b)} = (2^a\beta + \alpha)x^{(b)} + C - \tau 2^b$ ($\tau = \lfloor \frac{(2^a\beta+\alpha)x^{(b)}+C}{2^b} \rfloor$), $y^{(a)} = (\alpha + k)x^{(a)} + \tau \pmod{2^a}$, let $\lambda = \alpha + k \equiv 1 \pmod 2$, $\tau_i = \lfloor \frac{(2^a\beta+\alpha)x_i^{(b)}+C}{2^b} \rfloor$, we can get

$$\begin{cases} x_t^{(b)} = (2^a\beta + \alpha)x_{t-1}^{(b)} + C \pmod{2^b} \\ x_t^{(a)} = \lambda^t x_0^{(a)} + \sum\limits_{i=0}^{t-1} \lambda^{t-1-i}\tau_i \pmod{2^a} \end{cases}$$

Then we can conclude that

$$x_{T_aT_b}^{(a)} = \lambda^{T_aT_b}x_0^{(a)} + \sum_{j=0}^{T_a-1} \lambda^{jT_b} \sum_{i=0}^{T_b-1} \lambda^{T_b-1-i}\tau_i \pmod{2^a}$$

Since λ is odd, according to lemma 2, $\sum\limits_{i=0}^{T_b-1} \lambda^{T_b-1-i}\tau_i \equiv 1 \pmod 2$. So

$x_{T_aT_b}^{(a)} = x_0^{(a)} \Rightarrow \lambda^{T_aT_b} \equiv 1 \pmod{2^a}$ and $2^a | \sum\limits_{j=0}^{T_a-1} \lambda^{jT_b}$.

Assume $T_a = 2^l q$ where q is an odd integer,

$$\sum_{j=0}^{T_a-1} \lambda^{jT_b} = \frac{\lambda^{T_aT_b} - 1}{\lambda^{T_b} - 1} = \prod_{i=0}^{l-1}((\lambda^{q2^b})^{2^i} + 1)(\frac{\lambda^{q2^b} - 1}{\lambda^{2^b} - 1})$$

Since λ is odd, we have $\lambda^{2^b} \equiv 1 \pmod 4$ for $b \geq 1$ and $\frac{\lambda^{q2^b}-1}{\lambda^{2^b}-1} = \sum_{i=0}^{q-1} \lambda^{2^b i} \equiv q \equiv 1 \pmod 2 \Rightarrow 2^l | \sum_{j=0}^{T_a-1} \lambda^{jT_b}$ and $2^{l+1} \nmid \sum_{j=0}^{T_a-1} \lambda^{jT_b} \Rightarrow l \geq a$ and $T_a \geq 2^a$. We can check that $x^{(a)}_{2^a \times 2^b} = x^{(a)}_0$, so the cycle length of $g(x)$ is 2^n when $a < b$.

For the case of $a \geq b$, we omit the details. □

3. Analysis of the New Construction

Though there are multiplication in the construction, when use $f(x)$ as in theorem 2, we chose α, β and γ which have low hamming weight for example if $a \geq 2$, we chose $\alpha = 5$, $\beta = 16$ and $\gamma = -16$, then

$$\begin{aligned} f(x) &= x + (x << 2) + (x <<< a) + ((x <<< a) << 4) \\ &\quad - (x >> b) - ((x >> b) << 4) + C \end{aligned}$$

is a single cycle invertible function which uses only simple operations of addition, shift and rotation.

Now, let's analyze the randomness of the new construction in the following aspect[10]: if two of the triple (x_0, t, x_t) are known, we compare the complexity of finding the 3rd number to the complexity of exhaustive search. Since the methods are similar to each other, we only write the algorithm to compute x_t, under the assumption that $a < b$ and we know (x_0, t). We only take $f(x) = x + (x >>> b) - (x >> b) + C$ as an example. The correctness of the algorithm is based on the theorem.

Step 1 Pre-compute $\delta_{(a,b)} = \sum_{i=0}^{2^b-1} \lfloor \frac{(2^a+1)i+C}{2^b} \rfloor \pmod{2^a}$;

Step 2 According to (7), we compute $t_0 = t \pmod{2^b}$, $x^{(b)}_t = x^{(b)}_{t_0}$ and $\lambda = \sum_{i=0}^{t_0-1} \lfloor \frac{(2^a+1)x^{(b)}_i + C}{2^b} \rfloor \pmod{2^a}$;

Step 3 Compute $t_1 = \lfloor \frac{t}{2^b} \rfloor$, then $x^{(a)}_t = x^{(a)}_0 + t_1 \delta_{(a,b)} + \lambda \pmod{2^a}$ and $x_t = x^{(a)} | x^{(b)}$.

For fixed (a, b), $\delta_{(a,b)}$ is a constant, so we omit the complexity of step 1; to compute $x^{(b)}_t = x^{(b)}_{t_0}$, we only need to do (7) t_0 times; the complexity of computing λ is at most t_0 times multiplication and t_0 times addition, sometimes we can also make a lookup table for λ's. Therefore, the total complexity is depend on t_0, and we don't need to compute x_is one by one.

The complexity of knowing (x_0, x_t) to compute t and knowing (t, x_t) to compute x_0 is smaller than 2^n which is the complexity of exhaustive search.

4. New Way to Generate Longer State

In this section, we give some new methods to construct longer cycles based on single cycle invertible functions.

Theorem 4.1. *Given x_0 and a sequence $a_0a_1a_2\cdots$ $(a_i \in \mathbb{N})$ whose period is T_a. Let $x_t = f(x_{t-1})$, and period of $x_0x_1x_2\cdots$ is T_x, let $y_0 = x_0$, $y_{t+1} = f^{a_t}(y_t)$. If $0 \le a_i \le T_x - 1$ and $(\sum_{i=0}^{T_a-1} a_i, T_x) = 1$, then, period of $y_0y_1y_2\cdots$ is T_aT_x.*

Proof. We can check that $y_t = f^{\sum_{j=i}^{t-1} a_j}(y_i)$. Let T denote the period of $\{y_t\}$, then $y_i = y_{T+i} = f^{\sum_{j=i}^{T+i-1} a_j}(y_i)$. Since $(\sum_{i=0}^{T_a-1} a_i, T_x) = 1$, $\{y_t\} = \{x_t\}$. For $x_{T_x} = f^{T_x}(x_0) = x_0$, we have $T_x | \sum_{j=i}^{T+i-1} a_j$ which means to

$$\sum_{j=0}^{T-1} a_j \equiv \sum_{j=1}^{T} a_j \equiv \cdots \equiv \sum_{j=T_a-1}^{T+T_a-2} a_j \pmod{T_x}$$

Thus

$$\sum_{j=1}^{T} a_j - \sum_{j=0}^{T-1} a_j \equiv \sum_{j=2}^{T+1} a_j - \sum_{j=1}^{T} a_j \equiv \cdots \equiv 0 \pmod{T_x}$$

we have

$$a_T \equiv a_0 \pmod{T_x}$$
$$a_{T+1} \equiv a_1 \pmod{T_x}$$
$$\vdots$$
$$a_{T+T_a-1} \equiv a_{T_a-1} \pmod{T_x}$$

Which means that T is also a period of $\{a_t\}$, thus $T_a | T$. Assume $T = kT_a$, we have $\sum_{j=i}^{T+i-1} a_j = k \sum_{i=0}^{T_a-1} a_i$, thus $y_i = y_{T+i} = f^{(k \sum_{i=0}^{T_a-1} a_i)}(y_i)$. Take $T_x | (k \sum_{i=0}^{T_a-1} a_i)$ into consideration and since $(\sum_{i=0}^{T_a-1} a_i, T_x) = 1$, the minimal integer k is T_x. □

Note: The stop-and-go sequence[11] is a special case of the theorem.

Corollary 4.1. *If $y = f(x)$ is a single cycle invertible function, let d be an odd integer, $y = f^d(x)$ is also a single cycle invertible function.*

5. Proposal: Construction for Stream Ciphers

Since $LFSR$ is a widely used cryptographic primitives, we will combine $LFSR$ and $SCIF$ to construct a new stream cipher.

Theorem 5.1. *If $f(x)$ is an n-stage single cycle invertible function and $a_0a_1a_2\cdots$ is a k-stage m-sequence, then period of $y_0y_1y_2\cdots$ is $2^n(2^k-1)$ where $y_{i+1} = f^{a_i+1}(y_i)$. Further more, each element of Z_{2^n} appears exactly $2^k - 1$ times in a period.*

We can use more LFSRs to control the $f(x)$ in a wide way. As an example, if the output of k LFSRs are $a_0a_1\cdots a_{k-1}$, we simply iterate $f(x)$ $\sum\limits_{i=0}^{k-1} 2^i a_i + 1$ times. In this case, we can prove that:

Theorem 5.2. *Let $F : \mathbb{Z}_{2^n} \to \mathbb{Z}_{2^n}$,if we use k LFSRs to control $f(x)$ and output l bits of $f(x)$ as the key stream, then $l \leq min\{\frac{n}{2}, k\}$.*

Proof. Assume $l > min\{\frac{n}{2}, k\}$. We compute the probability of $x_{i+1}^{(l)}$ under the assumption that $x_i^{(l)}$ is known. Let $x_i = (x_i^{(l)}, x_i^{(u)})$ where $x_i^{(u)}$ is unknown. Now we compute $x_{i+1}^{(l)}$:

$$(x_{i+1}^{(l)}, x_{i+1}^{(u)}) = f^m(x_i^{(l)}, x_i^{(u)})$$

If we know $x_i^{(l)}$, we can compute at most 2^k different $x_{i+1}^{(l)}$s; and if we know m, we can compute at most 2^l different $x_{i+1}^{(l)}$s. In both cases, $2^u \geq 2^k$ or $2^u \geq 2^l$. Take $l + u = n$ into consideration, we can find it is contradict with $l > min\{\frac{n}{2}, k\}$. Thus $l \leq min\{\frac{n}{2}, k\}$. □

If output of f is a word, usually, $n = 32$ or $n = 64$, so $l \leq 16$ or $l \leq 32$ respectively. In the following, we only consider the case $n = 32$, thus $l \leq 16$. From this theorem, the number of output bits depends on how many LFSRs are used. But if too many LFSRs are used, f will be iterated many times which will take too much time for example if we use 5 LFSRs, f will be iterated $(\sum\limits_{i=1}^{32} i)/32 = 16.5$ times in average. So large k's are rejected. Now, we assume that f is a T-function and take $k = l = 1$ as an example just as figure 1 shows.

Proposition 5.1. *Let f be a T-function, and an s-stage LFSR is used to control the iterated time of f, if the least significant bit x_0 of f is outputted as the key streams $k_0k_1k_2\cdots$, we can re-construct LFSR and recover the initial state of both LFSR and f if we know $k_0k_1k_2\cdots k_{\alpha-1}(\alpha \geq 2s)$.*

Proof. Since f is a T-function, the least significant bits of f must be $010101\cdots$ or $101010\cdots$. Assume the initial state of f is x_0, $x_{t+1} = f(x_t)$ and the least significant bit of x_t is $k_t = f(x_{t-1})_0 = g(k_{t-1})$, from the property of T-function, we have $k_t = g(k_{t-1}) = \overline{k_{t-1,0}}$, thus $k_t = k_{t+2}$.

Assume the output of LFSR is $a_0a_1a_2\cdots$, if $a_t = 0$, $k_t = f^{(a_t+1)}(x_{t-1})_0 = g(k_{t-1}) = \overline{k_{t-1}}$, and if $a_t = 0$, we have $k_t = f^{(a_t+1)}(x_{t-1})_0 = g^2(k_{t-1}) = k_{t-1}$, since $x_{t,0}$ is the output key streams, we have

$$\begin{cases} k_{i+1} = k_i & if \;\; a_i = 1 \\ k_{i+1} = \overline{k}_i & if \;\; a_i = 0 \end{cases}$$

So we only need to know k_{i+1} and k_i for different i's. By using *Berlekamp–Massey* algorithm, we can re-construct the s-stage LFSR, furthermore, k_is only depend on the least significant bits of x_is, we only need to re-construct the least significant bit of x_0. □

Example 5.1. If the key streams are $010011010\cdots$, we can re-construct the m-sequence according to proposition 1: $00101000\cdots$.

Proposition 5.2. *Let f be a T-function, and an s-stage LFSR is used to control the iterated number of f, if the second least significant bit x_1 of f is outputted as the key streams, we can distinguish the key stream from a truly random sequence.*

Proof. We compute the probability of the appearance of $*10*$ and $*11*$.

$$p(*10*) = \frac{1}{2}(1 + \frac{2^{s-1}}{2^s - 1})$$

$$p(*11*) = \frac{1}{2}(1 - \frac{2^{s-1}}{2^s - 1})$$

For large $s \to \infty$,

$$\frac{p(*10*)}{p(*11*)} \to 3 \qquad (13)$$

From this ratio, we can can distinguish the key stream from a truly random sequence. □

As a proposal, a 128 stage primitive LFSR is used to control the iterated number of $f(x) = x + (x^2 \vee 5)$ which is a single cycle invertible function and output the most significant bit of $f(x)$ as the keystream.

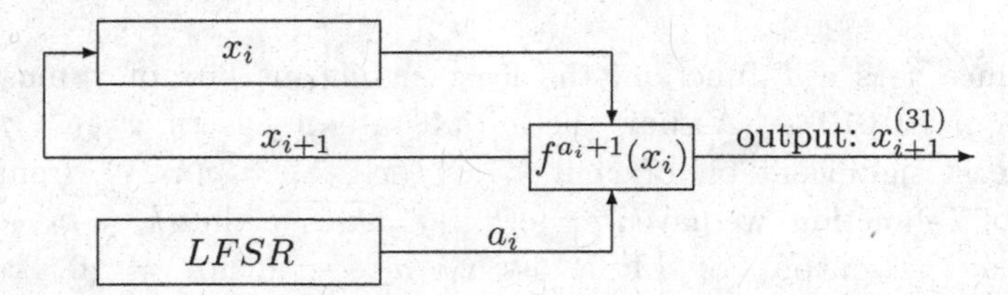

Figure 1: A bit-oriented stream cipher

From theorem 3, we know that:

Proposition 5.3. *Period of the output keystream is* $2^{32}(2^{128} - 1)$*, and 0-1 are strictly balanced.*

6. Conclusion

A new class of single cycle invertible functions is given in this paper. From the process of our proof, we can find that this new class of single cycle invertible function is not "strong" in cryptographic view, because $y^{(b)}$ is T-function and the statistic property of T-function's lower bits is very weak[2]. This is why [4] says that the smaller a is, the better statistic property of the new functions will be. From this aspect, we can't use the new class of single cycle invertible function to generate key streams, but we can use it as a cryptographic primitive to generate other random numbers. Later on, we give a new way to generate longer period state that can be used as the state transition function in stream ciphers. As an example, we give a proposal of new stream cipher based on T-function and $LFSR$. Some cryptography properties are studied.

Acknowledgments

The work in this paper is partially supported by the Natural Science Foundation of China (No:60573028) and Open Foundation of University Key Laboratory of Network Security and Cryptography Technology of Fujian(No:07A003).

References

1. D. J.Bernstein, Salsa20 Specification. `http://www.ecrypt.eu.org/stream/salsa20p2.html`.
2. A.Klimov and A.Shamir, A New Class of Invertible Mappings, in *CHES2002*, (Springer-Verlag, 2003) pp. 470–483.
3. V.Anashin, Uniformly Distributed Sequences of p-adic integers,II. `http://www.arxiv.org/ps/math.NT/0209407`.
4. I. O.Levin, On Single Cycle Functions. `http://www.literatecode.com/get/scfn.pdf`.
5. U.Dudley, *Elementary Number Theory.* (W.H. Freeman and Company, 1972).
6. Y. J.Hong, D.H.Lee and D.Han, T-function Based Stream Cipher TSC-3. ECRYPT Stream Cipher Project, Report 2005/031,2005.
7. Y. J.Hong, D.H.Lee and D.Han, New Class of Single Cycle T-functions., in *FSE2005*, (Springer-Verlag, 2005) pp. 68–82.
8. F.Muller and T.Peyrin, Linear Cryptanalysis of TSC Stream Ciphers Applications to the ECRYPT proposal TSC-3. `http://www.cosic.esat.kuleuven.ac.be/ecrypt/stream/`, 2005.
9. P. J. S. Künzli and W. Meier, Distinguishing attacks on T-functions., in *Mycrypt 2005*, (Springer-Verlag, 2005) pp. 2–15.
10. A.Klimov and A.Shamir, Cryptographic Applications of T-functions, in *SAC2003*, (Springer-Verlag, 2004) pp. 248–261.
11. T.Beth and F. Piper, The Stop-and-go-generator., in *EUROCRYPT'84*, (Springer-Verlag, 1985) pp. 88–92.

Concurrent Signatures without a Conventional Keystone

Yi Mu[1], Duncan Wong[2], Liqun Chen[3], Willy Susilo[1], Qianhong Wu[4]

[1] *Center for Computer and Information Security Research*
School of Computer Science and Software Engineering
University of Wollongong, Wollongong NSW 2522, Australia
Email: {ymu, wsusilo}@uow.edu.au
[2] *Department of Computer Science,*
City University of Hong Kong, Hong Kong, China
Email: duncan@cityu.edu.hk
[3] *Hewlett-Packard Laboratories, Bristol, UK*
Email: liqun.chen@hp.com
[4] *College of Computer Science, Wuhan University, China*
Email: qhwu@xidian.edu.cn

The notion of concurrent signatures was introduced by Chen, Kudla, and Paterson in Eurocrypt 2004. A concurrent signature scheme allows two parties to concurrently exchange their signatures. A concurrent signature scheme has two phases. In the first phase, two parties exchange their ambiguous signatures, which are verifiable only by themselves. In the second phase, the ambiguity is removed by throwing out a "keystone", a secret known to one of involved parties. However, concurrent signatures are still one step short from fully solving the fair exchange problem and they require a normal public key infrastructure. Nevertheless, concurrent signatures have many useful applications. We present a variant of the notion of concurrently exchanging digital signatures and propose a generic construction and a concrete construction to show how to exchange signatures between two parties as fair as possible without using a conventional keystone.

Keywords: Concurrent signature, Chameleon trapdoor function, Chameleon two-party ring signature, Fair exchange

1. Introduction

In Eurocrypt 2004 Chen, Kudla, and Paterson (CKP)[1] introduced a notion of "concurrent signatures". A concurrent signature scheme consists of two phases. In the first phase, two parties, say, $\mathcal{A}$ and $\mathcal{B}$, are able to produce two signatures such that from any third party's point of view, both signatures are ambiguous in terms of the authorship. In the second phase, witnesses are produced by (one of) the two parties so that the ambiguity of both signatures are revoked and the signatures become binding to their true signers concurrently. Although the concept of concurrent signatures has its own independence of interest, it is generally believe that it is only one-step short from solving the problem of fair exchange digital signatures[2–4].

One may note that before the very last step of one participant revealing the keystone in the second phase, both signatures are still ambiguous. Hence, the notion of concurrent signatures is not fully fair for both parties because the party, say, $\mathcal{A}$, who executes the last step can decide whether to run it to revoke the ambiguity finally and thus, has an additional privilege. This is easy to see in the following example.

Suppose Alice wants to obtain a contract from Bob with one million dollars. If a conventional signature is used by letting Alice sign the message m, "Alice agrees to obtain the contract from Bob with one million dollars", and send the signature σ to Bob, Bob may maliciously terminate the protocol and leave. Later, if Carol also wants this contract, Bob can convince Carol that Alice has agreed to obtain it with one million dollars. Carol has to bid more than one million dollars to win it and she does not know what Bob's bottom line with respect to the contract. However, using the concurrent signatures, if Bob wants to convince Carol the fact, he also shows to Carol that he agreed to let Alice win the contract with one million dollars. If Bob sells Carol the contract with more than one million dollars, it is risky to Bob because Carol may forward his binding argument to Alice so that Alice can ask compensation for Bob's nonfulfillment of the signed contract.

In particular, the notion of currently signature applies to the applications where a partially fair exchange of digital signatures is sufficient: (1) there is no sense in $\mathcal{A}$ terminating before last step because she needs it to obtain a service from $\mathcal{B}$; (2) there is no possibility of $\mathcal{A}$ keeping $\mathcal{B}$'s signature private in the long term; (3) there is a single third party $\mathcal{C}$ who verifies both $\mathcal{A}$ and $\mathcal{B}$'s signature. Finally, concurrent signatures also provide a novel solution to the old problem of fair tendering of contracts.

In the original definition of current signatures[1], the second stage is completed by releasing a keystone of party $\mathcal{A}$. It seems that concurrent signatures must be based on a keystone mechanism in order to ensure a fair signature binding moment. Actually, all currently known current signature schemes[1,5,6] are based on the keystone technique. However, we observe that the keystone methodology is merely an approach to current signatures. Furthermore, since the keystone is controlled by one of committed signers, it could be the case that the keystone is not thrown out due to some reason such as loss of the keystone. Moreover, if the keystone is stolen, then it could be thrown out in any moment making one of signers disadvantaged. Hence, it is interesting to investigate whether there exist concurrent signature schemes that are built *without* using the keystone methodology; that is, the signature binding process (in the second stage) is achieved by other means. However, how to construct a concurrent signature scheme without a keystone remains an open problem till now.

1.1. *Our Contribution*

Due to its interesting applications where partially fair exchanges of digital signatures is sufficient, we further investigate the notion of concurrent signatures. Especially, our main contribution includes the following aspects.

- *Revisiting the definition of concurrent signatures.* As we remarked earlier, the original notion of concurrent signatures is defined via an approach to the implementation of this notion, which leads to a serious reliance on the keystone methodology. In this paper, we reconsider this notion from the final goal and functionalities of concurrent signatures. This reconsideration allows us a refined definition to captures more approaches to realization of concurrent signatures.
- *A new generic construction without a keystone.* We present a new notion of chameleon one-way function. Using this notion and two-party ring signatures, we propose a generic construction of concurrent signatures. Our construction does not rely on the keystone methodology.
- *An efficient instantiation.* We realize an efficient chameleon one-way function. Based on this scheme and known ring signature schemes, we implement an efficient concurrent signature scheme. The scheme is proven secure in the random oracle model.

1.2. *Technical approach*

We briefly explain how a concurrent signature protocol without a keystone can also be built on the general notion of ambiguity of two-party ring signatures. This introduction outlines the basic ideas of our concurrent signatures in this paper.

Suppose that there are two signers, U_0, U_1. In the first stage, they interactively sign two messages m_0, m_1 in such a way such that they become publicly committed their respective messages at the same time without the aid of a keystone and a third party. Each of the signers holds a valid private key and public key pair. Without losing generality, let us assume that U_0

is the initial signer. U_0 starts the signing process by signing her ambiguous partial signature on a message m_0. This partial signature can be verified as a normal signature, but it is ambiguous to the third party about its authorship. If U_1 is happy with the U_0's partial ambiguous signature, he will create his partial (half) ambiguous signature on m_1 in the same way. At this point of time, both signers obtain the partial signature of their counterpart. However, because both partial signatures are ambiguous to any third party, no one can claim that a valid signature from the other side is obtained.

If both signers are happy with the signing run at this point of time, either signer can compute the other half of the signature and the protocol proceeds to the second stage. Suppose that U_0 wants to move first. U_0 creates the other half of partial signature on message m_0. Combining both half pieces of her signature, a complete signature can be formed. U_0 then sends her second half of the signature to U_1. In this point of time, all signatures are still not binding, namely, the full signature from U_0 and the partial signature from U_1 are still ambiguous to any third party. U_1 can check if the signature from U_0 is valid. If the check outputs *accept*, then U_1 has two options: (1) creating the second half of his signature and sending it to U_0; or (2) showing that he is indeed not the signer of U_0's signature by proving that his private key is indeed different from that of U_0's in zero knowledge. We will show that both options make both signatures binding concurrently. Notice that in option (2), U_0's full signature and U_1's partial signature are binding concurrently. We assume that if the authorship of a partial signature becomes clear, then it will be regarded as a full signature.

We note that our approach requires four rounds, which seems to be less efficient than the original CKP approach. Nonetheless, the aim of this paper to show a different concept than the original CKP approach, i.e. to remove the necessity of relying on the keystone methodology. Therefore, we do *not* aim to outperform the efficiency of the CKP scheme. Nonetheless, an efficient scheme without the keystone methodology is an interesting research problem which is worth to pursue.

Our scheme has two distinct properties. (1) U_0 and U_1 do not have to decide at the beginning of the protocol who is going to create the keystone and to control the keystone release. The order, i.e. who is the initiator and who is the responder, of the first phase is separated from the order of the second phase. After the first phase, they can come together to decide who is going to start the second phase. (2) If we consider the very last message still playing the role of the keystone, then in the proposed scheme, the message is not totally controlled by either of the two signers. But the value of a conventional keystone is completely controlled by one signer.

We further note that the CKP concurrent signature scheme and its variants are based on the ring signature scheme[7]. The CKP concrete scheme is indeed built on [7] with an adaption of the Schnorr signature scheme and the keystone methodology. Our scheme is also based on the ring signature scheme due to [7] with a novel construction where the signature binding is achieved with the natural exchange of signatures and there is no need to use any keystone.

Paper Organization

The rest of this paper is organized as follows. In Section 2, we revisit the definition of concurrent signatures. In Section 3, we present a generic construction without a keystone. In Section 4, we propose an efficient instantiation followed by a security analysis. The final section is the conclusion.

2. Revisiting definition of concurrent signatures

In this section, we revisit the definition of concurrent signatures[1]. The main difference is that we define this notion via the functionalities of a concurrent signature protocol while it is defined via the implementation approach in the CKP scheme[1]. For the terminology, we follow the CKP scheme[1] unless additional terms are required.

2.1. *Concurrent signature algorithms*

We first define algorithms used in a concurrent signature scheme.

Definition 2.1. A concurrent signature scheme is a digital signature scheme comprised of the following algorithms:

- ***Setup***: A probabilistic algorithm that on input a security parameter ℓ, outputs descriptions of: the set of participants $\mathcal{U}$, the message space $\mathbb{M}$, the signature space $\mathbb{S}$. The algorithm also outputs the public keys $\{X_i\}$ of all the participants, each participant retaining their private key x_i, and any additional system parameters π.
- ***ASign***: A probabilistic algorithm that on inputs $\{X_i, X_j, x_i, M, \pi\}$, where X_i and $X_j \neq X_i$ are public keys, x_i is the private key corresponding to X_i, and $M \in \mathbb{M}$, outputs an ambiguous signature σ on M.
- ***AVerify***: An algorithm which takes as input $S = \{\sigma, X_i, X_j, M, \pi\}$, where $\sigma \in \mathbb{S}$, X_i and X_j are public keys, and $M \in \mathbb{M}$, outputs *accept* or *reject*. We also require that if $\sigma' = \textit{ASign}(X_j, X_i, x_j, M, \pi)$ and

$S' = \{\sigma', X_j, X_i, M, \pi\}$, then $\mathit{AVerify}(S) = \mathit{AVerify}(S')$. We call this the symmetry property of *AVerify*.

- *BSign*: An algorithm which takes as input $S = \{\sigma, X_i, X_j, M, \pi\}$ and x_j, outputs a binding arguement ϱ on an ambiguous signature σ of message M.
- *BVerify*: An algorithm which takes as input S, ϱ and outputs *accept* or *reject*. We also require that if $\varrho' = \mathit{BSign}(S', x_i)$, then $\mathit{BVerify}(S, \varrho) = \mathit{BVerify}(S', \varrho')$. We call this the symmetry property of *BVerify*.

We call a signature σ an ambiguous signature and any pair $\{\sigma, \varrho\}$, where ϱ is a valid binding witness for σ, a concurrent signature. A concurrent signature scheme is said to be correct if both parties who holds X_i, X_j follow the protocol honestly, it holds that $\mathit{AVerify}(S) = \mathit{AVerify}(S') = \mathit{BVerify}(S, \varrho) = \mathit{BVerify}(S', \varrho') =$*Accept*.

2.2. *Concurrent signature protocol*

We next describe a concurrent signature protocol between two parties $\mathcal{A}$ and $\mathcal{B}$ (or Alice and Bob). Since one party needs to create and send the first ambiguous signature, we call this party the initial signer. A party who responds to this initial signature by creating another ambiguous signature we call a matching signer. Without loss of generality, we assume $\mathcal{A}$ to be the initial signer, and $\mathcal{B}$ the matching signer. From here on, we will use subscripts A and B to indicate initial signer $\mathcal{A}$ and matching signer $\mathcal{B}$. The signature protocol works as follows:

$\mathcal{A}$ and $\mathcal{B}$ run `Setup` to determine the public parameters of the scheme. We assume that $\mathcal{A}$'s public and private keys are X_A and x_A, and $\mathcal{B}$'s public and private keys are X_B and x_B.

(1) $\mathcal{A}$ takes her own public key X_A and $\mathcal{B}$'s public key X_B and picks a message $M_A \in \mathbb{M}$ to sign. $\mathcal{A}$ then computes her ambiguous signature to be

$$\sigma_A = \mathtt{Asign}(X_A, X_B, x_A, M_A, \pi),$$

and sends this to $\mathcal{B}$.

(2) Upon receiving $\mathcal{A}$'s ambiguous signature σ_A, $\mathcal{B}$ verifies the signature by checking that $\mathtt{Averify}(\sigma_A, X_A, X_B, M_A) = \mathit{accept}$. If not $\mathcal{B}$ aborts, otherwise $\mathcal{B}$ picks a message $M_B \in \mathbb{M}$ to sign and computes his ambiguous signature

$$\sigma_B = \mathtt{Asign}(X_A, X_B, x_B, M_B, \pi),$$

and sends this back to $\mathcal{A}$.

(3) Upon receiving $\mathcal{B}$'s ambiguous signature σ_B, $\mathcal{A}$ verifies the signature by checking that $\texttt{Averify}(\sigma_A, X_B, X_A, M_A) = accept$. If not $\mathcal{A}$ aborts, otherwise $\mathcal{A}$ computes his binding argument

$$\varrho_A = \texttt{Bsign}(\sigma_B, X_A, X_B, x_A, M_A, \pi),$$

and sends this back to $\sigma_{\mathcal{B}}$. Note that given $\sigma_A, \sigma_B, \varrho_A$, the authorship of these signatures is still ambiguous for any third party.

(4) Upon receiving $\mathcal{A}$'s binding argument ϱ_A, $\mathcal{B}$ verifies the argument by checking that

$$\texttt{Bverify}(\varrho_A, X_B, X_A, M_A) = accept.$$

If not $\mathcal{B}$ aborts, otherwise $\mathcal{B}$ computes his binding argument

$$\varrho_B = \texttt{Bsign}(\sigma_A, X_A, X_B, x_B, M_A, \pi),$$

and sends this back to $\sigma_{\mathcal{A}}$.

(5) Upon receiving $\mathcal{B}$'s binding argument ϱ_B, $\mathcal{A}$ verifies the argument by checking that

$$\texttt{Bverify}(\varrho_B, X_B, X_A, M_A) = accept.$$

If not $\mathcal{A}$ aborts, otherwise the ambiguity of concurrent signatures are revoked.

2.3. *Unforgeability*

We give the formal security model of the existential unforgeability of our scheme under a chosen message attack. We extend the definition of existential unforgeability under a chosen message attack[1] and design a game to simulate such attacks against the proposed scheme. It is defined using the following game between an adversary $\mathcal{AD}$ and a challenger $\mathcal{CH}$.

Setup: $\mathcal{CH}$ runs $\texttt{Setup}$ for a given security parameter ℓ to obtain descriptions of $\mathbb{U}, \mathbb{M}, \mathbb{S}$. $\texttt{Setup}$ also outputs the public and private keys $\{X_i\}$ and $\{x_i\}$ and any additional public parameters π. $\mathcal{AD}$ is given all the public parameters and the public keys $\{X_i\}$ of all participants. $\mathcal{CH}$ retains the private keys $\{x_i\}$. $\mathcal{AD}$ can make the following types of query to the challenger $\mathcal{CH}$:

KReveal Queries: $\mathcal{AD}$ can request the private keys corresponding to the public keys of participants. In response, $\mathcal{CH}$ outputs $\{x_i\}$ requested by $\mathcal{AD}$.

ASign Queries: $\mathcal{AD}$ can request an ambiguous signature for any input of the form (X_i, X_j, M) where X_i and $X_j \neq X_i$ are public keys and $M \in \mathbb{M}$. $\mathcal{CH}$ responds with an ambiguous signature $\sigma = \mathsf{Asign}(X_i, X_j, x_i, M, \pi)$. Note that using ASign queries, $\mathcal{AD}$ can obtain concurrent signatures σ for messages and pairs of users of his choice.

BSign Queries: $\mathcal{AD}$ can request a binding argument for any input of the form (σ, X_i, X_j, M) where $\sigma = \mathsf{Asign}(X_i, X_j, x_i, M, \pi)$. $\mathcal{CH}$ responds with a binding argument $\varrho = \mathsf{Bsign}(\sigma, X_i, X_j, x_i, M, \pi)$. Using BSign queries, $\mathcal{AD}$ can obtain binding arguments of σ on messages of his choice.

AVerify and BVerify Queries: Answers to these queries are not provided by $\mathcal{CH}$ since $\mathcal{AD}$ can compute them for himself using the Averify and BVerify algorithms.

Output: Finally $\mathcal{AD}$ outputs a tuple $(\sigma', \varrho, X_c, X_d, m)$ for public keys $X_c, X_d \in \{X_i\}$, and a message $m \in \mathbb{M}$. The adversary wins the game if

$$\mathsf{Averify}(\sigma', X_c, X_d, m, \pi) = \mathsf{Bverify}(\sigma', \varrho', X_c, X_d, m, \pi) = accept,$$

and if one of the following three cases hold:

(1) (X_c, X_d, m) has never been queried to ASign or (σ', X_c, X_d, m) has never been queried to Bsign; and no KReveal query was made by A on either X_c or X_d.

(2) (X_c, X_i, m) has never been queried to ASign or (σ', X_c, X_i, m) has never been queried to Bsign for any $X_i \neq X_c, X_i \in \mathcal{U}$; and no KReveal query was made by A on either X_c.

(3) (X_d, X_i, m) has never been queried to ASign or (σ', X_d, X_i, m) has never been queried to Bsign for any $X_i \neq X_d, X_i \in \mathcal{U}$; and no KReveal query was made by A on either X_d.

Definition 2.2. We say that a concurrent signature scheme is existentially unforgeable under a chosen message attack in the multi-party model if the probability of success of any polynomially bounded adversary in the above game is negligible (as a function of the security parameter ℓ).

2.4. *Ambiguity*

Ambiguity for a concurrent signature is defined by the following game between an adversary $\mathcal{AD}$ and a challenger $\mathcal{CH}$.

Setup: This is as before in the unforeability game.

Phase 1: $\mathcal{AD}$ makes a sequence of KReveal, ASign, Bsign queries. These are answered by $\mathcal{CH}$ as in the unforgeability game.

Challenge: Then $\mathcal{AD}$ selects a tuple (X_i, X_j, M) where $M \in \mathbb{M}$ is the message to be signed and X_i and X_j are public keys while the corresponding secret keys x_i, x_j have never been queried for. $\mathcal{CH}$ randomly selects a bit $b \in \{0,1\}$. If $b = 0$, $\mathcal{CH}$ outputs challenge $\sigma_0 = \texttt{Asign}(X_i, X_j, x_i, M, \pi)$, $\sigma_1 = \texttt{Asign}(X_i, X_j, x_j, M, \pi)$, $\varrho = \texttt{Bsign}(\sigma_0, X_i, X_j, x_j, M, \pi)$; otherwise $\mathcal{CH}$ computes $\sigma_0 = \texttt{Asign}(X_i, X_j, x_i, M, \pi)$, $\sigma_1 = \texttt{Asign}(X_i, X_j, x_j, M, \pi)$, $\varrho = \texttt{Bsign}(\sigma_1, X_i, X_j, x_i, M, \pi)$. $\mathcal{CH}$ Sends $(\sigma_0, \sigma_1, \varrho)$ to $\mathcal{AD}$.

Phase 2: $\mathcal{AD}$ may make another sequence of queries as in Phase 1 except that the KReveal cannot be queried on X_i and X_j and neither σ_0 nor σ_1 can be queried to `Bsign`; these are handled by $\mathcal{CH}$ as before.

Output: Finally $\mathcal{AD}$ outputs a guess bit $b' \in \{0,1\}$. $\mathcal{AD}$ wins if $b' = b$.

Definition 2.3. We say that a concurrent signature scheme is ambiguous if no polynomially bounded adversary has advantage that is non-negligibly greater than 1/2 of winning in the above game.

2.5. *Partial fairness*

We note that the definition of fairness in the CKP scheme[1] is indeed a partial fairness due to the slight privilege of the signer who holds the keystone can choose whether to release the keystone to revoke the ambiguity of the concurrent signatures. However, if signer chooses to release the keystone, it shows that both he/she and the other signer agree on the signed message.

In our definition, after the initial signer publishes the binding argument of the matching signer's ambiguous signature, the signature is still ambiguous. But if the matching signer also publishes the binding argument, then their signatures's ambiguity is revoked and become conventional signatures, which implies that both parties agree on the signed message. As a result, the notion of the concurrent signatures makes a step in the design of fair exchange of signatures without a trusted party.

We require the concurrent signature scheme and protocol to be fair for both the initial signer **A** and the matching signer $\mathcal{B}$. Now we formalize the partial fairness of concurrent signatures. This concept is defined via the following game between an adversary $\mathcal{AD}$ and a challenger $\mathcal{CH}$:

Setup: This is as before in the unforgeability game.

KReveal, ASign and Bsign Queries: These queries are answered by $\mathcal{CH}$ as in the unforgeability game.

Output: Finally $\mathcal{AD}$ outputs a tuple $(\sigma_0, \varrho_0, \sigma_1, \varrho_1, X_c, X_d, M)$, where X_c and X_d are public keys, and $M \in \mathbb{M}$, such that $\texttt{Averify}(\sigma_0, X_c, X_d, M, \pi) = \textit{accept}$, $\texttt{Averify}(\sigma_1, X_d, X_c, M, \pi) = \textit{accept}$, $\texttt{Bverify}(\sigma_1, \varrho_0, X_c, X_d, M, \pi) = \textit{accept}$. The adversary wins the game if one of the following cases hold:

(1) If σ_0, ϱ_1 were previous outputs from Asign and Bsign queries, no KReveal query on X_d was made, and

$$\texttt{Bverify}(\sigma_1, \varrho_0, X_d, X_c, M, \pi) = \textit{accept}.$$

(2) $\texttt{Bverify}(\sigma_1, \varrho_0, X_d, X_c, M, \pi) = \textit{reject}$.

Definition 2.4. We say that a concurrent signature scheme is fair if a polynomially bounded adversary's probability of success in the above game is negligible.

This definition of fairness captures the intuitive of the partial fairness for $\mathcal{B}$ in the concurrent signature protocol (in case 1 of the output conditions), since it guarantees that only the matching party $\mathcal{B}$ who holds the secret key can use it to create a binding argument. It also formalizes the fairness for **A** (in case 2 of the output conditions), since it guarantees that any valid ambiguous signatures from the two parties will all become binding. Thus $\mathcal{B}$ cannot be left in a position where **A**'s ambiguous signature is binding to **A** while $\mathcal{B}$'s own ambiguous signature is not binding to $\mathcal{B}$. However note that the definition does not guarantee that **A** will ever receive the binding argument from $\mathcal{B}$ and the notion only provides a partial fairness which is sufficient for many applications as we remarked earlier.

2.6. *Security of concurrent signatures*

Definition 2.5. We say that a correct concurrent signature scheme is secure if it is existentially unforgeable, ambiguous, and partially fair under a chosen message attack in the multi-party setting.

3. Generic construction

We present a generic construction based on a new notion of chameleon trapdoor function $\mathcal{T}$ and a chameleon two-party ring signature scheme. Before introducing our generic concurrent signature protocol, we formalize the building blocks.

Definition 3.1. (Chameleon Trapdoor Function $\mathcal{T}$) Let $y = f(x)$ be a public key, where $f : \mathbb{X} \rightarrow \mathbb{Y}$ be a one-way function. Let $\otimes : \mathbb{Y} \times \mathbb{Y} \rightarrow \mathbb{Y}$

and $\odot : \mathbb{X} \times \mathbb{X} \rightarrow \mathbb{X}$ be two efficient reversible operators. The chameleon trapdoor function associated with y is then defined as $y = u \otimes v$, where $u = f(a)$ and $v = f(b)$ for $a \odot b = x$. (u, v) is called an image pair of $\mathcal{T}$. (a, b) is called an image pair of the secret x. a or b is called partial image of x.

Like Chameleon functions[8], to construct a $\mathcal{T}$ function, one must know a secret, in this case, x. However, the additional feature of $\mathcal{T}$ is ambiguity where any one without the secret can construct an image pair (u, v) from a $\mathcal{T}$ function y such that $y = u \otimes v$. It is indistinguishable to any third party whether or not the image pair was created by a party who knows the secret or a party who does not know the secret. The only difference is that the party who knows the secret can know the values (a, b) such that $x = a \odot b$, $u = f(a)$, and $v = f(b)$.

Definition 3.2. (Chameleon Two-Party Ring Signature $\mathcal{R}$) Let $x_i = a_i \odot b_i$ be the private key of party $i \in \{0, 1\}$, $y_i = u_i v_i$ be the corresponding public key, and m_i be the message to be signed. A chameleon two-party ring signature pair $(\sigma(a_i, m_i), \sigma(b_i, m_i)))$ is a knowledge signature pair

$$\sigma(a_i, m_i) = KS\{f(\alpha\beta)|u_0 = f(\alpha) \vee u_1 = f(\alpha)\}(m_i),$$

$$\sigma(b_i, m_i) = KS\{f(\alpha\beta)|v_0 = f(\alpha) \vee v_1 = f(\alpha)\}(m_i).$$

Let U_0, U_1 be two polynomial time parties who want to fairly exchange their signatures. Let $\{m_0, m_1\} \in \mathbb{M}$ be the agreed messages to be signed by them, respectively. Let x_i be the private key and y_i be the public key of party U_i for $i \in \{0, 1\}$. The two parties compute a_i, b_i, u_i, v_i such that $y_i = u_i \otimes v_i$, $u_i = f(a_i)$, $v_i = f(b_i)$ and $x_i = a_i \odot b_i$. For simplicity, we omit the system parameters π.

Setup:

- U_0 and U_1 exchange u_i, v_i and other system parameters including their public keys.

ASign:

- U_0 computes $\sigma(a_0, m_0)$ and sends it to U_1. U_1 can verify its correctness. At this point of time, U_1 is sure that U_0 has created $\sigma(a_0, m_0)$, since he has not done so, but the authorship of $\sigma(a_0, m_0)$ is ambiguous to any third party; to whom, the signer could be either U_0 or U_1.

– If the verification of $\sigma(a_0, m_0)$ by U_1 outputs *accept*, U_1 will compute its signature $\sigma(a_1, m_1)$ and send it to U_0; otherwise abort. U_0 can in turn verify the correctness of $\sigma(a_1, m_1)$ and is sure that it is indeed created by U_0 since she has not done so. The authorship of $\sigma(a_1, m_1)$ is ambiguous to any third party; to whom, either U_0 or U_1 could be the signer.

BSign:

– Once both participants are happy with the signatures, one of them can start the binding process. Let U_0 move first. She computes $\sigma(b_0, m_0)$ without any message as her binding commitment and sends it to U_1.
– U_1 verifies $\sigma(b_0, m_0)$. If it outputs *accept*, he knows that U_0 has indeed committed to her signature. However, to any third party, $\sigma(b_0, m_0)$ could be created by either U_0 or U_1. Therefore, with $\sigma(a_0, m_0)$ and $\sigma(b_0, m_0)$, the third party is still not able to distinguish the signer. Then U_1 can make both $\sigma(a_0, m_0)$ and $\sigma(a_1, m_0)$ binding by computing $\sigma(b_1, m_1)$ as his binding commitment and sending it to U_0.
– U_0 verifies $\sigma(b_1, m_1)$. If it outputs *accept*, then both parties's signatures are binding; it aborts, otherwise.

Before the binding phase, the above protocol holds prefect authorship ambiguity. Any one (say, U) can select a random number a and compute $u = f(a)$ and find v such that $y = u \otimes v$. However, if U does not have the knowledge on the private key corresponding to y, he must break the one-way property of $f(\cdot)$ in order to show his knowledge on 'b' from v.

A signing process is correctly complete as long as both signatures are bound to their signers. This is because only the owner of the private key can prove its knowledge on private key images from u_i and v_i such that $u_i \otimes v_i = y_i$. We suppose that U_1 show his knowledge on b_1 in any way he chooses. As long as he has proved his knowledge on b_1, his signature is binding and can be used as his full signature. Therefore, U_1 cannot present an attack by excluding himself from U_0's signature.

3.1. *Signature of knowledge proof of discrete logarithm*

We first recall the zero knowledge proofs of discrete logarithms in finite cyclic group $\mathbb{G} = \langle g \rangle$ of a prime order p. We consider the following proof.

Given $u = g^z$, $w_0 = g_0^{z_0}$, $w_1 = g_1^{z_1}$, where $g_0 = g^{z_1}$ and $g_1 = g^{z_0}$, prove $\log_g u = \{\log_{g_0} w_0 \vee \log_{g_1} w_1\}$ in zero knowledge. The proof is defined as

$$\mathsf{SKP}[\alpha : \{u = g^\alpha\} \wedge \{w_0 = g_0^\alpha \vee w_1 = g_1^\alpha\}](m). \tag{1}$$

Here, SKP denotes Signature Knowledge Proof. m denotes the signed message. The proof can be referred to the literature[9–11].

3.2. *The scheme*

- $\mathsf{Setup}(1^\ell)$. Choose a suitable finite cyclic group $\mathbb{G} = \langle g \rangle$ of large prime order p. For each participant, select a private key $x_i \in \{0,1\}^\ell$ and compute his public key as $y_i = g^{x_i}$. Select a cryptographic hash function $H : \{0,1\}^* \to \mathbb{Z}_p^*$.
- `ASign`. U_i sets $y_i = u_i v_i = g^{a_i} g^{b_i}$ for $i \in \{0,1\}$ and products the ambiguous signature knowledge proof on message $m_i \in \mathbb{M}$:

$$\sigma(a_0, m_0) \leftarrow \mathsf{SKP}[\alpha : \{g^{a_0 a_1} = (g^{a_1})^\alpha\} \wedge \{\frac{y_0}{v_0} = g^\alpha \vee \frac{y_0}{v_1} = g^\alpha\}](m_0).$$

$$\sigma(a_1, m_1) \leftarrow \mathsf{SKP}[\alpha : \{g^{a_1 a_0} = (g^{a_0})^\alpha\} \wedge \{\frac{y_1}{v_1} = g^\alpha \vee \frac{y_1}{v_0} = g^\alpha\}](m_1).$$

- `AVerify`. Check if $\sigma(a_i, m_i)$ is correct.
- `BSign`. U_i computes the binding arguments as follows:

$$\sigma(b_0, m_0) \leftarrow \mathsf{SKP}[\beta : \{g^{b_0 b_1} = (g^{b_1})^\beta\} \wedge \{\frac{y_0}{u_0} = g^\beta \vee \frac{y_0}{u_1} = g^\beta\}](m_0).$$

$$\sigma(b_1, m_1) \leftarrow \mathsf{SKP}[\beta : \{g^{b_1 b_0} = (g^{b_0})^\beta\} \wedge \{\frac{y_1}{u_1} = g^\beta \vee \frac{y_1}{u_0} = g^\beta\}](m_1).$$

- `BVerify`. Check if $\sigma(b_i, m_i)$ is also correct.

3.3. *Discussions*

In this section, we explain the ambiguity of the signatures prior to the binding moment and two cases for signature binding.

Ambiguity case 1: After both $\sigma(a_0, m_0)$ and $\sigma(a_1, m_1)$ were exchanged, they are ambiguous, since both parties can create $\sigma(a_0, m_0)$ and $\sigma(a_1, m_1)$. Notice that $y = g^x = uv = g^a g^b$. Any one can select a random number b and compute $y = ug^b$.

$$\sigma(a_0, m_0) \leftarrow \mathsf{SKP}[\alpha : \{g^{a_0 a_1} = (g^{a_1})^\alpha\} \wedge \{u_0 = g^\alpha \vee u_1 = g^\alpha\}](m_0).$$

$$\sigma(a_1, m_1) \leftarrow \mathsf{SKP}[\alpha : \{g^{a_1 a_0} = (g^{a_0})^\alpha\} \wedge \{u_0 = g^\alpha \vee u_1 = g^\alpha\}](m_1).$$

Ambiguity case 2: After both signatures $\sigma(a_0, m_0)$ and $\sigma(a_1, m_1)$ were exchanged and U_0 sent $\sigma(b_0, m_0)$ to U_1, both signatures are still ambiguous. Observe that U_1 can also create $\sigma(b_0, m_0)$ by himself:

$$\sigma(b_0, m_0) \leftarrow \mathsf{SKP}[\beta : \{g^{b_0 b_1} = (g^{b_1})^\beta\} \wedge \{ v_0 = g^\beta \vee v_1 = g^\beta\}](m_0).$$

$\{g^{b_0 b_1} = (g^{b_1})^\beta\}$ can be replaced by $\{g^{b_1 b_1} = (g^{b_1})^\beta\}$.

Binding case 1: After $\sigma(a_0, m_0)$, $\sigma(a_1, m_1)$, and $\sigma(b_0, m_0)$ have been exchanged, U_1 makes both signatures binding by revealing $\sigma(b_1, m_1)$ to U_0:

$$\sigma(b_1, m_1) \leftarrow \mathsf{SKP}[\beta : \{g^{b_1 b_0} = (g^{b_0})^\beta\} \wedge \{ v_0 = g^\beta \vee v_1 = g^\beta\}](m_1).$$

where U_1 proves his knowledge on b_1 which matches one of $\{v_0 = g^{b_0}, v_1 = g^{b_1}\}$. Recall that U_0 has proved her knowledge on b_0 which also matches one of $\{v_0 = g^{b_0}, v_1 = g^{b_1}\}$. Both signatures are therefore binding concurrently.

Binding case 2: After $\sigma(a_0, m_0)$, $\sigma(a_1, m_1)$, and $\sigma(b_0, m_0)$ have been exchanged, U_1 shows that he is not U_0 by utilizing

$$\sigma(b_0, m_0) \leftarrow \mathsf{SKP}[\beta : \{g^{b_0 b_1} = (g^{b_1})^\beta\} \wedge \{ v_0 = g^\beta \vee v_1 = g^\beta\}](m_0).$$

U_1 proves his knowledge on b_1 which matches one of g^{b_1} such that $u_1 g^{b_1} = y_1$ and $g^{b_0 b_1} \neq g^{b_1 b_1}$. While doing this, $\sigma(a_1, m_1)$ is binding too:

$$\sigma(a_1, m_1) \leftarrow \mathsf{SKP}[\alpha : \{g^{a_0 a_1} = (g^{a_0})^\alpha\} \wedge \{ u_0 = g^\alpha \vee u_1 = g^\alpha\}](m_1).$$

Therefore, both signatures are binding concurrently.

3.4. *Security Analysis*

Lemma 3.1. *The proposed scheme is existentially unforgeable under a chosen message attack in the random oracle model, assuming the hardness of the discrete logarithm problem.*

Proof: We prove the lemma by using the game defined earlier in this paper. $\mathcal{CH}$ preforms the **Setup** and generates all necessary parameters. $\mathcal{AD}$ chooses a public key by himself without letting $\mathcal{CH}$ know which one he has chosen. $\mathcal{CH}$ guesses that $\mathcal{AD}$ will choose y_a in input. For each $i \neq a$, x_i is chosen at random and sets $y_i = g^{x_i}$. All public keys $\{y_i\}$ are given to $\mathcal{AD}$. $\mathcal{CH}$ simulates the challenger by simulating all the oracles which $\mathcal{AD}$ can query.

- Random Oracle Queries. $\mathcal{AD}$ can query the random oracle at any time. $\mathcal{CH}$ simulates it by keeping a list of tuple (m_i, h_i) called the H_i-list. When the oracle is queried with an input $m \in \mathbb{M}$, $\mathcal{CH}$ responds: If the query m is already on the H_i-list in tuple (m_i, h_i), then $\mathcal{CH}$ outputs h_i. Otherwise, $\mathcal{CH}$ selects a random $h \in \mathbb{Z}_p^*$ and adds (m, h) to the H_i-list.

- KReveal Queries. $\mathcal{AD}$ can request the private key for any public key y_i. If $y_i = y_a$, then $\mathcal{CH}$ terminates the simulation. Otherwise, $\mathcal{CH}$ returns the corresponding private key x_i.
- ASign Queries. $\mathcal{CH}$ simulates the signing oracle. $\mathcal{AD}$ requests an ambiguous signature knowledge proof from $\mathcal{CH}$ on input (m, g^z, Y), where Y is the set of all public keys. If $y_i \neq y_a$, then $\mathcal{CH}$ responds with a normal ambiguous signature proof $\sigma = (c_0, s_0, s_1)$.
  If $y_i = y_a$, then $\mathcal{CH}$ answers the query as follows.
  (a) $\mathcal{CH}$ picks at random (h, s_0, s_1), computes $B_1 = g^{s_i}(g^z)^{h_i}$, $B_2 = g^{s_i}(g^z)^{h_i}$, $B_3 = g^{s_i}(y_i/g^z)^{h_i}$, and forms two bit-strings $B_1\|B_2\|Y\|m$ and $B_1\|B_3\|Y\|m$.
  (b) If h_i is equal to some previous output for the H_i oracle, or either $B_1\|B_2\|Y\|m$ or $B_1\|B_3\|Y\|m$ is the same as some previous input, then return to step (a).
  (c) Otherwise, add $(B_1\|B_2\|Y\|m, h_i)$ and $(B_1\|B_3\|Y\|m, h_i)$ to the H_i-list.
  (d) $\mathcal{CH}$ outputs $\sigma_i = (h_i, s_0, s_1)$ as the signature for message m for g^z.
- BSign Queries. They are similar to the ASign Queries.
- AVerify and BVerify Queries. Answers to these queries can be done by $\mathcal{AD}$ himself.

Finally, $\mathcal{AD}$ outputs a tuple $(\sigma', \varrho, y, g^z, m)$ for public keys $y, g^z \in \{y_i\}$, and a message $m \in \mathbb{M}$, where $\sigma' = (m, h_i, s_0, s_1)$ for g^z and Verify(σ') $\rightarrow$ *accept*. $\mathcal{AD}$ wins the game if the conditions defined in section 2.3 are satisfied.

It is easy to show that output conditions can occur only with negligible probability δ. This follows immediately from the unforgeability of the underlying ring signature[7], assuming the hardness of the discrete algorithm problem. Here, we omit the detail of the unforgeability proof, which is similar to that in Abe et. al. [7].

In the game, if $y_a \neq y_i$, then $\mathcal{CH}$ aborts, having failed to guess the correct challenge public key. The probability of $y_a = y_i$ is $1/\lambda(\ell)$, where λ is a polynomial function[1].

Recall that we can rewrite the signature above in the form $(c_0, s_0, s_1, R_0, R_1)$, where c_0 implies (h_0, h_1) which are associated with the values stored in H_i-list. In the game, $\mathcal{CH}$ aborts, if one of $\{h_i\}$ matches some value in H_i-list. Therefore, if $\mathcal{AD}$ made μ_s queries to the signing oracles ASign and BSign, and μ_i to the H_i oracle, the probability for a match is at most $2\mu_1\mu_i/2^k = \mu_1\mu_i/2^{k-1}$.

As a result, the probability that $\mathcal{CH}$ does not have to abort at some point in the game is at least

$$\gamma = (1-\delta)\frac{1}{\lambda(\ell)}(1-\frac{\mu_s\mu_i}{2^{k-1}}).$$

According to the forking lemma[12,13], B can repeat its simulation so that $\mathcal{AD}$ produces another such signature knowledge proof $(c_0', s_0', s_1', R_0, R_1)$, where $c_0 \neq c_0'$ or $\{h_i\} \neq \{h_i'\}$. There is one seal in the ring: (h_j', s_j', R_j'). Similarly, there is one in the other solution $(h_k, s_k, R_k,)$, where $h_j' \neq h_j$, $s_j' \neq s_k$, and $R_j' = R_k$. The probability finding these pair of seals is at most 1/4. We then have two equations: $s_j' = r - h_j' z$, $s_k = r - h_k z$. Solving these two equations, we have $z = (s_j' - s_k)/(h_k - h_j')$, where $z \in (a_i, b_i)$. Notice that $\mathcal{AD}$ does not need to solve the discrete logarithm problem twice in order to break the scheme. Recall that $\mathcal{AD}$ can randomly choose a number z to compute the partial public key u (for $y = uv$) and then find $v = y/u$. $\mathcal{AD}$ only needs to solve the discrete logarithm problem with respect to v. The forking lemma states that if $\mathcal{AD}$ is a polynomial time Turing machine with input only public data, which produces, in time τ and with probability $\eta \geq 10(\mu_s+1)(\mu_s+\mu)/2^\ell$ a valid signature (m, s, h, R), where μ is the number of hash queries, and μ_s is the number of signature queries. In our case, it should be $\eta/4 = 10(\mu_s+1)(\mu_s+\mu_i)/2^{\ell+2}$, where μ_i is the number of queries to H_i and μ_s is the number of queries made to ASign and BSign. If (m, s_i, R_i) are simulated with indistinguishable probability distribution without knowledge of the secret key, then there exists an algorithm $\mathcal{AD}$, which controls $\mathcal{CH}$ and replaces A's interaction with the signer by the simulation, and which produces two valid signatures in expected time at most $\tau' = 120686\mu_s\tau/\eta$. In our case, the expected time is at most $\tau'/\gamma = 120686\mu_s\tau/(\eta\gamma)$.

Lemma 3.2. *The proposed scheme is ambiguous in the random oracle model.[14]*

We omit the proof and refer the reader to the security analysis of ring signatures[7] plus the the deniability of our deniable ring based on the trapdoor construction.

Lemma 3.3. *The proposed scheme is fair in the random oracle model.*

The proof is omitted due to page limit.

Theorem 3.1. *The proposed scheme is secure in the random oracle model, assuming the hardness of the discrete logarithm problem.*

From the proofs of Lemmas 1-3.

4. Conclusion

We presented a new notion of chameleon one-way function. Using this notion and two-party ring signatures, we were able to construct a novel generic concurrent signature scheme without relying on the keystone methodology. We also presented a concrete construction of the chameleon one-way function. Based on this concrete scheme and a known ring signature scheme, we construct a concrete signature scheme without a keystone.

References

1. L. Chen, C. Kudla and K. G. Paterson, Concurrent signatures, in *Advances in Cryptology, Proc. EUROCRYPT 2004,* LNCS 3027, (Springer-Verlag, Berlin, 2004).
2. N. Asokan, V. Shoup and M. Waidner, *IEEE Journal on Selected Areas in Communications* **18** (2000).
3. F. Bao, R. Deng and W. Mao, Efficient and practical fair exchange protocols, in *Proceedings of 1998 IEEE Symposium on Security and Privacy, Oakland*, 1998.
4. Y. Dodis and L. Reyzin, Breaking and repairing optimistic fair exchange from podc 2003., in *Proc. the 3rd ACM workshop on Digital rights management*, 2003.
5. W. Susilo, Y. Mu, F. Zhang, Perfect concurrent signature schemes, in *Proceedings of International Conference on Information and Communications Security*, (Springer-Verlag, Berlin, 2004).
6. W. Susilo and Y. Mu, Tripartite concurrent signatures, in *Proc. of the 20th IFIP International Information Security Conference (IFIP/SEC 2005)*, 2005.
7. M. Abe, M. Ohkubo and K. Suzuki, 1-out-of-n signatures from a variety of keys, in *Advances in Cryptology–ASIACRYPT 2002,* LNCS 2501, (Springer-Verlag, Berlin, 2002).
8. H. Krawczyk and T. Rabin, Chameleon hashing and signatures, in *Proc. of Network and Distributed System Security Symposium*, (The Internet Society, 2000).
9. D. Chaum, J.-H. Evertse and J. V. de Graaf, An improved protocol for demonstrating possession of discrete logarithms and some generalizations, in *Advances in Cryptology, Proc. EUROCRYPT 87,* LNCS 304, (Springer-Verlag, 1988).
10. C. P. Schnorr, Efficient identification and signatures for smart cards, in *Adances in cryptology - CRYPT0'89, Lecture Notes in Computer Secience 435*, (Springer-Verlag, Berlin, 1990).
11. J. Camenisch and M. Stadler, Efficient group signature schemes for large groups, in *Advances in Cryptology, Proc. CRYPTO 97,* LNCS 1296, (Springer-Verlag, Berlin, 1997).

12. D. Pointcheval and J. Stern, Security proofs for signature schemes, in *Advances in Cryptology, Proc. EUROCRYPT 96,* LNCS 1070, (Springer-Verlag, 1996).
13. D. Pointcheval and J. Stern, *Journal of Cryptology* **13**, 361 (2000).
14. M. Naor, Deniable ring authentication, in *Advances in Cryptology, Proc. CRYPTO 2002,* LNCS 2442, (Springer-Verlag, Berlin, 2002).

Authentication Codes in the Query Model

R. Safavi-Naini

Department of Computer Science, University of Calgary, 2500 University Drive, NW Calgary T2N 1N4, Canada
E-mail: rei@ucalgary.ca
cisac.ucalgary.ca

D. Tonien

Faculty of Informatics, University of Wollongong, Wollongong, New South Wales 2522, Australia
E-mail: dong@uow.edu.au

P. R. Wild

Department of Mathematics, Royal Holloway, Egham, Surrey TW20 0EX, United Kingdom
E-mail: p.wild@rhul.ac.uk
www.isg.rhul.ac.uk

We consider unconditionally secure authentication systems in which a sender communicates a source state to a receiver by encoding it as an authenticated message under a key agreed with the receiver. An authentication code is a triple $(\mathcal{S}, \mathcal{M}, \mathcal{E})$ where $\mathcal{E}$ is a collection of *encoding rules*, *i.e.* mappings from the set $\mathcal{S}$ of *source states* into the set $\mathcal{M}$ of *messages*. A probability distribution on $\mathcal{E}$ models the key agreement process by which the encoding rule is chosen by the sender and receiver. In the usual model the adversary observes messages transmitted between the sender and the receiver before introducing to the channel a spoofing message, chosen according to some strategy. The adversary is successful if the spoofed message is accepted by the receiver as a valid (authenticated) message. In this paper we consider the extension to the model in which the adversary interacts with the sender and the receiver. In this query model the adversary may send messages to the receiver and observe a response to determine whether or not they are accepted or the adversary may provide the sender with the source state and observe the corresponding authenticated message that the sender transmits. We discuss the nature of an optimal strategy for such an adversary and derive bounds on the probability of deception for an authentication code in this model. This also leads to combinatorial characterisations of optimal authentication codes.

Keywords: authentication code; query model; combinatorial characterisation.

1. Introduction

Authentication schemes that offer unconditional security for sending authenticated messages were introduced by Gilbert, MacWilliams and Sloane[1]. An information theoretic model was developed by Simmons[7,8], while Massey[2] took a combinatorial approach.

In these models there are three participants, a sender, a receiver and an adversary. The sender communicates a sequence of distinct *source states* from a finite set $\mathcal{S}$ to the receiver by encoding them using one from a finite set $\mathcal{E}$ of *encoding rules*. We do not consider splitting in this paper, so each encoding rule is an injective mapping from $\mathcal{S}$ into a set $\mathcal{M}$ of *messages*. The triple $(\mathcal{S}, \mathcal{M}, \mathcal{E})$ is called an authentication code or *A*-code. The set sizes $k = |\mathcal{S}|, v = |\mathcal{M}|$ and $b = |\mathcal{E}|$ are called the parameters of the authentication code. The receiver recovers the source states from the received messages by determining their (unique) pre-images under the agreed encoding rule. The receiver accepts a message as authentic if it lies in the image of the agreed encoding rule and decodes to a source state that has not been received before. The adversary observes the resulting sequence of messages and attempts to determine another message which will be accepted by the receiver as authentic, thereby deceiving the receiver. This message is called the *spoofing message*. As discussed by Rosenbaum[4], in determining the probability of success of the adversary it is sufficient to consider the impersonation attack in which the spoofing message is transmitted in addition to the observed message rather than as a substitution for them.

The agreed encoding rule constitutes a key and is chosen according to a probability distribution $P_{\mathcal{E}}$ on $\mathcal{E}$. We assume that the sequence of distinct source states that is to be transmitted is generated with a known probability distribution. The adversary uses knowledge of these distributions to formulate a strategy for choosing a spoofing message given a sequence of observed transmitted messages. Many authors (for example, Walker[14], Rosenbaum[4], Pei[3], Stinson[9–11], Rees and Stinson[12]) have contributed to the study of this model. Both information theoretic and combinatorial bounds on the probability of success of the adversary have been given as well as bounds on the number of encoding rules of authentication codes for which these bounds on the probability of deception are achieved. Such authentication codes that have a minimum number of encoding rules are called optimal and combinatorial characterisations of optimal codes have been given.

Shikata *et al*[6] and Safavi-Naini and Wild[5] have considered authentication systems in a more general model that admits an adversary who, as well as passively observing transmitted messages, can be proactive in gathering information about the authentication code. We call this the query model. The adversary may submit queries to the sender or the receiver and observe the responses. In a signature query the adversary provides the sender with a source state which the sender, in response, authenticates and transmits to the receiver. The adversary observes this transmitted message. In a verification query the adversary transmits a message to the receiver for verification and observes the receiver's response in accepting or rejecting it (as authentic).

We are interested in the probability of success of an adversary in this query model and, following the method of analysis of authentication codes with passive adversaries, consider combinatorial bounds on the probability of success and the number of encoding rules. In section 2 we review the known results for the model with a passive adversary. In section 3 we consider the query model – we establish bounds and characterizations for the signature query model and review results for the verification query model.

2. Passive Adversaries

We begin by recalling some results (see Massey[2], Stinson[10,11], Rosenbaum[4]) for the standard model of an authentication system in which the adversary is passive and observes a sequence of authenticated messages before spoofing.

Let $(\mathcal{S}, \mathcal{M}, \mathcal{E})$ be an authentication code with parameters (k, v, b). We write $p(E = e)$ to denote the probability that the agreed encoding rule is $e \in \mathcal{E}$ and $p(S_i = s)$ to denote the probability that the ith source state that the sender encodes is $s \in \mathcal{S}$. We also write $p(S^i = s^i)$ to denote the probability that the sequence of source states that are encoded by the sender is the sequence $s^i = s_1, \ldots, s_i$. Note that this probability is non-zero only if $s_1, \ldots, s_i$ are distinct. These distributions determine the probability distribution on the message sequences transmitted by the sender. The probability $p(M_i = m)$ that the ith transmitted message is $m \in \mathcal{M}$ is given by

$$p(M_i = m) = \sum_{e \in \mathcal{E}} \sum_{s \in \mathcal{S}: e(s) = m} p(E = e) p(S_i = s).$$

If the adversary has observed a sequence $m^i = m_1, m_2, \ldots, m_i$ of i messages then the agreed coding rule must belong to the set

$$\mathcal{E}_{m^i} = \{e \in \mathcal{E} | e^{-1}(m_1), \ldots, e^{-1}(m_i) \in \mathcal{S} \text{ and are distinct}\}.$$

If e is the agreed encoding rule then the ith transmitted message is m when the ith source state is $e^{-1}(m)$. The probability $p(M^i = m^i)$ that the sequence of transmitted messages is $m^i = m_1, \ldots, m_i$ is given by

$$p(M^i = m^i) = \sum_{e \in \mathcal{E}_{m^i}} p(E = e)p(S^i = e^{-1}(m_1), \ldots, e^{-1}(m_i)).$$

For each $e \in \mathcal{E}_{m^i}$ there is a conditional probability $p(e|m^i)$ that the agreed encoding rule is e given that the adversary has observed m^i. The adversary is successful in spoofing with $m \in \mathcal{M}$ if there exists $s \in \mathcal{S}$ with $s \neq e^{-1}(m_1), \ldots, e^{-1}(m_i)$ and $e(s) = m$ where e is the agreed encoding rule. Thus the adversary is successful if

$$e \in \mathcal{E}_{m^i,m} = \{e \in \mathcal{E}_{m^i} | e(s) = m, \text{ for some } s \neq e^{-1}(m_1), \ldots, e^{-1}(m_i)\}.$$

The probability of success in spoofing with m given that the sequence m^i has been observed is $\sum p(e|m^i)$ where the sum is over $e \in \mathcal{E}_{m^i,m}$. This is usually denoted Payoff$(m; m^i)$. An adversary's best strategy is to spoof with a message m for which this is a maximum. If we denote this maximum by $P(m^i)$ then the adversary's expected probability of success by spoofing after observing i messages using the best strategy is $P_i = \sum_{m^i} p(M^i = m^i)P(m^i)$.

Now $\sum_{m \neq m_1, \ldots, m_i}$ Payoff$(m; m^i) = \sum_{e \in \mathcal{E}} \sum p(e|m^i)$ where the inner sum is over $m \in \mathcal{M}$ with $m = e(s)$ for $s \neq s_1, \ldots, s_i$. For each e there are exactly $|\mathcal{S}| - i$ possibilities for such an m and so $\sum_{m \neq m_1, \ldots, m_i}$ Payoff$(m; m^i) = |\mathcal{S}| - i$. It follows that

$$\max_{m \neq m_1, \ldots, m_i} \text{Payoff}(m; m^i) \geq \frac{|\mathcal{S}| - i}{|\mathcal{M}| - i}$$

and so $P_i \geq \frac{|\mathcal{S}|-i}{|\mathcal{M}|-i}$. Note that this is a combinatorial bound and depends only on the authentication code $(\mathcal{S}, \mathcal{M}, \mathcal{E})$ and not on the probability distributions on $\mathcal{E}$ and $\mathcal{S}$.

Since the adversary learns information about the encoding rule with each observed message, the sender and receiver adopt an upper bound (which we denote by $L+1$) on the number of messages to be encoded using a given encoding rule before a new encoding rule is agreed. If $P_i = \frac{|\mathcal{S}|-i}{|\mathcal{M}|-i}$ for $i = 0, \ldots, L$ then for each sequence m^{L+1} of distinct messages the set $\mathcal{E}_{m^{L+1}}$ is non-empty and so $|\mathcal{E}| \geq \frac{\binom{|\mathcal{M}|}{L+1}}{\binom{|\mathcal{S}|}{L+1}}$ with equality when $|\mathcal{E}_{m^{L+1}}| = 1$ for each sequence m^{L+1} of distinct messages. If equality holds then it follows that the design with point set $\mathcal{M}$ and block set $\mathcal{E}$ is a $(L+1) - (|\mathcal{M}|, |\mathcal{S}|, 1)$ design where incidence is defined by $(m, e) \in \mathcal{I}$ if and only if $e \in \mathcal{E}_m$.

3. The Query Model

In the query model we assume that the adversary is not restricted to observing messages transmitted by the sender to the receiver. The adversary can send messages of the adversary's choice to the receiver and observe the receiver's response in either accepting or rejecting them. The adversary can also have the sender authenticate source states of the adversary's choice and may observe the corresponding transmitted messages. We refer to the former as verification queries and to the latter as signature queries.

In the query model we are not concerned with the probability distribution on $\mathcal{S}$ but with the adversary's strategy for choosing the queries. We model the adversary's strategy by probability distributions on the query set, $\mathcal{S}$ for signature queries and $\mathcal{M}$ for verification queries.

3.1. *Signature Queries*

In the signature query model the adversary chooses a source state $s \in S$ and observes the corresponding message $m = e(s)$ for the unknown encoding rule e that the sender and receiver share. We model the adversary's strategy as a family of probability distributions $\tau_{(s,m)^j}$, indexed by $(s,m)^j \in (\mathcal{S} \times \mathcal{M})^j, j = 0, \ldots, L-1$ on $\mathcal{S}$. Here $\tau_{(s,m)^j}(s_{j+1})$ denotes the probability that the adversary chooses s_{j+1} as the next query given that m_l was the response to query s_l for $l = 1, \ldots j$. Since a repeated query will produce the same response we may assume that $\tau_{(s,m)^j}(s) = 0$ for $s \in \{s_1, \ldots, s_j\}$.

An adversary who has observed a sequence $(s,m)^j$ of query and response pairs may decide to spoof with a message m. The adversary's strategy for choosing m is represented as a probability distribution denoted $\sigma_{(s,m)^j}$, one from a family of probability distributions on $\mathcal{M}$. Let $\mathcal{E}_{(s,m)^j} = \{e \in \mathcal{E} | e(s_1) = m_1, \ldots, e(s_j) = m_j\}$. The adversary may determine the conditional probability distribution on $\mathcal{E}$ given the the sequence $(s,m)^j$. This conditional probability is non-zero only for encoding rules belonging to $\mathcal{E}_{(s,m)^j}$. A spoofing message m is successful if $m = e(s)$ for some $s \in \mathcal{S}, s \neq s_1, \ldots, s_j$ where e is the agreed encoding rule, that is, if e belongs to the subset $\mathcal{E}_{(s,m)^j,(m,1)} = \{e \in \mathcal{E}_{(s,m)^j} | e(s) = m \text{ for some } s \in \mathcal{S}, s \neq s_1, \ldots, s_j\}$. The probability that the adversary's spoofing message m is successful is $\text{Payoff}(m; (s,m)^j) = \sum_{e \in \mathcal{E}_{(s,m)^j,(m,1)}} p(e|(s,m)^j)$. The expected probability of success of an adversary using these strategies and spoofing after j signature queries is

$$P_j^{\tau,\sigma} = \sum_{s_1} \tau(s_1) \sum_{m_1} p(m_1|s_1) \sum_{s_2 \neq s_1} \tau_{(s_1,m_1)}(s_2) \sum_{m_2} p(m_2|(s_1,m_1), s_2) \\ \cdots \sum_{s_j \neq s_1,\ldots,s_{j-1}} \tau_{(s,m)^{j-1}}(s_j) \sum_{m_j} p(m_j|(s,m)^{j-1}, s_j)$$

$$\sum_{m \notin m^i} \sigma_{(s,m)^j}(m)\text{Payoff}(m; (s,m)^j).$$

Let P_j be the maximum of $P_j^{\tau,\sigma}$ taken over all possible strategies of the adversary. An adversary's strategy is said to be optimal if $P_j^{\tau,\sigma} = P_j$. We are interested in bounding P_j. We do this by considering a special class of (simple) strategies. We say that an adversary's strategy is *deterministic* if, for $l = 0, \ldots, j-1$, whenever a sequence $(s,m)^l$ has arisen from the strategy, it holds that, for some $s \in \mathcal{S}$, $\tau_{(s,m)^l}(s) = 1$, and $\tau_{(s,m)^l}$ takes the value 0 otherwise, and whenever a sequence $(s,m)^j$ has arisen from the strategy, it holds that, for some $m \in \mathcal{M}$, $\sigma_{(s,m)^j}(m) = 1$, and $\sigma_{(s,m)^j}$ takes the value 0 otherwise. We show that an adversary need only consider deterministic strategies in order to adopt an optimal strategy.

Proposition 3.1. *Let $(\mathcal{S}, \mathcal{M}, \mathcal{E})$ be an authentication code. In the signature query model, for each $j = 0, \ldots, |\mathcal{S}|$, there always exists an optimal adversary strategy for spoofing after j queries that is deterministic.*

Proof. For each sequence $(s,m)^j$ of query and response pairs there exists m such that $\text{Payoff}(m; (s,m)^j)$ is maximal. Let $m_{(s,m)^j}$ be such a message m. Define $\sigma_{(s,m)^j}$ so that $\sigma_{(s,m)^j}(m)$ equals 1 for $m = m_{(s,m)^j}$ (and equals 0 otherwise). Then $\sum_{m \neq m_1, \ldots, m_j} \sigma_{(s,m)^j}(m)\text{Payoff}(m; (s,m)^j)$ is greater than $\sum_{m \neq m_1, \ldots, m_j} \sigma'_{(s,m)^j}(m)\text{Payoff}(m; (s,m)^j)$ for any other probability distribution $\sigma'_{(s,m)^j}$.

For each sequence $(s,m)^{j-1}$ of query and response pairs there exists $s_j \neq s_1, \ldots, s_{j-1}$ such that

$$\sum_{m_j} p(m_j | (s,m)^{j-1}, s_j) \sum_{m \neq m_1, \ldots, m_j} \sigma_{(s,m)^j}(m))\text{Payoff}(m; (s,m)^j)$$

is maximal. Let $s_{(s,m)^{j-1}}$ be such a source state s_j. Define $\tau_{(s,m)^{j-1}}$ so that $\tau_{(s,m)^{j-1}}(s_j)$ equals 1 for $s_j = s_{(s,m)^{j-1}}$ (and equals 0 otherwise). Now

$$\sum_{s_j \neq s_1, \ldots, s_{j-1}} \tau_{(s,m)^{j-1}}(s_j) \sum_{m_j} p(m_j | (s,m)^{j-1}, s_j)$$
$$\sum_{m \neq m_1, \ldots, m_j} \sigma_{(s,m)^j}(m))\text{Payoff}(m; (s,m)^j)$$

is greater than

$$\sum_{s_j \neq s_1, \ldots, s_{j-1}} \tau'_{(s,m)^{j-1}}(s_j) \sum_{m_j} p(m_j | (s,m)^{j-1}, s_j)$$
$$\sum_{m \neq m_1, \ldots, m_j} \sigma'_{(s,m)^j}(m)\text{Payoff}(m; (s,m)^j)$$

for any other $\tau'_{(s,m)^{j-1}}$ and $\sigma'_{(s,m)^j}$.

In general, for each sequence $(s,m)^{j-l}$, $l = 1,\ldots,j$ we define $\tau_{(s,m)^{j-l}}$ so that $\tau_{(s,m)^{j-l}}(s_{j-l+1}) = 1$ for s_{j-l+1} such that

$$\sum_{s_{j-l+1}} \tau_{(s,m)^{j-l}}(s_{j-l+1}) \sum_{m_{j-l+1}} p(m_{j-l+1}|(s,m)^{j-l}, s_{j-l+1})$$

$$\cdots \sum_{s_j} \tau_{(s,m)^{j-1}}(s_j) \sum_{m_j} p(m_j|(s,m)^j, s_j) \sum_{m \neq m_1,\ldots,m_j} \sigma_{(s,m)^j}(m)\mathrm{Payoff}(m;(s,m)^j)$$

is maximal. With these definitions we have a deterministic strategy which is optimal. □

For a deterministic strategy τ, σ we have

$$P_j^{\tau,\sigma} = \sum_{m_1} p(m_1|s_1) \sum_{m_2} p(m_2|(s_1,m_1),s_2) \cdots \sum_{m \neq m_1,\ldots,m_j} \mathrm{Payoff}(m;(s,m)^j)$$

where, for $l = 1,\ldots,j$, $\tau_{(s,m)^{l-1}}(s_l) = 1$ and $\sigma_{(s,m)^j}(m) = 1$. Thus $P_j^{\tau,\sigma} = \sum_{m^j} \Pr[\mathcal{E}_{(s,m)^j,(m,1)}]$ where $s^j = (s_1,\ldots,s_j)$ satisfies $\tau(s_1) = 1$, $\tau_{(s,m)^l}(s_{l+1}) = 1$ for $l = 1,\ldots,j-1$ and $\sigma_{(s,m)^j}(m) = 1$.

3.2. *A Bound on the Probability of Success*

We establish a bound on the probability of success of an adversary in the signature query model by finding the average value of $P_j^{\tau,\sigma}$ for deterministic strategies. Let $(\mathcal{S},\mathcal{M},\mathcal{E})$ be an authentication code with parameters (k,v,b). Let Δ_j be the set of deterministic strategies in the signature query model where the adversary spoofs after j queries. First we count the number of deterministic strategies.

Lemma 3.1.

$$\begin{aligned}|\Delta_j| &= k(k-1)^v(k-2)^{v(v-1)}\ldots(k-j+1)^{v\ldots(v-j+2)}(v-j)^{v\ldots(v-j+1)}\\ &= \left(\prod_{l=0}^{j-1}(k-l)^{\prod_{w=0}^{l-1}(v-w)}\right)(v-j)^{\prod_{w=0}^{j-1}(v-w)}\end{aligned}$$

Proof. There are k choices for s_1 to satisfy $\tau(s_1) = 1$. For each such choice there are v possibilities for m_1 and for each of these there are $k-1$ choices for s_2 (being distinct from s_1) to satisfy $\tau_{(s_1,m_1)}(s_2) = 1$. Thus there are $(k-1)^v$ possibilities for the choices of s_2 values in a deterministic strategy.

In general, for each choice of $s_1,\ldots,s_l$, $l = 1,\ldots,j-1$, there are $v(v-1)\ldots(v-l+1)$ possibilities for (the distinct) $m_1,\ldots m_l$ and for each of these there are $k-l$ choices for s_{l+1} (being distinct from $s_1,\ldots,s_l$) to satisfy $\tau^l_{(s,m)}(s_{l+1})=1$. This gives $(k-l)^{v(v-1)\ldots(v-l+1)}$ possibilities for the choices of s_l values.

Finally, for each choice of $s_1,\ldots,s_j$ there are $v(v-1)\ldots(v-j+1)$ possibilities for (the distinct) $m_1,\ldots,m_j$ and for each of these there are $v-j$ choices for m (being distinct from $m_1,\ldots,m_j$) to satisfy $\sigma_{(s,m)^j}(m)=1$, yielding $(v-j)^{v\ldots(v-j+1)}$ possibilities for the choices of m values.

Taking the product of these numbers of choices establishes the result □

We now show that the sum of $P_j^{\tau,\sigma}$ over all deterministic strategies is a value dependent only on the parameters k and v of the code $(\mathcal{S},\mathcal{M},\mathcal{E})$.

Lemma 3.2. $\sum P_j^{\tau,\sigma}=|\Delta_j|\frac{k-j}{v-j}$ *where the sum is over all deterministic strategies* τ,σ.

Proof. For a deterministic strategy τ,σ we have $P_j^{\tau,\sigma}=\sum_{m^i}\Pr[\mathcal{E}_{(s,m)^i,(m,1)}]$ where $s^j=(s_1,\ldots,s_j)$ satisfies $\tau(s_1)=1$, $\tau_{(s,m)^l}(s_{l+1})=1$ for $l=1,\ldots,j-1$ and $\sigma_{(s,m)^j}(m)=1$. For a given m^j let $\Delta_j((s,m)^j,m)$ denote the subset of deterministic strategies with $\tau(s_1)=1,\tau_{(s,m)^{l-1}}(s_l)=1$, $l=2,\ldots,j$ and $\sigma_{(s,m)^j}(m)=1$. By a similar argument to that given in the proof of Lemma 3.1, there are $(k-1)^{v-1}(k-2)^{v(v-1)-1}\ldots(k-j+1)^{v\ldots(v-j+2)-1}(v-j)^{v\ldots(v-j+1)-1}$ deterministic strategies in $|\Delta_j((s,m)^j,m)|$ independent of $(s,m)^j$ and m.

Now summing over deterministic strategies we obtain (where, for $\mathcal{E}_{(s,m)^j,(m,1)}$ below, s^j and m satisfy $\tau_{(s,m)^{l-1}}(s_l)=1$, for $l=1,\ldots,j$ and $\sigma_{(s,m)^j}(m)=1$).

$$\begin{aligned}
\sum_{\tau,\sigma\in\Delta_j}P_j^{\tau,\sigma} &= \sum_{\tau,\sigma\in\Delta_j}\sum_{m^j}\Pr[\mathcal{E}_{(s,m)^j,(m,1)}]\\
&=\sum_{m^j}\sum_{s^j,m}\sum_{\tau,\sigma\in\Delta((s,m)^j,m)}\Pr[\mathcal{E}_{(s,m)^j,(m,1)}]\\
&=|\Delta_j((s,m)^j,m)|\sum_{m^j}\sum_{s^j,m}\sum_{e\in\mathcal{E}_{(s,m)^j,(m,1)}}p(e)\\
&=|\Delta_j((s,m)^j,m)|\sum_{e\in\mathcal{E}}\sum_{s^j}\sum_{m\in\mathcal{M}:m=e(s);s\neq s_1,\ldots,s_j}p(e)\\
&=|\Delta_j((s,m)^j,m)|k(k-1)\ldots(k-j+1)(k-j)\sum_{e\in\mathcal{E}}p(e)\\
&=|\Delta_j|\frac{k-j}{v-j}
\end{aligned}$$

□

Corollary 3.1. $P_j \geq \frac{k-j}{v-j}$ *and if equality holds then* $P_j^{\tau,\sigma} = \frac{k-j}{v-j}$ *for every deterministic strategy* τ, σ.

Proof. Since P_j is greater than the average of $P_j^{\tau,\sigma}$ over all deterministic strategies τ, σ, the result follows immediately from Lemmas 3.1 and 3.2. □

3.3. *A Bound on the Size of* $\mathcal{E}$

We do not have a characterization of optimal codes for the signature query model as complete as that for the passive adversary model. However we make the following observations.

By Corollary 3.2 $P_j^{\tau,\sigma} = P_j^{\tau,\sigma'}$ for deterministic strategies τ, σ and τ, σ' which only differ in the spoofing strategy. This means that, given $(s,m)^j$, $\Pr[\mathcal{E}_{(s,m)^j,(m,1)}]$ is independent of m. Therefore $\Pr[\mathcal{E}_{(s,m)^j,(m,1)}] = \frac{k-j}{v-j}\Pr[\mathcal{E}_{(s,m)^j}]$. Thus whenever $\Pr[\mathcal{E}_{(s,m)^j}] \neq 0$, we have that, for all $m \neq m_1, \ldots, m_j$, $\Pr[\mathcal{E}_{(s,m)^j,(s,m)}] \neq 0$ for some $s \neq s_1, \ldots, s_j$. It follows that $|\mathcal{E}_{(s,m)^j}| \geq |\mathcal{E}_{(s,m)^j,(m,1)}|\frac{v-j}{k-j}$. Hence if $P_j = \frac{k-j}{v-j}$ for $j = 0, \ldots, i$ then we have, for all distinct $m_1, \ldots, m_{i+1}$, that $\Pr[\mathcal{E}_{(s,m)^{i+1}}] \neq 0$ for some distinct $s_1, \ldots, s_{i+1}$, That is $\mathcal{E}_{m^{i+1}} \neq \emptyset$. It follows that $|\mathcal{E}| \geq \frac{v(v-1)\ldots(v-i)}{k(k-1)\ldots(k-i)}$ as in the case for authentication codes with passive adversaries. We show, however, that equality cannot occur if $i > 0$ and $v > k$.

For $i = 0$ the code consisting of $\frac{v}{k}$ encoding rules which map the k source states to disjoint subsets of messages achieves equality. Indeed this is the only way equality may be achieved.

For $i > 0$, if $|\mathcal{E}| = \frac{v(v-1)\ldots(v-i)}{k(k-1)\ldots(k-i)}$ then, for any sequence m^{i+1} of distinct messages there is a unique sequence s^{i+1} of source states with $\mathcal{E}_{(s,m)^{i+1}} = \mathcal{E}_{m^{i+1}}$ consisting of the unique encoding rule that maps this sources state sequence to the message sequence. This also means that, as we noted in the case of passive observers, that the code corresponds to a $(i+1)-(v,k,1)$ design. Further, the code has the following property: for any sequence m^i of distinct messages there is a unique sequence s^i of source states with $\mathcal{E}_{(s,m)^i} = \mathcal{E}_{m^i}$ consisting of $\frac{v-i}{k-i}$ encoding rules. Indeed, the encoding rules of $\mathcal{E}_{(s,m)^i}$ restricted to the source states and messages not belonging to $(s,m)^i$ determine a code with structure similar to that of the code described above for the case $i = 0$. Now, by the property noted above, for all the encoding rules that map s^{i-1} to m^{i-1} there is only one preimage of m_i, (namely s_i), while there are $\frac{v-i}{k-i}$ messages with preimage s_{i+1}. This is impossible if $v > k$ as then $\frac{v-i}{k-i} \neq 1$ and this contradicts the fact that the design with parameters $(i+1)-(v,k,1)$ to which the code corresponds is also a design with parameters $i-(v,k,\frac{v-1}{k-1})$.

By considering codes with

$$|\mathcal{E}| > \frac{v(v-1)\dots(v-i)}{k(k-1)\dots(k-i)}$$

we can still obtain authentication codes in the signature query model with $P_i = \frac{k-j}{v-j}$ for $j = 0, \dots, i$. Put

$$\mathcal{M}^i(\tau, \sigma) = \{m^i | \mathcal{E}_{(s,m)^i} \neq \emptyset, \text{ where } \tau_{(s,m)^{l-1}}(s_l) = 1, l = 1, \dots, i\}.$$

So $\mathcal{M}^i(\tau, \sigma)$ is the collection of all possible responses (arising from all possible choices of encoding rule) to the strategy τ. Since $\mathcal{E}$ is the disjoint union of $\mathcal{E}_{(s,m)^i}$ for $m^i \in \mathcal{M}^i(\tau, \sigma)$ it follows that $|\mathcal{E}| \geq |\mathcal{M}^i(\tau, \sigma)| \frac{v-i}{k-i}$. When every sequence m^i of distinct messages arises under some encoding rule for this strategy then $|\mathcal{M}^i(\tau, \sigma)| = v(v-1)\dots(v-i+1)$ and $|\mathcal{E}| = v(v-1)\dots(v-i+1)\frac{v-i}{k-i}$. An authentication code $(\mathcal{S}, \mathcal{M}.\mathcal{E})$ with these parameters may be constructed from an authentication code $(\mathcal{S}, \mathcal{M}, \mathcal{E}')$ corresponding to a $(i+1) - (v, k, 1)$ design and a subset Π of permutations on $\mathcal{S}$ which is transitive on sequences s^i of distinct source states. The elements of $\mathcal{E}$ are pairs $e = (e', \pi) \in \mathcal{E}' \times \Pi$ where $e(s)$ for $s \in \mathcal{S}$ is defined by $e(s) = e'(\pi(s))$. For such a code, whichever deterministic strategy the adversary adopts is irrelevant as no matter what choice of source state the adversary makes there is an encoding rule that produces the same message. Thus the adversary learns no more from the queries than from passive observation of the authentication code $(\mathcal{S}, \mathcal{M}, \mathcal{E}')$.

3.4. *Verification queries*

In the verification query model the adversary chooses $m \in \mathcal{M}$ to transmit to the receiver and observes the response, accept or reject, by the receiver. Tonien *et al*[13] consider two scenarios which they call *on-line* and *off-line* attacks. These depend on the state of the receiver. In the on-line attack it is assumed that the receiver is reacting to all in-coming messages and so the adversary succeeds as soon as a message is sent that is accepted by the receiver. In the off-line attack, although the adversary obtains a response that indicates whether the message is accepted or rejected, an accept response does not represent a successful attack as it is assumed that the receiver is off-line and is not taking action on received messages. Instead the adversary must wait until the receiver is on-line and use the information gathered from the responses to submit a spoofing message at this later time.

Tonien *et al*[13] have considered these scenarios and used analysis of deterministic strategies to establish the following bounds. In the on-line attack a deterministic strategy determines the next message to be sent to the receiver should the previous one be rejected. There are $v(v-1)\dots(v-i)$ such strategies for sending $i+1$ messages. In the off-line attack a deterministic strategy determines the next message for each possible sequence of accept or reject responses to the previous messages. There are $v(v-1)^2(v-2)^{2^2}\dots(v-i)^{2^i}$ of these. In the on-line case, failure only occurs if all messages are rejected and the bound on the probability of success is $P_i \geq 1 - \frac{\binom{v-k}{i+1}}{\binom{v}{i+1}}$ while the bound on the probability of success in the off-line case is $P_i \geq \frac{k}{v}$ for each i, as this is a weighted average of the success probabilities for the various combinations of numbers of acceptance and rejection responses to the previous $i-1$ messages.

The conditions for equality for the two cases are equivalent and occur when, for all sequences of i distinct messages, all i messages are accepted with the same probability. This condition implies that $|\mathcal{E}| \geq \frac{\binom{v}{i+1}}{\binom{k}{i+1}}$ with equality exactly when the messages and the encoding rules determine a $(i+1)-(v,k,1)$ design as described for the passive adversary case.

4. Conclusion

We have considered the extension of the standard model of authentication to include an active adversary that can make queries to the sender and receiver. We have given a bound on the probability of success of the adversary that makes signature queries to the sender and considered the number of encoding rules in an authentication code that attains this bound. The probability of success of the active attacker is kept to that of a passive attacker by using an authentication code with more encoding rules. In both cases such authentication codes are related to combinatorial designs.

The bound on the probability of success of an attacker that makes verification queries depends on whether the receiver is on-line or off-line. However, in either case an authentication code that attains the bound is the same as in the case of a passive attacker.

References

1. E.N. Gilbert, F.J. MacWilliams and N.J.A. Sloane, Codes which detect deception, *Bell System Technical Journal* **53** (1974), 405–424.
2. J.L. Massey, Cryptography a selective survey, In *Digital Communications*, North Holland, pp. 321, 1983.
3. D. Pei, Information-theoretic bounds for authentication codes and block designs, *Journal of Cryptology* **8** (1995), 177–188.

4. U. Rosenbaum, A lower bound on authentication after having observed a sequence of messages, *Journal of Cryptology* **6** (1993), 135–156.
5. R. Safavi-Naini and P. Wild, Bounds on authentication systems in query model, *Proceedings of the 2005 IEEE Information Theory Workshop on Theory and Practice in Information-Theoretic Security*, pp. 85–91, 2005.
6. J. Shikata, G. Hanaoka, Y. Zheng, T. Matsumoto and H. Imai, Unconditionally Secure Authenticated Encryption , *IEICE Trans Fundam Electron Commun Comput Sci*, **E87-A** (2004), 1119-1131.
7. G.J. Simmons, Message authentication: a game on hypergraphs, *Congressus Numerantium* **45** (1984), 161–192.
8. G.J. Simmons, Authentication theory / coding theory, *CRYPTO'84, LNCS* **196**, pp. 411–432, 1995.
9. D.R. Stinson, Some constructions and bounds for authentication codes, *Journal of Cryptology* **1** (1988), 37–51.
10. D.R. Stinson, The combinatorics of authentication and secrecy codes, *Journal of Cryptology* **2** (1990), 23–49.
11. D.R. Stinson, Combinatorial characterizations of authentication codes, *Designs, Codes, and Cryptography* **2** (1992), 175–187.
12. R.S. Rees and D.R. Stinson, Combinatorial characterizations of authentication codes II, *Designs, Codes, and Cryptography* **7** (1996), 239–259.
13. D. Tonien, R. Safavi-Naini and P.R. Wild, Combinatorial characterizations of authentication codes in verification oracle model, *ASIACCS '07: Proceedings of the 2nd ACM Symposium on Information, Computer and Communications Security*, ACM, New York, pp. 183–193. 2007.
14. M. Walker, Information-theoretic bounds for authentication schemes, *Journal of Cryptology* **2** (1990), 257–263.

Collision in the DSA Function

Igor E. Shparlinski and Ron Steinfeld
Department of Computing, Macquarie University
Sydney, NSW 2109, Australia
{igor,rons}@ics.mq.edu.au

We study possible collisions among the values of the DSA function $f(s) = (g^s \operatorname{rem} p) \operatorname{rem} t$ where g is order t modulo a prime p and $n \operatorname{rem} k$ denotes the remainder of n on division by k. In particular, in a certain range of p and t we guarantee the existence of collisions and also give a nontrivial algorithm for inverting this function.

Keywords: DSA function, collisions

1. Motivation

Some cryptographic applications of digital signature schemes (such as authenticated key exchange protocols[7] and chosen ciphertext secure public key encryption schemes[2]) require the signature scheme to be *strongly* unforgeable. A randomised signature scheme is called strongly unforgeable if it is difficult for an adversary, given a set of valid message/signature pairs $(m_1, \sigma_1), \ldots, (m_n, \sigma_n)$, to compute a valid 'new' forgery message/signature pair (m^*, σ^*), that is, $(m^*, \sigma^*) \neq (m_i, \sigma_i)$ for $i = 1, \ldots, n$. Note that it is possible that such a forgery would have an 'old' message ($m^* = m_j$ for some $j \in \{1, \ldots, n\}$) but a 'new' signature ($\sigma^* \neq \sigma_i$ for $i = 1, \ldots, n$). This is contrast to the notion of a *weakly* unforgeable signature scheme we only allow forgeries where the message is 'new' ($m^* \neq m_i$ for $i = 1, \ldots, n$).

The Digital Signature Algorithm[12] (DSA) is an example of a signature scheme whose strong unforgeability may be easier to break than its (standard) weak unforgeability. We recall that a DSA public key is of the form (p, q, g, y), where p is a prime, q is a prime divisor of $p-1$, g is an element of order q in $\mathbb{Z}_p^*$ and $y = g^x \operatorname{rem} p$, where $n \operatorname{rem} k$ denotes the remainder of n on division by k. A valid DSA signature on a message m with respect to public key (p, q, g, y) consists of a pair $(\sigma_1, \sigma_2) \in \mathbb{Z}_q \times \mathbb{Z}_q$ satisfying $\sigma_2 = ((g^{H(m)} y^{\sigma_2})^{1/\sigma_1} \operatorname{rem} p) \operatorname{rem} q$, where $H(m)$ is the chosen cryptographic hash function.

We now explain why the strong unforgeability of DSA might be easier to break than its weak unforgeability. Given a valid DSA message/signature pair (m, σ_1, σ_2), let us define $h = g^{H(m)} y^{\sigma_2} \operatorname{rem} p$ and consider the mapping $f : \mathbb{Z}_q \to \mathbb{Z}_q$ which sends s to

$$f(s) = (h^s \operatorname{rem} p) \operatorname{rem} q,$$

so we have $f(1/\sigma_1) = \sigma_2$.

Heuristically, we expect that the mapping f behaves as a random function on $\mathbb{Z}_q$. Modelling f as a random function, we have for any given $s = 1/\sigma_1 \in \mathbb{Z}_q$, with high probability over the choice of f (namely with probability $1 - (1 - 1/q)^{q-1} \approx 0.63$, as q is large enough), that there exists a second distinct preimage $\widetilde{s} \neq s$ such that $f(\widetilde{s}) = f(s) = \sigma_2$. Evidently, if an attacker can find such a second preimage $\widetilde{s}$ for the mapping f then the attacker can break the strong unforgeability of DSA, by returning the 'new' valid message/signature forgery pair $(m, \widetilde{\sigma}_1, \sigma_2)$ with $\widetilde{\sigma}_1 = 1/\widetilde{s} \operatorname{rem} q$. Note that finding such *second preimages* for f does not seem sufficient for breaking the *weak* unforgeability of DSA, due to the cryptographic hash function applied to the message. We also note that the existence of collisions (two or more distinct preimages) in the mapping f means that DSA is not a *partitioned* signature scheme as defined in[1]. As a consequence, the transform presented in[1] for converting weakly unforgeable signatures into strongly unforgeable signatures (whose security proof assumes the weakly unforgeable signature scheme is partitioned) does not apply to DSA (instead, one can use the more general transforms in[3,15,16]).

2. Our Results

Motivated by the above cryptographic implications, we attempt to study the properties of the DSA mapping rigorously, without any heuristic assumptions. We actually look at a slightly more general function $f(s) = (g^s \operatorname{rem} p) \operatorname{rem} t$ where $g \in \mathbb{Z}_p^*$ has order t, and t is not necessarily prime. We prove that, under the condition $p^{3/4+\varepsilon} \leq t \leq (p-1)/2$ and for sufficiently large p, collisions in the DSA mapping f exist. We also show that the mapping f can be inverted using $O(p^{1/2})$ time and space using a variant of the baby-step giant-step method. Unfortunately, our assumed conditions on t mean that our results do not apply to DSA with typical parameter values $t \approx 2^{160}$ and $p \approx 2^{1024}$. Obtaining non-trivial results for such parameters is an open problem.

3. Preparations

Let p be a prime and let g be an integer with $\gcd(p, g) = 1$ of multiplicative order t modulo p.

Throughout the paper, we use the Landau symbols 'O' and 'o' as well as the Vinogradov symbols '$\ll$' and '$\gg$' with their usual meanings. We recall that $U = O(V)$, $U \ll V$ and $V \gg U$ are all three equivalent to the inequality $|U| \leqslant cV$ with some constant $c > 0$. All the implied constants are absolute.

For an integer n we denote by r_n the smallest positive residue modulo p of g^n, that is,

$$r_n = g^n \operatorname{rem} p.$$

We denote

$$\mathbf{e}_p(z) = \exp(2\pi i z/p).$$

and recall the identity

$$\sum_{|c| \leqslant (p-1)/2} \mathbf{e}_p(cu) = \begin{cases} 0, & \text{if } u \not\equiv 0 \pmod p, \\ p, & \text{if } u \equiv 0 \pmod p, \end{cases} \tag{1}$$

which follows from the formula for the sum of a geometric progression.

For a divisor $d|p-1$ we denote by $\mathcal{X}_d$ be the set of all d multiplicative characters modulo p of order d, that is the set of characters χ with $\chi^d = \chi_0$, where χ_0 is the principal character. Also, let $\mathcal{G}_d$ be the group of dth residue modulo p, that is $u \in \mathcal{G}_d$ if and only if $u \equiv v^d \pmod p$ for some integer $v \not\equiv 0 \pmod p$. We recall that

$$\sum_{\chi \in \mathcal{X}_d} \chi(u) = \begin{cases} d, & \text{if } u \in \mathcal{G}_d, \\ 0, & \text{if } u \notin \mathcal{G}_d, \end{cases} \tag{2}$$

see [6, Theorem 5.4].

We also need the following bound of incomplete exponential sums, which also follows from the formula for the sum of a geometric progression, see [4, Bound (8.6)].

Lemma 3.1. *The bound*

$$\left| \sum_{u=1}^{h} \mathbf{e}_p(au) \right| \ll \min\{h, p/|a|\},$$

holds for any integer a with $1 \leqslant |a| \leqslant (p-1)/2$ and any integer h.

The following bound is a special case of the Weil bound, see, for example, [11, Chapter 6, Theorem 3] and also [6, Comments to Chapter 5].

Lemma 3.2. *For any multiplicative character $\chi \neq \chi_0$ and integers a and c with* $\gcd(a,p) = 1$, *we have*

$$\left| \sum_{v=1}^{p-1} \chi(v^d + a)\mathbf{e}_p(cv^d) \right| \ll dp^{1/2}.$$

The following bound is a special case of several well-known and more general results which can be found in[8–10,13,14].

Lemma 3.3. *For any integer c with* $\gcd(c,p) = 1$, *we have*

$$\left| \sum_{n=1}^{t} \mathbf{e}_p(cg^n) \right| \leqslant p^{1/2}.$$

We now need to estimate the number of $n = 1, \ldots, t$ for which r_n belongs to a prescribed interval and $r_n + q \equiv r_m \pmod p$ for some $n = 1, \ldots, t$.

Lemma 3.4. *For any integers k and h with* $0 \leqslant k < k + h \leqslant p - 1$

$$\#\{n = 1, \ldots, t \mid r_n \in [k, k+h]\} = \frac{th}{p} + O(p^{1/2} \log p).$$

Proof. Using (1) we write

$$\begin{aligned}
\#\{n = 1, \ldots, t \mid r_n \in [k+1, k+h]\} \\
&= \frac{1}{p} \sum_{n=1}^{t} \sum_{u=1}^{h} \sum_{|c| \leqslant (p-1)/2} \mathbf{e}_p\left(c(g^n - u - k)\right) \\
&= \frac{1}{p} \sum_{|c| \leqslant (p-1)/2} \mathbf{e}_p(-ck) \sum_{n=1}^{t} \mathbf{e}_p\left(cg^n\right) \sum_{u=1}^{h} \mathbf{e}_p\left(-cu\right) = \frac{th}{p} + E,
\end{aligned}$$

where

$$|E| = \frac{1}{p} \sum_{0 < |c| \leqslant (p-1)/2} \left| \sum_{n=1}^{t} \mathbf{e}_p\left(cg^n\right) \right| \left| \sum_{u=1}^{h} \mathbf{e}_p\left(-cu\right) \right|. \qquad (3)$$

Substituting the bounds of Lemmas 3.1 and 3.3 in (3), we have

$$|E| \leqslant 2p^{1/2} \left(\sum_{c=1}^{(p-1)/2} \frac{1}{c} \right) \ll p^{1/2} \log p,$$

as required. □

4. Existence of Collisions

Theorem 4.1. *Let us fix an arbitrary* $\varepsilon > 0$. *If* $p^{3/4+\varepsilon} \leqslant t \leqslant (p-1)/2$ *and* p *is sufficiently large, then there are integers* n *and* m *with* $1 \leqslant n < m \leqslant t$ *and such that* $r_n \equiv r_m \pmod t$.

Proof. Let $\mathcal{S}$ be the sets of $n = 1, \dots, t$ such that $r_n < p - t$.

It is enough to show that the congruence $g^n + t \equiv g^m \pmod p$ has a solution in $n \in \mathcal{S}$ and $m = 1, \dots, t$.

Let N be the number of solutions to the above congruence.

Putting $d = (p-1)/t$, we see that N is equal to the number of $n \in \mathcal{S}$ for which $g^n + t \pmod p$ belongs to $\mathcal{G}_d$. Therefore, using (2), we write

$$N = \frac{1}{d} \sum_{n \in \mathcal{S}} \sum_{\chi \in \mathcal{X}_d} \chi(g^n + t) = \frac{\#\mathcal{S}}{d} + \frac{1}{d} \sum_{\substack{\chi \in \mathcal{X}_d \\ \chi \neq \chi_0}} \sum_{n \in \mathcal{S}} \chi(g^n + t).$$

By Lemma 3.4 we see that $\#\mathcal{S} = t(p-t)/p + O(p^{1/2} \log p)$, thus

$$N - \frac{t^2(p-t)}{p(p-1)} \ll \frac{t \log p}{p^{1/2}} + \frac{1}{d} \sum_{\substack{\chi \in \mathcal{X}_d \\ \chi \neq \chi_0}} \left| \sum_{n \in \mathcal{S}} \chi(g^n + t) \right|. \tag{4}$$

Furthermore, using (1), we write

$$\begin{aligned}
\sum_{n \in \mathcal{S}} \chi(g^n + t) &= \frac{1}{p} \sum_{n=1}^{t} \chi(g^n + t) \sum_{|c| \leqslant (p-1)/2} \sum_{u=1}^{p-t} \mathbf{e}_p\left(c(g^n - u)\right) \\
&= \frac{1}{p} \sum_{n=1}^{t} \chi(g^n + t) \sum_{|c| \leqslant (p-1)/2} \sum_{u=1}^{p-t} \mathbf{e}_p\left(c(g^n - u)\right) \\
&= \frac{1}{p} \sum_{|c| \leqslant (p-1)/2} \sum_{n=1}^{t} \chi(g^n + t) \mathbf{e}_p\left(cg^n\right) \sum_{u=1}^{p-t} \mathbf{e}_p\left(-cu\right).
\end{aligned}$$

Therefore

$$\left| \sum_{n \in \mathcal{S}} \chi(g^n + t) \right| \leqslant \frac{1}{p} \sum_{|c| \leqslant (p-1)/2} \left| \sum_{n=1}^{t} \chi(g^n + t) \mathbf{e}_p\left(cg^n\right) \right| \left| \sum_{u=1}^{p-t} \mathbf{e}_p\left(-cu\right) \right|. \tag{5}$$

By Lemma 3.2 we have

$$\left| \sum_{n=1}^{t} \chi(g^n + t) \mathbf{e}_p\left(cg^n\right) \right| = \frac{1}{d} \left| \sum_{v=1}^{p-1} \chi(v^d + a) \mathbf{e}_p(cv^d) \right| \leq p^{1/2}.$$

Plugging this bound in (5) and using Lemma 3.1, we get

$$\left|\sum_{n=1}^{t}\chi(g^n+t)\mathbf{e}_p\left(cg^n\right)\right| \le p^{1/2}\sum_{|c|\le(p-1)/2}\frac{1}{|c|} \ll p^{1/2}\log p.$$

Therefore, from (4) we conclude

$$N-\frac{t^2(p-t)}{p(p-1)} \ll \frac{t\log p}{p^{1/2}}+p^{1/2}\log p \ll p^{1/2}\log p.$$

Since $p^{3/4+\varepsilon}\leqslant t\le(p-1)/2$ we have

$$\frac{t^2(p-t)}{p(p-1)}\geqslant\frac{t^2}{2p}\geqslant p^{1/2+2\varepsilon}.$$

Hence $N>0$ for sufficiently large p, as required. □

5. Finding Collisions

Theorem 5.1. *The mapping $f(s)=(g^s \text{ rem} p) \text{ rem} t$ can be inverted using time and space $O(p^{1/2}(\log p)^{O(1)})$.*

Proof. We may assume that $t>p^{1/2}$ since otherwise, the result follows trivially from an exhaustive search algorithm. The inversion algorithm is given $r\in\mathbb{Z}_t$. Suppose there exists a preimage $s\in\mathbb{Z}_t$ such that $r=f(s)=(g^s \text{ rem} p) \text{ rem} t$. Hence, there exists an integer $0\le k<d$ (where as before $d=(p-1)/t$) such that $g^s \text{ rem} p=a+kt$. Letting $m=\lceil p^{1/2}\rceil$, write $s=u+mv$ for $0\le u<m$ and $0\le v<t/m$. We therefore have

$$g^u \text{ rem} p=(a+k\cdot q)g^{-mv} \text{ rem} p. \tag{6}$$

The inversion algorithm computes and stores a list $\mathfrak{L}$ at most $m=O(p^{1/2})$ pairs $(x,g^x \text{ rem} p)$ over over all possible candidates $0\le x<m$ for u. This can be done in time and space $O(p^{1/2}(\log p)^{O(1)})$.

Then it computes the at most $d\cdot t/m=O(p^{1/2})$ possible values of $(a+z\cdot q)g^{-my} \text{ rem} q$ over all possible candidates $0\le y<t/m$ and $0\le z<d$ for (u,k). Then, for each such pair (y,z) it uses a binary search in the list $\mathfrak{L}$ to find a possible match between $g^x \text{ rem} p$, $0\le x<m$, and $(a+z\cdot q)g^{-my} \text{ rem} q$. Note that this stage also can be done in time and space $O(p^{1/2}(\log p)^{O(1)})$, as required.

When such a match is found the algorithm recovers $s=x+my$. □

References

1. D. Boneh, E. Shen, and B. Waters, 'Strongly unforgeable signatures based on computational Diffie-Hellman', *Lect. Notes in Comp. Sci.*, Springer-Verlag, Berlin, **3958** (2006), 229–240.
2. R. Canetti, S. Halevi, and J. Katz, 'Chosen-Ciphertext security from identity-based encryption', *Lect. Notes in Comp. Sci.*, Springer-Verlag, Berlin, **3027** (2004), 207–222.
3. Q. Huang, D.S. Wong, and Y. Zhao, 'Generic transformation to strongly unforgeable signatures', *Lect. Notes in Comp. Sci.*, Springer-Verlag, Berlin, (to appear).
4. H. Iwaniec and E. Kowalski, *Analytic number theory*, American Mathematical Society, Providence, RI, 2004.
5. W.-C. W. Li, *Number theory with applications*, World Scientific, Singapore, 1996.
6. R. Lidl and H. Niederreiter, *Finite fields*, Cambridge University Press, Cambridge, 1997.
7. J. Katz and M. Yung, 'Scalable protocols for authenticated group key exchange', *Lect. Notes in Comp. Sci.*, Springer-Verlag, Berlin, **2729** (2003), 110–125.
8. S. V. Konyagin and I. E. Shparlinski, *Character sums with exponential functions and their applications*, Cambridge Univ. Press, Cambridge, 1999.
9. N. M. Korobov, 'On the distribution of digits in periodic fractions', *Matem. Sbornik*, **89** (1972), 654–670 (in Russian).
10. N. M. Korobov, *Exponential sums and their applications*, Kluwer Acad. Publ., Dordrecht, 1992.
11. W.-C. W. Li, *Number theory with applications*, World Scientific, Singapore, 1996.
12. A. J. Menezes, P. C. van Oorschot and S. A. Vanstone, *Handbook of applied cryptography*, CRC Press, Boca Raton, FL, 1997.
13. H. Niederreiter, 'Quasi-Monte Carlo methods and pseudo-random numbers', *Bull. Amer. Math. Soc.*, **84** (1978), 957–1041.
14. H. Niederreiter, *Random number generation and Quasi–Monte Carlo methods*, SIAM Press, 1992.
15. R. Steinfeld, J. Pieprzyk, and H. Wang, 'How to strengthen any weakly unforgeable signature into a strongly unforgeable signature', *Lect. Notes in Comp. Sci.*, Springer-Verlag, Berlin, **4377** (2007), 357–371.
16. I. Teranishi, T. Oyama, and W. Ogata, 'General conversion for obtaining strongly existentially unforgeable signatures', *Lect. Notes in Comp. Sci.*, Springer-Verlag, Berlin, **4329** (2006) 191–205.

The Current Status in Design of Efficient Provably Secure Cryptographic Pseudorandom Generators

R. Steinfeld*

Department of Computing, Macquarie University,
North Ryde, New South Wales 2109, Australia
** E-mail: rons@ics.mq.edu.au*
http://www.ics.mq.edu.au/~rons

Cryptographic Pseudorandom Generators (PRGs) are fundamental to symmetric key cryptography, and are widely used in practical applications. Unfortunately, the majority of PRG designs currently used in practice are 'ad hoc' constructions, whose security is not related to the difficulty of a well studied hard computational problem, and many such designs have been found to be insecure. In contrast, 'provably secure' PRG designs have a security reduction which shows that any efficient attack on the PRG can be converted to an efficient attack on a well studied hard computational problem. The reason why 'ad hoc' PRG designs are currently preferred over 'provably secure' designs in practice is that the former are currently much faster than the latter. However, several recent research results on efficient provably secure PRG design have narrowed this speed 'gap' significantly. In this paper we survey some of these recent constructions and their proven security reductions.

Keywords: pseudorandom bit generator, symmetric key cryptography, provable security, stream cipher design.

1. Introduction

PRGs and their Applications. The secrets used in cryptographic systems must be efficiently computable by legitimate system users, while at the same time being unpredictable by entities attacking the system. Consider, for example, the classical problem of symmetric key encryption: Two users, Alf and Bonnie, sharing a secret key $\kappa \in \{0,1\}^n$, want to communicate privately over an insecure channel which may be eavesdropped by an attacker Clive. To communicate a message $m \in \{0,1\}^n$, Alf can use the classical one-time pad encryption scheme, computing a ciphertext $c = m \oplus \kappa$ as the bitwise XOR of the message and key. Alf sends c over the insecure channel to Bonnie, who then can recover $m = c \oplus \kappa$. As first shown by Shannon[1], as long as κ is uniformly random and unknown to Clive, the ciphertext c gives Clive no information on the message m. Note that this scheme assumes that the message m is of the same length as the key κ. However, in typical applications, practical constraints on exchanging and storing the secret κ typically mean that the secret length n is quite small, whereas the length of the message ℓ that Alf and Bonnie want to communicate later may be much longer (for example, κ may be a short secret stored in a mobile phone, whereas m may be a sequence of phone calls made over the lifetime of the phone). What Alf and Bonnie need here is an efficient way of stretching the short random secret $\kappa \in \{0,1\}^n$ into a much longer random looking 'pseudorandom' string $\kappa' \in \{0,1\}^\ell$, which can then be XORed to the long message to form the ciphertext.

A cryptographic PseudoRandom bit Generator (PRG for short) is an algorithm for achieving the above goal. Informally, a PRG G is a *length expanding* function $G : \{0,1\}^n \to \{0,1\}^\ell$, taking a short *random* bit string called a *seed* $x \in \{0,1\}^n$ (typically n is in the range 80 to 5000 bits), and producing a long output bit string $s = G(x) \in \{0,1\}^\ell$ (Typically, ℓ can be chosen by the user to be arbitrarily large, in the range of Gigabits and beyond). What is the security ('pseudorandomness') requirement on G? Informally, we would like the output $G(x)$ to be a 'random looking' ℓ-bit string when the input seed $x \in \{0,1\}^n$ is a uniformly random n-bit string (throughout this paper we will use the notation $x \in_R \{0,1\}^n$ to denote that x is chosen uniformly at random from $\{0,1\}^n$). However, observe that, since G is a *deterministic* function with an n-bit input, the set $\{G(x) : x \in \{0,1\}^n\}$ contains at most 2^n ℓ-bit strings. Hence, since $n < \ell$, the probability distribution of $s = G(x)$ in $\{0,1\}^\ell$ will always be highly non-uniform, regardless of the particular choice of function for G. So we cannot hope to define 'pseudorandomness' as 'close to uniform'. Fortunately, in applications this non-uniformity of the distribution of $G(x)$ is not a problem as long as detecting this non-uniformity is *computationally intractable*. This is the basis for the complexity-based cryptographic definition of pseudorandomness, first proposed by Blum and Micali in the early 1980's[2], and the one adopted since then in the cryptographic literature: *A PRG is secure if it is computationally infeasible to distinguish between the output $s = G(x) \in \{0,1\}^\ell$ of the PRG (when fed with a uniformly random seed $x \in \{0,1\}^n$) and a uniformly random ℓ-bit string* (see Sec. 2 for a quantitative security definition).

PRGs have many cryptographic applications. Besides the particular application to symmetric key encryption described above (which is known as a *stream cipher* scheme), PRGs can be used to construct pseudorandom functions[3] and permutations[4] which in turn can be used to construct Message Authentication Codes (MACs) and block ciphers. PRGs are also widely used to generate secure pseudorandom strings for randomised public-key cryptographic schemes and protocols such as digital signatures and key exchange protocols.

PRG Construction Approaches: Provable vs. Heuristic. There are currently two distinct approaches to the design of a PRG: the 'Heuristic Approach' and the 'Provable Approach'.

The 'Heuristic' approach is the older and currently the dominant one in practice. In this approach, the PRG is based typically on 'custom' Boolean functions chosen to satisfy a list of necessary properties for avoiding certain types of known attacks (e.g. balance for avoiding statistical bias, high non-linearity for avoiding linear and differential cryptanalysis techniques, high algebraic immunity for resisting algebraic attacks). The resulting functions (and PRG) are usually very computationally efficient, and this is the main reason for their current adoption in practice. However, the security of such PRGs remains uncertain, and as new attack techniques are discovered, existing PRGs are rendered insecure and more necessary conditions must be added to the list to avoid the new attacks in new designs.

The more recent 'Provable Security' approach to design of PRGs originated in the early 1980's with the work of Blum and Micali[2]. In this approach, the designer bases the security of the PRG on a natural computationally intractable mathematical problem Π whose computational complexity has been studied by the mathematics/computer-science community for a significant period of time. The PRG is designed in such a way that a mathematical proof can be constructed for the pseudorandomness of the PRG assuming the intractability of the problem Π. The proof consists of a computational reduction, showing how any efficient distinguisher attack against the PRG can be converted into an efficient algorithm for solving the problem Π, thus contradicting the assumed intractability of Π. Unfortunately, the strong security guarantee provided by PRGs designed with this approach has traditionally come at the cost of relatively low computational efficiency: the first PRGs designed with this approach were slower than competing 'Heuristic Approach' designs by up to 5 orders of magnitude, and hence were not adopted in practice. Recent research has focused on design of more efficient provably secure PRGs, which has narrowed this efficiency 'gap' by roughly 2-3 orders of magnitude. We believe that closing this gap further forms an interesting and practically significant research challenge.

This Paper. In this paper we survey some recent efficient designs of provably secure cryptographic PRGs. Although we do not give full details of all proofs, we do attempt to explain the main ideas of central proofs to give the reader an understanding of the techniques employed. We begin in Sec. 2 with a quantitative definition of pseudorandomness used throughout this paper and review some general results useful in PRG design. In Sec. 3 we briefly review generic constructions of PRGs based on any given one-way permutation (or function), and explain the efficiency limitations of such constructions. The core of the survey is Sec. 4, where we review three efficient PRG constructions based on specific hard problems: a DL-based construction due to Gennaro[5], an RSA-based construction due to Steinfeld, Pieprzyk and Wang[6], and an MQ-based construction due to Berbain et al[7]. In Sec. 5, we compare the surveyed constructions. Finally, Sec. 6 concludes the paper.

2. Preliminary Definitions and Useful Lemmas

2.1. *Definition of PRG Families and their Security*

In many PRG constructions, the pseudorandomness security proof relies on the length expanding function G being chosen at random from a large *family* of functions $\{G_N\}_{N \in \mathcal{I}_n}$ (by specifying a function index N from index space $\mathcal{I}_n$). In principle, one could think of N as part of the input seed of the PRG. However, N can be revealed to the attacker without compromising the pseudorandomness of the PRG. Hence, we think of N as a public parameter (which in applications could be communicated over a public channel), and the input to G alone is considered the secret *seed.*

Definition 2.1 (PRG Family). *A PRG Family* G *is specified by the following:*

- *Bit length of random seed :* n
- *Bit length of output string:* ℓ *(where* $\ell > n$*).*
- *Family of efficiently computable PRG functions* $G_N : \mathcal{S}_N \rightarrow \{0,1\}^{\ell}$ *indexed by* $N \in \mathcal{I}_n$*, where:*
  - $\mathcal{I}_n \subseteq \{0,1\}^{poly(n)}$ *is PRG function index space*
  - $\mathcal{S}_N \subseteq \{0,1\}^n$ *is PRG seed domain*

The above definition does not say anything about security. The standard quantitative way to assess the pseudorandomness security of a PRG family, introduced by Blum and Micali[2], is to measure the efficiency and performance of the best algorithm for distinguishing between the pseudorandom string produced by the PRG and a truly uniformly random string of the same length. First we define a (T, δ) Distinguisher.

Definition 2.2 ((T, δ) Distinguisher). *A (T, δ) distinguisher algorithm* D *for PRG Family* G*:*

- *Runs in time at most T.*
- *Has distinguishing advantage* $\mathsf{Adv}_{\mathsf{D}}(\mathcal{D}_P, \mathcal{D}_R)$ *at least δ between the random distribution $\mathcal{D}_R$ and the pseudorandom distribution $\mathcal{D}_P$, where:*

$$\left[\mathcal{D}_P \stackrel{\text{def}}{=} \{(N, s) : N \in_R \mathcal{I}_n; x_0 \in_R \mathcal{S}_N; s = G_N(x_0)\},\right]$$
$$\left[\mathcal{D}_R \stackrel{\text{def}}{=} \{(N, s) : N \in_R \mathcal{I}_n; s \in_R \{0,1\}^\ell\}.\right]$$

The distinguishing advantage $\mathsf{Adv}_{\mathsf{D}}(\mathcal{D}_P, \mathcal{D}_R)$ *is defined as*
$$\left[\mathsf{Adv}_{\mathsf{D}}(\mathcal{D}_P, \mathcal{D}_R) \stackrel{\text{def}}{=} \left|\Pr_{(N,s) \leftarrow \mathcal{D}_P}[\mathsf{D}(N, s) = 1] - \Pr_{(N,s) \leftarrow \mathcal{D}_R}[\mathsf{D}(N, s) = 1]\right|\right]$$

This motivates the following measure of security for a PRG.

Definition 2.3 ((T, δ) secure PRG Family). *A PRG Family* G *is called (T, δ) secure if there does NOT exist a (T, δ) Distinguisher for* G.

Hence for a (T, δ) secure PRG Family, any distinguisher for G running in time at most T has distinguishing advantage at most δ. For example, a typical requirement might be $T = 2^{70}$ machine cycles on a specified computing machine, and $\delta = 1/100$ (a preferred stricter requirement might be (T, δ) security for *all* pairs (T, δ) with, say, $T/\delta < 2^{80}$).

Iterative PRG. Typical applications require the output length ℓ of a PRG G_N to be chosen arbitrarily by the user. To achieve this, the algorithm implementing G_N is constructed as a state machine which outputs the bits of $s = G_N(x)$ as a sequence of r-bit blocks, for some small r. The initial state of the machine x_0 is set equal to the PRG input seed $x \in \mathcal{S}_N$. When the machine is clocked, a deterministic state transition function $f_N : \mathcal{S}_N \to \mathcal{S}_N$ is applied to the current state x_i to compute the next state $x_{i+1} = f(x_i)$. Upon entering state x_i, the PRG outputs the r-bit block $s_i = g_N(x_i)$ using some output function $g_N : \mathcal{S}_N \to \{0,1\}^r$. Hence to generate an $\ell = m \cdot r$-bit psuedorandom string, the machine is clocked $m - 1$ times, the output being $(g_N(x_0), g_N(x_1), \ldots, g_N(x_{m-1}))$.

Definition 2.4 (Iterative PRG). *An iterative PRG Family* G *with output parameter r is a PRG family in which the PRG output length ℓ can be chosen as any integer multiple m of r. The PRG family index $N \in \mathcal{I}_n$ indexes a* state transition function $f_N : \mathcal{S}_N \to \mathcal{S}_N$ *and an* output function

$g_N : \mathcal{S}_N \rightarrow \{0,1\}^r$. *Given a seed* $x_0 \in \mathcal{S}_N$, *and integer* $m \geq 1$, *the PRG function* $G_N^m : \mathcal{S}_N \rightarrow \{0,1\}^\ell$ *for generating* $\ell = m \cdot r$ *output bits runs as follows:*

- *For* $i = 0, \ldots, m-1$, *compute* $(x_{i+1}, s_i) = (f_N(x_i), g_N(x_i))$, *and return* $s = (s_0, s_1, \ldots, s_{m-1}) \in \{0,1\}^\ell$.

2.2. *Useful General Lemmas in PRG Design*

The following general result (originating in the work of Blum and Micali[2], see also Ref. 8) lower bounds the pseudorandomness of any Iterative PRG family running for $m > 1$ iterations in terms of a pseudorandomness requirement on the (single iteration) length expanding function sending $x \in \mathcal{S}_N$ to $(f_N(x), g_N(x)) \in \mathcal{S}_N \times \{0,1\}^r$, and the number of iterations m. The reduction holds for *any* choice of functions f_N and g_N.

Lemma 2.1 ($m = \ell/r$ iterations to 1 iteration.[2]). *Let* G *be an Iterative PRG Family with output parameter* r, *state transition function* f_N *(with evaluation time* T_f*) and output function* g_N *(with evaluation time* T_g*). Any* (T, δ) *distinguisher for the* m *iteration PRG* G_N^m *can be converted into a* $(T + O(m \cdot (T_f + T_g)), \delta/m)$ *distinguisher* D' *between the distributions*

$$\mathcal{D}'_P = \{(N, y = f_N(x), s = g_N(x)) : N \in_R \mathcal{I}_n; x \in_R \mathcal{S}_N\}$$

and

$$\mathcal{D}'_R = \{(N, y, s) : N \in_R \mathcal{I}_n; y \in_R \mathcal{S}_N; s \in_R \{0,1\}^r\}.\}.$$

Note that since the security (maximal distinguisher advantage at a given run-time) deteriorates only linearly with m, the above result essentially allows the PRG designer to focus on the behaviour of the functions f_N and g_N after just a *single* iteration (although in practice the deterioration factor m does impact efficiency to some extent).

Another general result which finds widespread use in design of provably secure PRGs is the result of Yao[9], which shows that any efficient distinguisher having non-negligible distinguishing advantage between a probability distribution $\mathcal{D}_P$ on $\{0,1\}^k$ and the uniformly random distribution $\mathcal{D}_R$ on $\{0,1\}^k$, can be converted into an efficient *bit predictor* for strings from the distribution $\mathcal{D}_P$. The bit predictor is given (for some $i \in \{1, \ldots, k\}$) the first $j-1$ bits of a string $s \in \{0,1\}^k$ sampled from $\mathcal{D}_P$ and predicts the value of the next jth bit of s with probability exceeding $1/2$ by a non-negligible amount ε. Once again, there is a linear deterioration factor k relating ε to the advantage δ of the given distinguisher.

Lemma 2.2 (Distinguisher to Bit Predictor[9]). *Let $\mathcal{D}_P$ and $\mathcal{D}_R$ denote two probability distributions on $\{0,1\}^k$. Then any (T,δ) distinguisher algorithm* D *for distinguishing these two distributions can be converted into a jth bit predictor algorithm O_j (for some $j = 1,\ldots,k$) such that O_j runs in time $T+O(k)$ and given the first $j-1$ bits of s (for s sampled from $\mathcal{D}_P$), outputs a prediction for the jth bit of s which is correct with probability at least $1/2+\delta/k$.*

3. PRG Constructions from General One-Way Permutations/Functions

We briefly review the classical generic construction of PRGs from a given one-way permutation due to Blum and Micali[2] and Goldreich-Levin[10]. We also discuss some extensions and efficiency limitations of these generic constructions.

Blum-Micali reduction to Hard-Core Bits. Blum and Micali[2] suggested constructing an iterated PRG with $r = 1$, state space $\{0,1\}^n$ and using a family of *one-way permutations* $f_N : \{0,1\}^n \to \{0,1\}^n$ (i.e. f_N is a one-to-one mapping on $\{0,1\}^n$) for the state transition function. We recall[8] that informally, a one-way function $f : \mathcal{S}_N \to \mathcal{S}_N$ can be efficiently evaluated (given x it is easy to compute $f(x)$) but there does not exist an efficient inversion algorithm that given $y = f(x)$ (for a uniformly random $x \in_R \mathcal{S}_N$) outputs a preimage x' such that $f(x') = y$ with non-negligible probability. We say that a one-way function is B secure if any inversion algorithm running in time T and success-probability ε has $T/\varepsilon \geq B$.

Observe that if f_N is a one-way *permutation* then $f_N(x)$ is uniformly distributed in $\mathcal{S}_N$ when x is uniformly distributed in $\mathcal{S}_N$. So in this case we obtain a relaxation of Lemma 2.1.

Corollary 3.1. *If f_N is a permutation, then Lemma 2.1 holds with distribution $\mathcal{D}'_R$ replaced with distribution*

$$[\mathcal{D}''_R = \{(N, y = f_N(x), s) : N \in_R \mathcal{I}_n; x \in_R \mathcal{S}_N; s \in_R \{0,1\}^r\}.]$$

Hence the security of the iterated PRG with a permutation transition function f_N and output function g_N is guaranteed as long as it is infeasible to distinguish between the distributions $\mathcal{D}'_P = \{(N, y = f_N(x), s = g_N(x)) : N \in_R \mathcal{I}_n; x \in_R \mathcal{S}_N\}$ and $\mathcal{D}''_R = \{(N, y = f_N(x), s) : N \in_R \mathcal{I}_n; x \in_R \mathcal{S}_N; s \in_R \{0,1\}^r\}$. An output function g_N which has this property relative to a given one-way function f_N is called a *hard core function* of f_N (if $r = 1$, the boolean function g_N is called a hard core bit of f_N). Goldreich and Levin[10] gave a simple construction for a hard core function (for small r) that can be applied to *any* given one-way function f_N, after a simple modification. The modification of f_N consists of simply adding an additional $r \cdot n$ bits to the domain of f_N, which are not modified by the new function; i.e. define $f'_N : \{0,1\}^n \times \{0,1\}^{r \cdot n} \rightarrow \{0,1\}^n \times \{0,1\}^{r \cdot n}$ by $f'_N(x, M) = (f_N(x), M)$. Here we state an improved result due to Håstad and Näslund[11]:

Theorem 3.1 (r-bit Golreich-Levin Hard-Core Function[10,11]). *Fix $N \in \mathcal{I}_n$ and $r < n$. For a $r \times n$ matrix M over $GF(2)$, the r-bit* Goldreich-Levin Hard-Core Function $GL_M^{(r)} : GF(2)^n \rightarrow GF(2)^r$ *sends an input vector $x \in GF(2)^n$ to the vector $GL_M^{(r)}(x) = M \cdot x$ (matrix-vector multiplication over $GF(2)$). Then any (T, δ) distinguisher between the distributions*
$\left[\mathcal{D}_{P,N} = \{(y = f_N(x), M, s = GL_M^{(r)}(x)) : x \in_R \mathcal{S}_N, M \in_R \{0,1\}^{r \times n}\}\right]$ *and*
$[\mathcal{D}_{R,N} = \{(y = f_N(x), M, s) : x \in_R \mathcal{S}_N; M \in_R \{0,1\}^{r \times n}; s \in_R \{0,1\}^r\}]$
(with $T = \Omega(n + \log(\delta^{-1}))$) can be converted to an inversion algorithm that runs in time T_I and given $y = f_N(x)$ for a uniformly random $x \in \mathcal{S}_N$, outputs a preimage of y under f with probability at least ε_I, where $[T_I = O(2^r n^2 \delta^{-2} \cdot T)$ *and $\varepsilon_I = 1/2$.*]

Given any one-way permutation f_N, we can apply Theorem 3.1 together with Corollary 3.1 to construct a provably secure iterative PRG outputting r bits per iteration for small r (we refer the reader to Ref. 11 for details). The limitation on the size of r comes from the exponential factor 2^r in the reduction time/success ratio $\frac{T_I}{\varepsilon_I}$; namely, if f_N is B secure, then Theorem 3.1 only guarantees (T, δ) security for the PRG if $T_I/\varepsilon_I < B$, which leads to the 'small r' condition $r < \log(\frac{B}{n^2 \delta^{-2} \cdot T}) + O(1)$. For example, for subexponentially secure functions f_N with $B = 2^{o(n)}$, Theorem 3.1 only allows $r = o(n)$, significantly limiting performance of the PRG.

Before we leave the topic of generic PRG constructions, we make a few final remarks. There has been much research in relaxing the need for f_N to be *permutation* in the above construction. A celebrated theoretical result[12] (recently improved in Ref. 13) works with any one-way function f_N but is significantly less efficient than the above construction. Alternatively, one can get better efficiency by making a slightly stronger requirement on f_N than just one-wayness; in Ref. 11 it is shown that the same construction above can be proven secure as long as f_N is hard to invert when the challenge value y to be inverted is distributed according to the output distribution of the i-times iterated transition function $f_N^i(x)$ (for some i) with uniformly random $x \in S_N$. As pointed out in Ref. 11, applying this result to fast and exponentially secure 'heuristic' candidate one-way functions (e.g. based on the AES) allows constructing quite efficient PRGs whose security is provable based on a reasonable inversion (one-wayness) assumption on f_N. This is a useful 'intermediate' approach between the 'heuristic' and 'provable' design approaches.

4. Efficient PRG Constructions from Specific One-Way Functions

In this section we review recent efficient PRG constructions which are based on specific well-studied one-way functions. These constructions significantly improve upon the efficiency of generic constructions based on the same one-way functions by a delicate combination of two methods: (1) Identifying efficient variants of well known one-way functions which have themselves been well studied, and (2) Exploiting specific properties of the one-way functions which allow for efficient security reductions.

4.1. *'Short Exponent' Discrete Log Based PRG*

We begin with the Discrete-Log (DL) based construction of Gennaro[5], which improves upon previous work by Patel and Sundaram[14]. The classical DL one-wayness problem is the following: given an n-bit prime p, a generator g of the multiplicative group $\mathbb{Z}_p^*$, and $y = f_{DL}(x) \stackrel{\text{def}}{=} [g^x]_p$ for a random $x \in \mathbb{Z}_{p-1}$, find x (for the remainder of the paper we use $[z]_p$ to denote $z \bmod p$). The DL problem has been extensively studied in cryptography since the seminal paper of Diffie and Hellman[15] and is widely believed to be hard; the best known attack is Number Field Sieve (NFS), which runs in subexponential time $2^{O(n^{1/3} \log^{2/3} n)}$. Consequently, due to the 2^r security reduction factor discussed in the previous section, the generic PRG constructed from f_{DL} can output at most $r = O(n^{1/3} \log^{2/3} n)$ bits per iteration while maintaining a security proof relative to the DL problem. The cost of each iteration is dominated by time T_f needed to evaluate the exponentiation function f_{DL}, which is proportional to the bit length n of the exponent x, taking $\Theta(n)$ n-bit modular multiplications. So the throughput of the generic PRG constructed from f_{DL} is at most $O(1/(n/\log n)^{2/3})$ output bits per n-bit Modular Multiplication (MM). In contrast, Gennaro's DL-based PRG to be described below achieves a rate up to $O((n/\log n)^{2/3})$ bits per n-bit MM.

The first step towards the improved DL-based PRG is the observation that variants of the DL problem in which the random exponent x is chosen to be 'small', i.e. $x < 2^c$ for some $c < n$, have also been well studied in cryptography. We call this problem the "Short Exponent" Discrete Log (c-DLSE), and the associated one-way function $f_{DLSE} : \mathbb{Z}_{2^c} \to \mathbb{Z}_p$ sending $x \in \mathbb{Z}_{2^c}$ to $f_{DLSE}(x) = [g^x]_p$. Apart from standard DL attacks, the best known attack against c-DLSE which makes use of the smallness of c is Shanks' 'Baby-Step/Giant-Step' which has a run time $O(2^{c/2})$ (some better attacks[16] are known for a random modulus p, but they are easily avoided by using a random *safe* prime p, for which $(p-1)/2$ is also prime). Hence, recalling the NFS run-time, we can choose $c = \Theta(n^{1/3} \log^{2/3} n)$ without reducing security relative to f_{DL} (according to known attacks). Thanks to the smallness of c, the function f_{DLSE} can be evaluated faster than f_{DL} by a speedup factor $n/c = \Theta((n/\log n)^{2/3})$.

A first efficient PRG construction from f_{DLSE} was given by Patel and Sundaram[14]. They exploited the homomorphic properties of f_{DL} to show that the string of $n - c$ Least Significant (LS) bits of x (excluding the least significant bit) is a hard core function of f_{DL}, assuming that f_{DLSE} is hard to invert. Moreover, the reduction is quite efficient. We now sketch the main ideas of this result. Below we use the notation $(x_n, \ldots, x_1)$ to denote the binary representation of an n-bit integer x (i.e. $x = \sum_{i=1}^{n} x_i \cdot 2^{i-1}$).

Theorem 4.1 (Patel-Sundaram[14]). *Consider the length expanding function* f_{PS} : $\mathbb{Z}_{p-1} \to \mathbb{Z}_p^* \times \{0,1\}^{n-c-1}$ *defined by* $[f'(x) = ([g^x]_p, (x_{n-c}, x_{n-c-1}, \ldots, x_2).]$ *Any* (T, δ) *distinguisher* D *between the distributions* $\mathcal{D}_P = \{f'(x) = ([g^x]_p, (x_{n-c}, x_{n-c-1}, \cdots, x_2)) : x \in_R \mathbb{Z}_{p-1}\}$ *and* $\mathcal{D}_R = \{(y, r) : y \in_R \mathbb{Z}_p^*; r \in_R \{0,1\}^{n-c-1}\}$ *can be converted into an inversion algorithm against the c-DLSE problem with run-time* T_I *and success probability* ε_I*, where* $[T_I = O(c \log(c) \delta^{-3}) \cdot (T + O(n^3))$ *and* $\varepsilon_I = 1/2.]$

Proof. (Sketch of Main Ideas) First, we apply Lemma 2.2 to convert the distinguisher D to an ith bit predictor O_i for some $i \in \{2, \ldots, n-c\}$, such that given $([g^x]_p, (x_2, x_3, \ldots, x_{i-1}))$, O_i predicts x_i with probability at least $1/2 + \delta'$ for $\delta' = \delta/(n-c-1)$ and runs in time $T + O(n)$.

Next, we convert O_i to a c-DLSE inversion algorithm A that, given c-DLSE instance $y = [g^{(0,\cdots,0,x_c,x_{c-1},\cdots,x_1)}]_p$, recovers x bit by bit. For each $j = 1, \ldots, c$, A computes an estimate $\widehat{x}_j$ for bit x_j as follows:

- If $j = 1$, compute $\widehat{x}_j = \left(\frac{y}{p}\right)$. If $j > 1$, assuming we already know $x_1, \ldots, x_{j-1}$, compute:

$$y[j] = [(y \cdot g^{-(0,\cdots,0,x_{j-1},x_{j-2},\cdots,x_1)})^{2^{i-j}}]_p$$

$$= [g^{(0,\cdots,0,x_c,x_{c-1},\cdots,x_j,0,\cdots,0)}]_p,$$

i.e. *zero out* known $j-1$ LS bits of x (by multiplying by $g^{-(0,\cdots,0,x_{j-1},x_{j-2},\cdots,x_1)}$) and *shift* bit x_j to the ith exponent bit position (by raising to the 2^{i-j}th power). Note that if $i-j<0$ this involves taking $i-j$ square-roots modulo p. Since p is prime, this can be done efficiently.

- Run ith bit predictor O_i on m *randomised inputs* of the form $(y[j,k],(r[k]_2,\ldots,r[k]_{i-1}))$, where $y[j,k]=[y[j]\cdot g^{r[k]}]_p$ for $m=\Theta(\log(c)\delta^{-2})$ independent uniformly random $r[k]$'s in $\mathbb{Z}_{p-1}$ ($k=1,\ldots,m$). Since $r[k]$ is uniform in $\mathbb{Z}_{p-1}$, each randomised $y[j,k]$ is uniformly random in $\mathbb{Z}_p^*$ (as expected by O_i). Furthermore, since we have zeroed out the $i-1$ LS bits in the exponent $x[j]$ of $y[j]=[g^{x[j]}]_p$, the bits $r[k]_2,\ldots,r[k]_{i-1}$ match the bits in positions $2,\ldots,i-1$ of $[x[j]+r[k]]_{p-1}$ (as expected by O_i), unless there was a 'wraparound' modulo $p-1$ (i.e. $x[j]+r[k]\geq p-1$). Assume that $i<n-c-u$ for $u=\log(1/(\delta'/2))$. Since the most significant $n-(i+c)>u$ bits of $x[j]$ are zero, the chance that $x[j]+r[k]\geq p-1$ is at most $1/2^u=\delta'/2$. Hence the prediction bit $b[k]$ returned by O_i on the kth query is correct with probability at least $1/2+\delta'-\delta'/2=1/2+\delta'/2$ and it that case (recalling again the zero $i-1$ LS bits of $x[j]$) $b[k]=x_j\oplus r[k]_j$, so the estimate $\widehat{x}_j[k]=b[k]\oplus r[k]_j$ for x_j is correct with probability $1/2+\delta'/2$. A standard probability argument now shows that taking for $\widehat{x}_j$ the majority vote over the $m=\Theta(\log(c)\delta^{-2})$ independent estimates $\widehat{x}_j[k]$ gives the correct value of x_j with probability at least $1-O(1/c)$.

Hence at the end of this process A obtains all c bits of x with constant probability, as required (the method for dealing with the case $i>n-c-u$ involves guessing u MS of x and zeroing them out before running the above procedure – this multiplies the run-time in the worst case by a factor of $2^u=O(1/\delta')$). □

The iterative PRG obtained from Theorem 4.1 (with state transition function f_{DL} and output function g_N sending $x\in\mathbb{Z}_{p-1}$ to $(x_{n-c},\ldots,x_2)\in\{0,1\}^{n-c-1}$) outputs $r=n-c=\Omega(n)$ bits per iteration (with $c=O(n^{1/3}\log^{2/3}n)$ as before), a major improvement over the $O(n^{1/3}\log^{2/3}n)$ bits output by the generic construction. However, the state transition function f_{DL} still takes $O(n)$ modular multiplications to evaluate. Gennaro[5] improved this construction significantly by observing that the portion of the computation in evaluating f_{DL} which depends on the output bits

$(x_{n-c}, \ldots, x_2)$ is redundant, i.e. one can replace the state transition function $f_{DL}(x) = [g^{(x_n, \ldots, x_{n-c+1}, x_{n-c}, \ldots, x_2, x_1)}]_p$ by the much more efficient function $f_{Gen}(x) = [g^{(x_n, \ldots, x_{n-c+1}, 0, \ldots, 0, x_1)}]_p$, in which the bits of x in positions $2, \ldots, n-c$ are zeroed out. Note that by precomputing $\bar{g} = [g^{2^{n-c}}]_p$, the function can be evaluated as $f_{Gen}(x) = [\bar{g}^{(x_n, \ldots, x_{n-c+1})} \cdot g^{x_1}]_p$, which takes only $\Theta(c) = \Theta(n^{1/3} \log^{2/3} n)$ MMs, while still outputing $n - c - 1$ bits per iteration, a speedup factor of $\Theta((n/\log n)^{2/3})$ over the Patel-Sundaram construction. The fact that PRG security is preserved by the switch from f_{DL} to f_{Gen} is shown by the following Lemma[5] (together with Lemma 2.1).

Lemma 4.1 (Gennaro[5]). *Any (T, δ) distinguisher* D *between the distributions* $[\mathcal{D}_{Gen} = \{(f_{Gen}(x) = [g^{(x_n, \ldots, x_{n-c+1}, 0, \ldots, 0, x_1)}]_p, (x_{n-c}, \cdots, x_2)) : x \in_R \mathbb{Z}_{p-1}\}]$ *and* $[\mathcal{D}_R = \{(y, x) : y \in_R \mathbb{Z}_p^*; x \in_R \{0,1\}^{n-c-1}\}]$ *can be converted into* $(T + O(n^3), \delta)$ *distinguisher* D′ *between the distributions* $\mathcal{D}_{PS} = \{(f_{DL}(x) = ([g^x]_p, (x_{n-c}, \cdots, x_2)) : x \in_R \mathbb{Z}_{p-1}\}$ *and* $\mathcal{D}_R$.

Proof. Given input $(y, (\widehat{x}_{n-c}, \ldots, \widehat{x}_2))$, distinguisher D′ computes $\left[y' = \left[\frac{y}{g^{(0, \cdots, 0, \widehat{x}_{n-c}, \cdots, \widehat{x}_2, 0)}} \right]_p \right]$ and runs D on input $(y', (\widehat{x}_{n-c}, \ldots, \widehat{x}_2))$, returning whatever D returns. It is easy to verify that the transformation computed by D′ maps distribution $\mathcal{D}_{PS}$ to $\mathcal{D}_{Gen}$ while mapping $\mathcal{D}_R$ to itself, as required. □

(n, c)-DLSEPRG: DL Based Iterative PRG Family[5]

Index Space	$\{(p, g) : p$ is n-bit safe prime, $g \in \mathbb{Z}_p^*\}$
State Space	$\mathbb{Z}_p^*$
Transition Function	$f_{Gen}(x) = [g^{(x_n, \ldots, x_{n-c+1}, 0, \ldots, 0, x_1)}]_p$
Output Function	$g_{Gen}(x) = (x_{n-c}, \ldots, x_2)$

Combining Theorem 4.1, Lemma 4.1 and Lemma 2.1 gives the following security result.

Theorem 4.2 (DLSEPRG Security[5]). *Any (T, δ) distinguisher* D *for (n, c, ℓ)*-DLSEPRG *can be converted into a (T_I, ε_I) inversion algorithm* A *for the c-DLSE problem with*

$$[T_I = O(c \log(c) \delta^{-3} \ell^3) \cdot (T + O(n^3)) \text{ and } \varepsilon_I = 1/2].$$

We conclude that using $c = \Theta(n^{1/3} \log^{2/3} n)$, the DLSEPRG can achieve an asymptotic rate up to $\Theta((n-c)/c) = O((n/\log n)^{2/3})$ output per multiply modulo an n-bit modulus, with security level $T/\delta = 2^{\Theta(n^{1/3} \log^{2/3} n)}$ comparable to security of DL with n-bit modulus, assuming $\ell = poly(n)$.

Note that this asymptotic estimate does not show the effect of the reduction cost factor $O(c \log c\delta^{-3}\ell^3)$, which for large ℓ and small δ can be significant, and influences efficiency for a given provable security level (the numerical efficiency comparison in Section 5 takes this cost factor into account). Finally, we remark that alternative constructions with comparable efficiency to DLSEPRG but based instead on the hardness of the *Decisional Diffie-Hellman* (DDH) problem in $\mathbb{Z}_p^*$ (which also seems to be as hard as the DL problem) have been presented recently in Ref. 17.

4.2. *'Small Solution' RSA Based PRG*

The RSA one-way permutation $f_{RSA}(x) = [x^e]_N$ for an RSA modulus $N = pq$ (with prime p, q and small e coprime to $\phi(n)$) and public exponent e is another attractive candidate for PRG generation. Similarly to the case of the DL problem, the problem has been well studied and the best known attack on the RSA problem (via factoring the modulus N) has subexponential run time $2^{O(n^{1/3}\log(n)^{2/3})}$, so the generic PRG construction can output at most $r = O(n^{1/3}\log(n)^{2/3})$ bits per iteration. So again we are motivated to find a specific PRG construction which has a security proof for larger values of r. In fact, our motivation is even more powerful in this case than for DL, since f_{RSA} with a small (constant) exponent e takes only $O(1)$ n-bit MMs to evaluate (versus $\Omega(n^{1/3}\log^{2/3} n)$ n-bit MMs for f_{DL} at a comparable security level).

We now describe the efficient RSA-Based PRG construction of Steinfeld, Pieprzyk and Wang[6]. The first observation leading to the efficient construction is analogous to the one by Patel and Sundaram for the DL problem - i.e. make use of the hardness of certain special variants of the RSA inversion problem, in which the input x is 'small'. In particular, we will use a problem called (δ, e)-'Small Solution RSA' ((δ, e)-SSRSA). In this problem, one is a given a random RSA modulus N, a small exponent e coprime to $\phi(N)$, a polynomial $f(x) = a_e x^e + a_{e-1}x^{e-1} + \ldots + a_1 x + a_0 \in \mathbb{Z}_N[x]$ of degree e, and the value $y = f(x)$ for a random 'small' integer $x < N^\delta$ for some constant $0 < \delta < 1$. The problem is to recover x. The (δ, e)-SSRSA problem has been well studied in cryptography for over a decade, beginning with the seminal work of Coppersmith[18], who showed how to use an efficient lattice reduction algorithm (such as LLL[19]) to solve (δ, e)-SSRSA in time polynomial in n when $\delta \leq 1/e$. But despite significant attempts[20–22], no efficient algorithm is known when $\delta > 1/e + \varepsilon$ for constant $\varepsilon > 0$. The best known attack which makes use of the 'smallness' of x is to guess the $\varepsilon \cdot n$ most signficant bits of x and apply the Coppersmith attack on the resulting

$(1/e, e)$-SSRSA instance, for each possible guess. This attack requires time $\Omega(2^{\varepsilon \cdot n})$ and hence choosing a sufficiently large $\varepsilon = O((\log n/n)^{2/3})$ makes this attack as hard as factoring N.

But how can we use the hardness of $(1/e + \varepsilon, e)$-SSRSA to obtain an efficient provably secure PRG? As it turns out, the construction[6] differs only in the setting of parameters and in the security proof, from the well known Blum Blum Shub and RSA based iterative PRGs analysed by many researchers in the last 25 years[23–27]. The RSA iterative PRG uses the RSA permutation f_{RSA} as the state transition function, while the output function g_{RSA} returns the r Least Significant (LS) bits of the state value. Prior to the work of Steinfeld et al[6], the best known reductions for the security of this PRG had, similarly to the security reductions for the generic constructions, a reduction cost factor $O(2^{2r})$ which is exponential in r, relative to the RSA problem, allowing at most $r = O(n^{1/3}(\log n)^{2/3})$. Taking a closer look at the reduction of Fischlin-Schnorr[27], Steinfeld et al[6] observed that the exponential cost factor 2^{2r} can be easily eliminated by reducing from a (not well studied) variant of the RSA problem called (n, e, r, k, l)-FSRSA, instead of the RSA problem. To define the problem, we introduce the following notation: for an integer $x \in \mathbb{Z}_N$, $\widehat{M}_{N,k}(x)$ denotes any approximation of x with additive error $|\widehat{M}_{N,k}(x) - x| \leq N/2^k$ (roughly speaking, this means that $\widehat{M}_{N,k}(x)$ provides the k most significant bits of x), while $L_r(x) \stackrel{\text{def}}{=} x \bmod 2^r$ denotes the r least significant bits of x.

(n, e, r, k, l)-FSRSA Problem. Recover $x \in_R \mathbb{Z}_N$ given an RSA modulus N and $y = [x^e]_N$, $a \in_R \mathbb{Z}_N$, $s_1 = L_r([ax]_N)$, $u_1 = \widehat{M}_{N,k}([ax]_N)$, $b \in_R \mathbb{Z}_N$, $s_2 = L_r([bx]_N)$, $u_2 = \widehat{M}_{N,l}([bx]_N)$.

The variant of the Fischlin-Schnorr reduction relative to FSRSA can be stated as follows Below we let $\ell_j(y)$ denote the j LS bit in the binary representation of integer y, and we let $L_j(y)$ denote the j LS bits of y.

Theorem 4.3 (Fischlin-Schnorr[27]). *Fix RSA modulus N. Consider the length expanding function $f_{RE} : \mathbb{Z}_N \to \mathbb{Z}_N \times \{0,1\}^r$ defined by $[f_{RE}(x) = ([x^e]_N, L_r(x)).]$ Any (T, δ) distinguisher D between the distributions $\mathcal{D}_P = \{f_{RE}(x) = ([x^e]_N, L_r(x)) : x \in_R \mathbb{Z}_N\}$ and $\mathcal{D}_R = \{(y, s) : y \in_R \mathbb{Z}_N; s \in_R \{0,1\}^r\}$ can be converted into an inversion algorithm against the (n, e, r, k, l)-FSRSA problem (with $k = 3\log(r/\delta) + 4$ and $l = \log(r/\delta) + 4$) having run-time T_I and success probability δ_I (over the random choice of a, b), where $[T_I = O(n \log n \delta^{-2}) \cdot (T + O(n^2))$ and $\varepsilon_I = 1/9.]$*

Proof. (Sketch of Main Ideas.) First, apply Lemma 2.2 to convert the distinguisher D to an jth bit predictor O_j for some $j \in \{1, \dots, r\}$, such that given $([x^e]_N, (r_1, r_2, \dots, r_{j-1}))$, O_j predicts r_j with probability at least $1/2 + \delta_p$ for $\delta_p = \delta/r$ and runs in time $T + O(n)$.

Next, we convert O_j to a (n, e, r, k, l)-FSRSA inversion algorithm A that, given FSRSA instance $y = [x^e]_N, a \in_R \mathbb{Z}_N, s_1 = L_r([ax]_N), u_1 = \widehat{M}_{N,k}([ax]_N), b \in_R \mathbb{Z}_N, s_2 = L_r([bx]_N), u_2 = \widehat{M}_{N,l}([bx]_N)$, recovers x. The reduction proceeds in n iterations $t = 1, \dots, n$ in a process called *binary division*. In this process, upon entering iteration t, we assume that we already know from the previous iteration the values $s_{t-1} = L_j([2^{-(t-1)}ax]_N)$ and an approximation u_{t-1} of $[2^{-(t-1)}ax]_N$ with absolute error at most $N/2^t$. In iteration t we compute the updated values (s_t, u_t) ready for the next iteration. Thus after n iterations we have u_n which approximates $[2^{-n}ax]_N$ within absolute error $1/2$. So rounding u_n to the nearest integer gives the exact value of $[2^{-n}ax]_N$, from which x can easily recovered (as n and a are known). We note that initially $t = 1$ and setting $(s_0, u_0) = (L_j(s_1), u_1)$ using the given FSRSA inputs s_1, u_1 satisfies the requirements. Then for $t = 1, \dots, n$, iteration t works as follows:

- Use $L_j([2^{-(t-1)}ax]_N)$ to compute $L_{j-1}([2^{-t}ax]_N)$ using the relation $L_{j-1}([2^{-t}ax]_N) = 1/2 L_j(L_j([2^{-(t-1)}ax]_N) + \ell_1([2^{-(t-1)}ax]_N)N)$.
- Estimate the jth bit of $[2^{-t}ax]_N$ (denoted $\ell_j([2^t ax]_N)$) with error probability $O(1/n)$ by querying O_j on $m = O(n\delta_p^{-2})$ pairwise independent randomised inputs of the form $(y_{t,i} = [(c_{t,i}x)^e]_N, L_{j-1}([c_{t,i}x]_N))$ with $c_{t,i} = (2i+1)[2^{-t}a]_N + b$ $(i = \pm 1, \dots, \pm m/2)$, and use majority vote over resulting m estimates for bit $\ell_j([2^{-t}ax]_N)$ (see below for more details on this step). This majority vote is correct except with error probability $O(1/n)$, hence all n iterations succeed (and the inversion attack succeeds) with at least a constant probability, as claimed.

We now sketch the main ideas used in the second step. Note this step requires providing O_j with $y_{t,i} = [(c_{t,i}x)^e]_N$. This is easily computed using the homomorphic RSA function property: $y_{t,i} = [c_{t,i}^e \cdot y]_N$. We also need to provide the string $L_{j-1}([c_{t,i}x]_N)$ to O_j on the ith query. This string is computed with error probability $O(\delta_p)$ by using the relation

$$[c_{t,i}x]_N = (1 + 2i)[2^{-t}ax]_N + [bx]_N - \omega_{t,i}N, \tag{1}$$

where

$$\omega_{t,i} \stackrel{\text{def}}{=} \left\lfloor \frac{(2i+1)[2^{-t}ax]_N + [bx]_N}{N} \right\rfloor \tag{2}$$

is a modular reduction coefficient. Reducing (1) modulo 2^{j-1}, we see that we can compute the $j-1$ LS bits of all terms on the right hand side except $\omega_{t,i}$. Namely, $L_{j-1}([2^{-t}ax]_N)$ was computed above, $L_{j-1}([bx]_N)$ can be computed as $L_{j-1}(s_2)$ from given FSRSA input s_2, and N is known. Finally $\omega_{t,i}$ can be estimated by replacing terms $[2^{-t}ax]_N$ and $[bx]_N$ in (2) by their known approximations u_{t-1} and s_2 respectively. Thanks to the floor function in (2), the resulting approximation for $\omega_{t,i}$ is clearly *exact* as along as the absolute error $\Delta_{t,i}$ in approximating the quantity $\Omega_{t,i} = \frac{(2i+1)[2^{-t}ax]_N+[bx]_N}{N}$ that appears inside the floor function is smaller than the distance of $\Omega_{t,i}$ to the nearest integer. Using the uniformly random choice of $a, b \in \mathbb{Z}_N$ and the known errors of u_{t-1} and s_2, it is not difficult to show that the exact value of $\omega_{t,i}$ is obtained with error probability at most $\delta_p/4$. Hence O_j given the correct answer for input $(y = [(c_{t,i}x)^e]_N, L_{j-1}([c_{t,i}x]_N))$ with probability at least $1/2 + \delta_p - (\delta_p/4) = 1/2 + 3\delta_p/4$. Using the pairwise independence of the inputs to O_j it can then be shown that the majority vote over the $m = O(n\delta_p^{-2})$ runs of O_j gives the correct value of jth bit $\ell_j([c_{t,i}x]_N)$ with error probability $O(1/n)$. Finally, the estimate for bit $\ell_j([c_{t,i}x]_N)$ is converted to an estimate for the desired bit $\ell_j([2^{-t}ax]_N)$ by the following relation, obtained from (1) by reducing modulo 2^j and solving for $\ell_j([2^{-t}ax]_N)$ in terms of other known quantities:

$$\ell_j([2^{-t}ax]_N) \equiv \ell_j([c_{t,i}x]_N) + \frac{1}{2^{j-1}}(L_{j-1}([c_{t,i}x]_N) \\ -(2i+1) \cdot L_{j-1}([2^{-t}ax]_N) - L_j([bx]_N) + \omega_{t,i} \pmod 2).$$

□

Although the modified Fischlin-Schnorr reduction to FSRSA is quite efficient, the improvement of efficiency came at the cost of introducing a new and not well studied problem (FSRSA). To close this gap, Steinfeld et al next proceeded to show that under certain choices of parameters, the FSRSA problem is in fact at least as hard as a specific type of SSRSA problem, where the degree e polynomial has the form $f(x) = (2^c x + s)^e$ for known c and s. More precisely, the specific SSRSA problem is called (n, e, r, w)-CopRSA and defined as follows. Here we use $M_c(x)$ to denote the c MS bits in the binary representation of x.

(n, e, r, w)-CopRSA Problem. Recover $x \in_R \mathbb{Z}_N$, given RSA modulus N and $y = [x^e]_N$, $s_L = L_r(x)$, $s_H = M_{n/2+w}(x)$.

Write $x = 2^{n/2-w}s_H + 2^r\bar{x} + s_L$, where $0 < \bar{x} < 2^{n/2-r}$ consists of the unknown portion of bits of x in positions $r, \ldots, n/2-1$. Hence we see that the (n, e, r, w)-CopRSA problem is equivalent to the specific SSRSA problem of recovering the random integer $\bar{x} < 2^{n/2-r}$ from $y = f(\bar{x}) = [(2^r \bar{x} + s)^e]_N$, given the polynomial coefficients 2^r and $s = 2^{n/2-w} s_H + s_L$. The missing link in the provable security of the RSA PRG is then filled by the following lemma[6].

Lemma 4.2 (Steinfeld-Pieprzyk-Wang). *Let* A' *be an attack algorithm against* $(n, e, r, w-1, w-1)$*-FSRSA with run-time* T' *and success probability* ε'*. Then we construct an attack algorithm* A *against* (n, e, r, w)*-CopRSA with run-time* T *and success probability at least* ε*, where* $[T = 4T' + O(n^2)$ *and* $\varepsilon = \varepsilon' - 4/2^{n/2}.]$

Proof. (Sketch) Attacker A is given a (n, e, r, w)-CopRSA instance $(N, y = [x^e]_N, s_L = L_r(x), s_H = M_{n/2+w}(x))$.

We need to somehow convert it to an FSRSA instance of the form (for some x' related in a known easily invertible way to x):

$$y' = [(x')^e]_N, a \in_R \mathbb{Z}_N, s_1 = L_r([ax']_N), u_1 = \widehat{M}_{N,k}([ax']_N)$$
$$b \in_R \mathbb{Z}_N, s_2 = L_r([bx']_N), u_2 = \widehat{M}_{N,l}([bx']_N).$$

We first observe that, thanks to the homomorphic property of RSA, we can set $x' = [m \cdot x]_N$ for any multiplier $m \in \mathbb{Z}_N^*$ of our choice, and still be able to efficiently compute $y' = [(x')^e]_N = [y \cdot m^e]_N$ (while being able to go back and efficiently recover $x = [m^{-1}x']_N$ from x').

Our second observation is that for "small" known multipliers $c \in \mathbb{Z}$ with $|c| < N^{1/2}$, we can efficiently compute (from the given r LS bits of x, i.e. s_L, and $n/2 + w$ MS bits of x, i.e. s_H), 2 candidates for: the r LS bits of $[c \cdot x]_N$ and an approximation to $[c \cdot x]_N$ with absolute error at most $N/2^{w-1}$. To see this, note that $[c \cdot x]_N = cx - \omega_c N$, where $\omega_c = \lfloor \frac{cx}{N} \rfloor$. Since $|c| < N^{1/2}$ and $\widehat{x} \stackrel{\text{def}}{=} 2^{n/2-w} s_H$ approximates x with error at most $N/2^{n/2+w-1}$, it follows that $\frac{c\widehat{x}}{N}$ approximates $\frac{cx}{N}$ with error at most $N^{1/2}/2^{n/2+w-1} < 1/2^{w-1} < 1$. Hence by rounding to the two nearest integers, we get only two candidates for ω_c. This gives us two candidates for $L_r([cx]_N) = L_r(c) \cdot L_r(x) - L_r(\omega_c N)$. Similarly, computing $c\widehat{x} - \omega_c N$ gives us two candidates for an approximation to $[cx]_N$ with error at most $N^{1/2} \cdot N/2^{n/2+w-1} < N/2^{w-1}$.

To make use of these observations, we first choose $a \in \mathbb{Z}_N^*$ and $b' \in \mathbb{Z}_N$ independently and uniformly at random. Then, using continued fractions (see, e.g. Lemma 16 in Ref. 28) we efficiently compute (in time $O(n^2)$) a small non-zero integer c with $|c| < N^{1/2}$ such that $|[b'c]_{\bar{N}}| < N^{1/2}$ is also small (here $[z]_{\bar{N}}$ denotes $z - N$ if $[z]_N > N/2$ or $[z]_N$ if $[z]_N < N/2$). We set $x' = [mx]_N$ with multiplier $m = [a^{-1}c]_N$. Hence $[ax']_N = [cx]_N$, so since c

is small we can apply the second observation above to efficiently compute 2 candidates for the desired $s_1 = L_r([ax']_N)$ and $u_1 = \widehat{M}_{N,k}([ax']_N)$ with $k = w - 1$. Defining $b = [ab']_N$, we have $[bx']_N = [b'cx]_N$, so since $[b'c]_{\bar{N}}$ is also small, we can apply the second observation again to compute 2 candidates for $s_2 = L_r([bx']_N)$ and $u_2 = \widehat{M}_{N,l}([bx']_N)$ with $l = w - 1$. Running the attacker A′ on the resulting 4 candidate FSRSA instance $(y', a, s_1, u_1, b, s_2, u_2)$ we get x' on one of those with probability at least $\varepsilon' - 4/2^{n/2}$ and hence recover the solution $x = [m^{-1}x']_N$ to FSRSA (the negligible term $4/2^{n/2}$ is due to the fact that a is chosen from $\mathbb{Z}_N^*$ here rather than $\mathbb{Z}_N$). □

In summary, we get the following construction and security result by combining Theorem 4.3, Lemma 4.2 and Lemma 2.1 (there is also an additional technical step related to moving from fixed N to a random N, refer to Steinfeld et al[6] for details).

(n, e, r)-RSAPRG: RSA Based Iterative PRG Family[6]

Index Space	$\{N = pq : p, q \text{ are } n/2\text{-bit primes}, \gcd(e, \phi(N)) = 1\}$
State Space	$\mathbb{Z}_N$
Transition Function	$f_{RSA}(x) = [x^e]_N$
Output Function	$g_{RSA}(x) = (x_r, \ldots, x_1)$

Theorem 4.4 (RSAPRG **Security[6]**). *Any (T, δ) distinguisher* D *for the iterated (n, e, r)-RSAPRG outputting ℓ bits can be converted into a (T_I, ε_I) inversion algorithm* A *for the (n, e, r, w)-CopRSA problem with $w = 3\log(2\ell/\delta) + 5$ and*

$$[T_I = O(n \log(n)\delta^{-2}\ell^2) \cdot (T + O(\ell/r \log(e)n^2)) \text{ and } \varepsilon_I = \delta/9 - 4/2^{n/2}.]$$

Recall that $(n, e, r, 3\log(2\ell/\delta) + 5)$-CopRSA is $(1/e + \varepsilon, e)$-SSRSA problem when

$$r/n = 1/2 - 1/e - \varepsilon - (3\log(2\ell/\delta) + 5)/n.$$

Hence for $e \geq 3$, we can use $r = \Theta(n)$ in the RSAPRG and achieve an asymptotic rate $\Theta(n)$ output bits per multiply modulo an n-bit modulus with security level $T_D/\delta_D = 2^{\Theta(n^{1/3}\log^{2/3} n)}$ comparable to security of factoring an n-bit modulus, assuming $\ell/\delta = poly(n)$. This improves on the rate of Gennaro DL-based PRG by a factor in the order of $n^{1/3}\log n$, for same modulus length n (again, this estimate does not show the effect of polynomial reduction cost factors, refer to numerical efficiency comparison in Section 5 for a comparison including this cost). We remark[6] that with suitable modifications the above reduction applies also for even exponents greater than 3 (e.g. $e = 8$).

4.3. *MQ-Based PRG*

The problem of solving a system of of m quadratic equations in n variables over the field $GF(2)$, known as the Multivariate Quadratics (MQ) problem, is another well known hard problem. This problem is known to be NP hard in the worst case[29], and also seems to be hard on average over random instances, when $\kappa \stackrel{\text{def}}{=} m/n = O(1)$ – the best attacks in this case run in time exponential in n. Here we describe a recent efficient PRG construction called QUAD[7] based on this problem.

Suppose we are given $m = \kappa \cdot n$ random quadratic polynomials $(Q^{(1)}), \ldots, Q^{(m)})$ in n variables $\mathbf{x} = (x_1, \ldots, x_n)$ over $GF(2)$, i.e. for $\mathbf{x} \in GF(2)^n$ we have:

$$Q^{(s)}(\mathbf{x}) = \sum_{1 \le i < j \le n} d_{i,j}^{(s)} x_i x_j + \sum_{1 \le i \le n} c_i^{(s)} x_i$$

for $s = 1, \ldots, m$, i.e. the $N = mn(n+1)/2$ coefficients $\mathbf{d} = (d_{1,2}^{(s)}, d_{1,3}^{(s)}, \ldots, d_{n-1,n}^{(s)}) \in GF(2)^{n(n-1)/2}$ and $\mathbf{c}^{(s)} = (c_1^{(s)}, \ldots, c_n^{(s)}) \in GF(2)^n$ are chosen independently and uniformly at random. We represent the m quadratic polynomials $Q^{(s)}$ $(s = 1, \ldots, m)$ as a $(m \times n(n+1)/2)$ matrix Q, whose sth row consists of the $n(n+1)/2$ coefficients of the sth quadratic polynomial $Q^{(s)}$, i.e. $\left[Q^{(s)} = (\mathbf{c}^{(s)}, \mathbf{d}^{(s)}) \in GF(2)^{n(n+1)/2}\right]$. Let $\mathcal{I}_n$ denote the set of all $(m \times n(n+1)/2)$ matrices with elements from $GF(2)$. Consider the family of functions $f_{MQ}^{Q} : GF(2)^n \to GF(2)^m$ indexed by $Q \in \mathcal{I}_n$, which maps n-bit input $\mathbf{x} \in_R GF(2)^n$ to m-bit output by evaluating the m polynomials in Q at $\mathbf{x}$, i.e.

$$f_{MQ}^{Q}(\mathbf{x}) = Q \cdot [\mathbf{x}, \mathbf{q}(\mathbf{x})]^T \in GF(2)^m,$$

where $\mathbf{x} = (x_1, \ldots, x_n) \in GF(2)^n$ is the vector of linear monomials evaluated at $\mathbf{x}$ and $\mathbf{q}(\mathbf{x}) = (x_1x_2, x_1x_3, \ldots, x_{n-1}x_n) \in GF(2)^{n(n-1)/2}$ is the vector of quadratic monomials evaluated at $\mathbf{x}$. When $\kappa = m/n > 1$, the mapping f_{MQ}^{Q} is length expanding, and, as mentioned above, if the expansion factor is not too big ($\kappa = O(1)$) then f_{MQ}^{Q} is one-way, assuming the average case hardness of the MQ problem. Moreover, the following Lemma[7] shows that the same one-wayness assumption also implies that the output of f_{MQ}^{Q} is indistinguishable from a random m-bit string and hence gives rise to an efficient iterative PRG. Namely, the Lemma shows that any efficient distinguisher can be converted into a Goldreich-Levin hardcore bit predictor for f_{MQ}^{Q}, and hence from the Goldreich Levin Theorem 3.1 with $r = 1$, such a distinguisher would contradict the assumed one-way property of f_{MQ}^{Q} (there is an intermediate technical step involved in applying Theorem 3.1 to the GL bit predictor from the lemma below due to the random choice of Q below; see Ref. 7 for details).

Lemma 4.3 (f_{MQ}^{Q} is Pseudorandom[7]). *Any* (T, δ) *distinguisher* D *for distinguishing the distributions*

$$\mathcal{D}_P = \{(Q, f_{MQ}^{Q}(\mathbf{x})) : \mathbf{x} \in_R GF(2)^n; Q \in_R \mathcal{I}_n\} \text{ and}$$
$$\mathcal{D}_R = \{(Q, y) : Q \in_R \mathcal{I}_n; y \in_R GF(2)^m\}$$

can be converted into a GL hardcore bit predictor A *for* f_{MQ}^{Q}*, that runs in time* $T' = T + O(n^3)$ *and given* $(Q, y = f_{MQ}^{Q}(\mathbf{x}), \mathbf{r})$ *for uniformly random* $Q \in \mathcal{I}_n$*,* $\mathbf{x} \in GF(2)^n$ *and* $\mathbf{r} \in GF(2)^n$*, outputs the value of the GL bit* $GL_{\mathbf{r}}^{(1)}(\mathbf{x}) = \mathbf{r} \cdot \mathbf{x}^T$ *with probability at least* $1/2 + \delta/4$*.*

Proof. (Sketch) First, we convert (T, δ) distinguisher D into a $(T, \delta/2)$ distinguisher D' having the property that on input coming from distribution $\mathcal{D}_R$, D' outputs 1 with probability $p_R = 1/2$, and on input coming from distribution $\mathcal{D}_P$, D' outputs 1 with probaility at least $p_P = 1/2 + \delta/2$. It is easy to see that such D' can be constructed as follows: on input $(Q, \mathbf{y})$, D' flips coin $c \in_R GF(2)$. If $c = 0$, D' returns $\mathsf{D}(Q, \mathbf{y})$, else D' returns $NOT(\mathsf{D}(Q, \mathbf{u}))$ for uniform and independent $\mathbf{u} \in GF(2)^m$ (we may have to apply a NOT to the output of D' to achieve $p_P > 1/2$).

On input $(Q, \mathbf{y} = f_{MQ}^{Q}(\mathbf{x}), \mathbf{r})$ for $Q \in_R \mathcal{I}_n$, $\mathbf{x} \in_R GF(2)^n$, $\mathbf{r} \in_R GF(2)^n$, GL bit predictor A works as follows:

- Choose a uniformly random bit $b \in_R GF(2)$ (guess for GL bit $\mathbf{r} \cdot \mathbf{x}^T$).
- Choose $m = \kappa \cdot n$ uniformly random bits $(a^{(1)}, \ldots, a^{(m)})$ from $GF(2)$.
- Replace matrix $Q = [(\mathbf{c}^{(1)}, \mathbf{d}^{(1)}), \ldots, (\mathbf{c}^{(m)}, \mathbf{d}^{(m)})]^T$ with matrix
$$Q' = [(\mathbf{c}^{(1)} + a^{(1)} \cdot \mathbf{r}, \mathbf{d}^{(1)}), \ldots, (\mathbf{c}^{(m)} + a^{(m)} \cdot \mathbf{r}, \mathbf{d}^{(m)})]^T$$
- Replace vector $\mathbf{y} = (y_1, \ldots, y_m)$ with vector
$$\mathbf{y}' = [y_1 + a^{(1)} \cdot b, \ldots, y_m + a^{(m)} \cdot b]^T$$
- If $\mathsf{D}(Q', \mathbf{y}') = 1$ return b else return $NOT(b)$.

It is easy to verify that conditioned on the case $b = \mathbf{r} \cdot \mathbf{x}^T$ (GL bit was guessed right) vector $\mathbf{y}'$ is indeed equal to $f^{Q'}(\mathbf{x})$ and the input to D' has the distribution $\mathcal{D}_P$, while conditioned on the case $b = \mathbf{r} \cdot \mathbf{x}^T + 1$ (GL bit was guessed wrong), vector $\mathbf{y}' = Q'(\mathbf{x}) + [a^{(1)}, \ldots, a^{(m)}]^T$ is uniformly random and independent of $\mathbf{x}, Q'$, and the input to D' has the distribution

$\mathcal{D}_R$. It follows that A guesses $\mathbf{r} \cdot \mathbf{x}^T$ correctly with probability at least $1/2 \cdot (1/2 + \delta/2) + 1/2 \cdot 1/2 = 1/2 + \delta/4$. □

We summarise the construction and security result as follows.

(n,m)-QUAD: MQ Based Iterative PRG Family[7]

Index Space	$\{Q = (Q_f, Q_L) : Q_f \in GF(2)^{n \times n(n+1)/2}, Q_g \in GF(2)^{(m-n) \times n(n+1)/2}\}$
State Space	$GF(2)^n$
Transition Function	$f^Q_{MQ}(\mathbf{x}) = Q_f \cdot [\mathbf{x}, \mathbf{q}(\mathbf{x})]^T$
Output Function	$g^Q_{MQ}(\mathbf{x}) = Q_g \cdot [\mathbf{x}, \mathbf{q}(\mathbf{x})]^T$

Theorem 4.5 (QUAD Security[7]).
Any (T, δ) distinguisher D *for $(n, m = \kappa n)$-QUAD PRG outputting ℓ bits can be converted into an inversion algorithm* A *for the (n,m)-MQ problem with run-time T_I and success probability at least ε_I, where* $\left[T_I = O(\delta^{-2}(\frac{\ell}{\kappa-1})^2) \cdot (T + O(\ell n^3 + \log(\delta^{-2}))) \text{ and } \varepsilon_I = \frac{\delta}{\ell/(m-n)}.\right]$

Using $\kappa = m/n = O(1)$ the QUAD PRG therefore attains an asymptotic rate $O(n)$ output bits per iteration, and each iteration takes time $O(n^3)$ bit operations.

5. Comparison of Constructions

To give a feeling for the state of the art in performance of provably secure PRGs, in Table 1 we compare the performance of the three efficient provably secure generators reviewed in the previous section, in both an asymptotic and concrete sense. We used the following assumptions in computing the estimates in Table 1:

- For asymptotic rates: We assume the goal is (T, δ) provable PRG security for $T = 2^k$, $\delta = 1/poly(k)$, where k is a security parameter, and express the output rate (output bits per bit operation) as a function of k, neglecting poly-logarithmic factors $O(\log^c(k))$ for simplicity. Also, we assume that a modular multiplication modulo an n-bit modulus takes $O(n^{1+\varepsilon})$ bit operations (for classical arithmetic we have $\varepsilon = 1$, but for large n improved algorithms exist with $\varepsilon < 1$).
- For concrete rates: We assume the goal is (T, δ) provable PRG security with $T = 2^{70}$ Pentium processor cycles and $\delta = 1/100$ for $\ell = 2^{30}$ output bits (note that this corresponds to a security level $T/\delta \approx 2^{77}$). For concrete complexity of big integer modular arithmetic, we use implementation figures from Ref. 30 and extrapolation assuming classical

arithmetic. For concrete complexity of QUAD we use implementation figures from Ref. 7 and extrapolation. We used the concrete security reduction costs of the constructions together with complexity estimates of best known attacks on the underlying hard problems to determine appropriate scheme parameters.

Table 1. Comparison of asymptotic and concrete PRG performance.

PRG	**Rate, Asymp.**	**Est. Rate, Abs.** (cyc/byte)
DLSEPRG	$\Theta(1/k^{1+3\varepsilon})$	1.0×10^5 ($n = 7648$)
RSAPRG	$\Theta(1/k^{3\varepsilon})$	5.4×10^3 ($n = 5920$)
QUADPRG	$\Theta(1/k^2)$	5.1×10^3 ($n = 212$)

We note in comparison with the above figures that rates of 'heuristic' PRG designs with 2^{80} security level are typically in the order of $5 - 10$ cycles per byte, and hence are still $2 - 3$ orders of magnitude faster than the above 'efficient' provably secure PRG designs.

6. Conclusions

We surveyed some recent efficient provably secure constructions based on the complexity of well studied hard computational problems. We have seen that the gap in performance between 'provable' and 'heuristic' designs has been reduced, but is still around $2 - 3$ orders of magnitude. Closing this gap further remains an interesting challenge for future research.

References

1. C. Shannon, *Bell Systems Technical Journal* **28**, 659 (1949).
2. M. Blum and S. Micali, *SIAM Journal on Computing* **13**, 850 (1984).
3. O. Goldreich, S. Goldwasser and S. Micali, *Journal of the ACM* **33**, 792 (1986).
4. M. Luby and C. Rackoff, *SIAM Journal on Computing* **17**, 373 (1988).
5. R. Gennaro, *Journal of Cryptology* **18**, 91 (2005).
6. R. Steinfeld, J. Pieprzyk and H. Wang, On the Provable Security of an Efficient RSA-Based Pseudorandom Generator, in *ASIACRYPT 2006*, LNCS Vol. 4284 (Springer-Verlag, Berlin, 2006).
7. C. Berbain, H. Gilbert and J. Patarin, QUAD: a Practical Stream Cipher with Provable Security, in *EUROCRYPT 2006*, LNCS Vol. 4004 (Springer-Verlag, Berlin, 2006).
8. O. Goldreich, *Foundations of Cryptography, Volume I* (Cambridge University Press, Cambridge, 2003).

9. A. Yao, Theory and Application of Trapdoor Functions, in *Proc. FOCS '82*, (IEEE Computer Society Press, 1982).
10. O. Goldreich and L. Levin, Hard-Core Predicates for Any One-Way Function, in *Proc. 21-st STOC*, (ACM Press, New York, 1989).
11. J. Håstad and M. Näslund, Practical Construction and Analysis of Pseudo-Randomness Primitives, in *ASIACRYPT 2001*, LNCS Vol. 2248 (Springer-Verlag, Berlin, 2001).
12. J. Håstad, R. Impagliazzo, L. Levin and M. Luby, *SIAM Journal on Computing* **28**, 1364 (1999).
13. I. Haitner, D. Harnik and O. Reingold, On the Power of the Randomized Iterate, in *CRYPTO 2006*, LNCS Vol. 4117 (Springer-Verlag, Berlin, 2006).
14. S. Patel and G. Sundaram, An Efficient Discrete Log Pseudo Random Generator, in *CRYPTO '98*, LNCS Vol. 1462 (Springer-Verlag, Berlin, 1998).
15. W. Diffie and M. Hellman, *IEEE Trans. on Information Theory* **22**, 644 (1976).
16. P. van Oorschot and M. Wiener, On Diffie-Hellman Key Agreement with Short Exponents, in *EUROCRYPT '96*, LNCS Vol. 1070 (Springer-Verlag, Berlin, 1996).
17. R. Farashahi, B. Schoenmakers and A. Sidorenko, Efficient Pseudorandom Generators Based on the DDH Assumption, in *PKC 2007*, LNCS Vol. 4450 (Springer-Verlag, Berlin, 2007).
18. D. Coppersmith, *J. of Cryptology* **10**, 233 (1997).
19. A. K. Lenstra, H. W. Lenstra and L. Lovász, *Mathematische Annalen* **261**, 515 (1982).
20. D. Coppersmith, Finding Small Solutions to Low Degree Polynomials, in *CALC '01*, LNCS Vol. 2146 (Springer-Verlag, Berlin, 2001).
21. P. Q. Nguyen and J. Stern, The Two Faces of Lattices in Cryptology, in *Cryptography and Lattices*, LNCS Vol. 2146 (Springer-Verlag, Berlin, 2001).
22. D. Catalano, R. Gennaro, N. Howgrave-Graham and P. Nguyen, Paillier's Cryptosystem Revisited, in *Proc. CCS '01*, (New York, 2001).
23. S. Goldwasser, S. Micali and P. Tong, Why and How to Establish a Private Code on a Public Network, in *Proc. FOCS '82*, (IEEE Computer Society Press, 1982).
24. M. Ben-Or, B. Chor and A. Shamir, On the Cryptographic Security of Single RSA Bits, in *Proc. 15-th STOC*, (ACM Press, New York, 1983).
25. U. Vazirani and V. Vazirani, Efficient and Secure Pseudo-Random Number Generation, in *Proc. FOCS '84*, (IEEE Computer Society Press, 1982).
26. W. Alexi, B. Chor, O. Goldreich and C. Schnorr, *SIAM Journal on Computing* **17**, 194 (1988).
27. R. Fischlin and C. Schnorr, *Journal of Cryptology* **13**, 221 (2000).
28. P. Q. Nguyen and I. E. Shparlinski, *J. Cryptology* **15**, 151 (2002).
29. M. Garey and D. Johnson, *Computers and Intractability: A Guide to the Theory of NP-Completeness* (W H Freeman, 1979).
30. W. Dai, *Crypto++ 5.2.1 Benchmarks*, (2006). `http://www.eskimo.com/~weidai/benchmarks.html`.

The Successive Minima Profile of Multisequences

Li-Ping Wang
Center for Advanced Study, Tsinghua University,
Beijing 100084, People's Republic of China
E-mail: wanglp@mail.tsinghua.edu.cn

Harald Niederreiter
Department of Mathematics, National University of Singapore,
2 Science Drive 2, Singapore 117543, Republic of Singapore
E-mail: nied@math.nus.edu.sg

In this paper we use the successive minima profile to assess the complexity of multisequences. We study the asymptotic behavior of the successive minima profile and establish an asymptotic formula for the expected value of the successive minima profile of random multisequences over a finite field. Furthermore, multisequences with almost perfect successive minima profile are defined. We investigate the relationship between almost perfect multisequences, strongly almost perfect multisequences, and multisequences with almost perfect successive minima profile.

Keywords: Multisequences; joint linear complexity profile; successive minima profile; almost perfect multisequences.

1. Introduction

Recent developments in stream ciphers point towards an interest in word-based or vectorized keystreams because of their efficiency gained through parallelization; see e.g. Dawson and Simpson [1], Hawkes and Rose [4], and the proposals DRAGON, NLS, and SSS to the ECRYPT stream cipher project [2]. The security of stream ciphers relies on good statistical randomness and complexity properties of keystreams. An important complexity measure for multisequences is the joint linear complexity, that is, the length of the shortest linear feedback shift register generating the constituent sequences of a multisequence simultaneously [9,10].

In this paper we use a more refined complexity measure for multisequences, the so-called successive minima (see Section 2 and [14]). In Section 3 we study the asymptotic behavior of the successive minima profile and establish an asymptotic formula for the expected value of the successive minima profile of random multisequences over a finite field. In Section 4 a multisequence with almost perfect successive minima profile is defined and we investigate the relationship between almost perfect multisequences, strongly almost perfect multisequences, and multisequences with almost perfect successive minima profile.

2. Successive minima profile

In [14] we introduced the successive minima profile to investigate the structure of multisequences. In this section we recall this concept and some of its properties as well as the fundamental definitions and notations in order to keep the article as self-contained as possible.

Let $\mathbb{F}_q$ be the finite field of order q, where q is an arbitrary prime power. For a positive integer m, consider m sequences $S^{(h)} = s_1^{(h)}, s_2^{(h)}, \ldots$, where $1 \leq h \leq m$, with terms $s_j^{(h)}$ in $\mathbb{F}_q$, i.e., an *m-fold multisequence* (or *m-dimensional vector sequence*)

$$\mathbf{S} = (S^{(1)}, \ldots, S^{(m)})$$

over $\mathbb{F}_q$. The joint linear complexity is a basic complexity measure for multisequences. The definition is as follows.

Definition 2.1. Let n be a nonnegative integer and let $\mathbf{S} = (S^{(1)}, \ldots, S^{(m)})$ be an m-fold multisequence over $\mathbb{F}_q$. Then the *nth joint linear complexity* $L_n^{(m)}(\mathbf{S})$ of $\mathbf{S}$ is the least length of a linear feedback register that simultaneously generates the first n terms of each sequence $S^{(h)}$, $h = 1, \ldots, m$. The *joint linear complexity profile* of $\mathbf{S}$ is the sequence $\{L_n^{(m)}(\mathbf{S})\}_{n \geq 0}$.

Here we define $L_0^{(m)}(\mathbf{S}) = 0$ for any multisequence $\mathbf{S}$. We refer to [10] for a recent survey of work on the joint linear complexity and the joint linear complexity profile of multisequences.

The linear feedback shift-register multisequence synthesis problem, that is, to compute the joint linear complexity, plays an important role in cryptanalysis. In [15,16] a multisequence synthesis algorithm called LBRMS algorithm was designed by means of the lattice basis reduction in function fields. As a by-product of the LBRMS algorithm, we introduced the concept of successive minima profile of a multisequence in [14].

For each $h = 1, 2, \ldots, m$, we identify the sequence $S^{(h)}$ with the formal power series $S^{(h)}(x) = \sum_{j=1}^{\infty} s_j^{(h)} x^{-j}$ which we view as an element of the Laurent series field $K = \mathbb{F}_q((x^{-1}))$. There is a standard (exponential) valuation v on K whereby for $\alpha = \sum_{j=j_0}^{\infty} a_j x^{-j} \in K$ we put $v(\alpha) = \max\{-j \in \mathbb{Z} : a_j \neq 0\}$ if $\alpha \neq 0$ and $v(\alpha) = -\infty$ if $\alpha = 0$. The *valuation* $v(\gamma)$ of an $(m+1)$-dimensional vector $\gamma = (\alpha_1, \ldots, \alpha_{m+1}) \in K^{m+1}$ is defined as $\max\{v(\alpha_i) : 1 \leqslant i \leqslant m+1\}$. We also use the *projection* $\theta : K^{m+1} \to \mathbb{F}_q^{m+1}$ such that $\gamma = (\alpha_i)_{1 \leq i \leq m+1} \mapsto (a_{1,-v(\gamma)}, \ldots, a_{m+1,-v(\gamma)})$, where $\alpha_i = \sum_{j=j_0}^{\infty} a_{i,j} x^{-j}$, $1 \leq i \leq m+1$.

A subset Λ of K^{m+1} is called an $\mathbb{F}_q[x]$-*lattice* if there exists a basis $\omega_1, \ldots, \omega_{m+1}$ of K^{m+1} such that

$$\left[\Lambda = \sum_{i=1}^{m+1} \mathbb{F}_q[x]\, \omega_i = \left\{ \sum_{i=1}^{m+1} f_i\, \omega_i : f_i \in \mathbb{F}_q[x],\ i = 1, \ldots, m+1 \right\}.\right]$$

In this situation we say that $\omega_1, \ldots, \omega_{m+1}$ form a *basis* for Λ and we often denote the lattice by $\Lambda(\omega_1, \ldots, \omega_{m+1})$. A basis $\omega_1, \ldots, \omega_{m+1}$ is *reduced* if $\theta(\omega_1), \ldots, \theta(\omega_{m+1})$ are linearly independent over $\mathbb{F}_q$.

There is an important notion of *successive minima* (see [7]). For $1 \leqslant i \leqslant m+1$, the ith successive minimum $M_i(\Lambda)$ is defined by

$$\begin{aligned} M_i(\Lambda) := \min\{&k \in \mathbb{Z} : \text{there are } i\ \mathbb{F}_q[x]\text{-linearly independent} \\ &\text{vectors } \gamma_1, \ldots, \gamma_i \text{ in } \Lambda \text{ such that } v(\gamma_j) \leq k, 1 \leqslant j \leqslant i\}. \end{aligned}$$

If the reduced basis $\omega_1, \ldots, \omega_{m+1}$ for Λ satisfies $v(\omega_1) \leqslant \cdots \leqslant v(\omega_{m+1})$, then $M_i(\Lambda) = v(\omega_i)$ for $1 \leqslant i \leqslant m+1$ (see [5]).

For any integer $n \geqslant 0$, we construct a special lattice $\Lambda(\varepsilon_1, \ldots, \varepsilon_m, \alpha_n)$ in K^{m+1} spanned by the vectors $\varepsilon_1 = (1, 0, \ldots, 0), \ldots, \varepsilon_m = (0, \ldots, 0, 1, 0)$, $\alpha_n = (S^{(1)}(x), \ldots, S^{(m)}(x), x^{-n-1})$.

By means of a lattice basis reduction algorithm [5,13], we can transform the initial basis $\varepsilon_1, \ldots, \varepsilon_m, \alpha_n$ into a reduced one, denoted by $\omega_{1,n}, \ldots, \omega_{m+1,n}$. The information about the joint linear complexity will appear in the basis, which is the fundamental idea of the LBRMS algorithm. By rearranging these elements such that $v(\omega_{1,n}) \leqslant \cdots \leqslant v(\omega_{m+1,n})$, we have $M_i(\Lambda(\varepsilon_1, \ldots, \varepsilon_m, \alpha_n)) = v(\omega_{i,n})$ for $1 \leqslant i \leqslant m+1$. It is clear that the lattice $\Lambda(\varepsilon_1, \ldots, \varepsilon_m, \alpha_n)$ is completely determined by the given multisequence $\mathbf{S}$ and the length n, and so these successive minima can be viewed as intrinsic parameters of multisequences. Therefore we also denote $M_i(\Lambda(\varepsilon_1, \ldots, \varepsilon_m, \alpha_n))$ by $M_{i,n}^{(m)}(\mathbf{S})$. Now we can introduce the following definition.

Definition 2.2. Let n be a nonnegative integer and let $\mathbf{S} = (S^{(1)}, \ldots, S^{(m)})$ be an m-fold multisequence over $\mathbb{F}_q$. The multiset

$\{M_{1,n}^{(m)}(\mathbf{S}), M_{2,n}^{(m)}(\mathbf{S}), \ldots, M_{m+1,n}^{(m)}(\mathbf{S})\}$, denoted by $\mathrm{SM}_n^{(m)}(\mathbf{S})$, is called the *successive minima* of the multisequence $\mathbf{S}$ at n, and the *successive minima profile* is the sequence $\{\mathrm{SM}_n^{(m)}(\mathbf{S})\}_{n\geq 0}$.

Note that the successive minima at n of a multisequence is just a multiset, and so there is no need to arrange the order of its $m+1$ elements. In addition, since the initial basis for any multisequence $\mathbf{S}$ with length $n=0$ is always reduced, we have

$$\mathrm{SM}_0^{(m)}(\mathbf{S}) = \{-1, \underbrace{0, \ldots, 0}_{m}\}.$$

We summarize an important relationship between the joint linear complexity and the successive minima that was proved in [14, Proposition 1].

Proposition 2.1. *For any m-fold multisequence $\mathbf{S}$ and any integer $n \geqslant 0$, there exists some integer k_n, $1 \leqslant k_n \leqslant m+1$, such that $M_{k_n,n}^{(m)}(\mathbf{S}) = L_n^{(m)}(\mathbf{S}) - n - 1$. Furthermore, we have*

$$\sum_{i=1}^{m+1} M_{i,n}^{(m)}(\mathbf{S}) = -n-1. \tag{1}$$

The following result in [14, Proposition 2] provides the dynamics of the successive minima profile and the joint linear complexity profile.

Proposition 2.2. *For any m-fold multisequence $\mathbf{S}$ and any integer $n \geqslant 1$, we have*

$$\left[\mathrm{SM}_n^{(m)}(\mathbf{S}) = (\mathrm{SM}_{n-1}^{(m)}(\mathbf{S}) \cup \{M_{h_{n-1},n-1}^{(m)}(\mathbf{S}) - 1\}) \setminus \{M_{h_{n-1},n-1}^{(m)}(\mathbf{S})\}\right]$$

and

$$\left[L_n^{(m)}(\mathbf{S}) = n + M_{h_{n-1},n-1}^{(m)}(\mathbf{S})\right]$$

for some integer h_{n-1}, $1 \leqslant h_{n-1} \leqslant m+1$.

It is clear that $M_{k_n,n}^{(m)}(\mathbf{S}) = M_{h_{n-1},n-1}^{(m)}(\mathbf{S}) - 1$ for k_n and h_{n-1} defined in Proposition 2.1 and Proposition 2.2, respectively.

3. The asymptotic behavior of the successive minima profile of multisequences

An important question, particularly for applications to cryptology, is that of the asymptotic behavior of the joint linear complexity profile of random multisequences over $\mathbb{F}_q$.

We use a canonical stochastic model with the following assumptions: (i) finite sequences over $\mathbb{F}_q$ of the same length are equiprobable; (ii) corresponding terms in the m constituent sequences making up an m-fold multisequence over $\mathbb{F}_q$ are statistically independent. Let $\mathbb{F}_q^m$ be the set of m-tuples of elements of $\mathbb{F}_q$ and let $(\mathbb{F}_q^m)^\infty$ be the sequence space over $\mathbb{F}_q^m$. It is obvious that the set $(\mathbb{F}_q^m)^\infty$ can be identified with the set of m-fold multisequences over $\mathbb{F}_q$, and henceforth we will use this identification. Let $\mu_{q,m}$ be the uniform probability measure on $\mathbb{F}_q^m$ which assigns the measure q^{-m} to each element of $\mathbb{F}_q^m$. Then we denote by $\mu_{q,m}^\infty$ the complete product measure on $(\mathbb{F}_q^m)^\infty$ induced by $\mu_{q,m}$. A property of multisequences $\mathbf{S} \in (\mathbb{F}_q^m)^\infty$ is said to be satisfied $\mu_{q,m}^\infty$-*almost everywhere* (abbreviated $\mu_{q,m}^\infty$-a.e.) if the property holds for a set of multisequences $\mathbf{S} \in (\mathbb{F}_q^m)^\infty$ of $\mu_{q,m}^\infty$-measure 1.

In [11] the authors determined the asymptotic behavior of the joint linear complexity profile of random multisequences over a finite field by showing the following result.

Proposition 3.1. *For any integer $m \geq 1$, we have $\mu_{q,m}^\infty$-almost everywhere*

$$\lim_{n\to\infty} \frac{L_n^{(m)}(\mathbf{S})}{n} = \frac{m}{m+1}. \tag{2}$$

The following refinement was proved in [12].

Proposition 3.2. *For any integer $m \geqslant 1$, we have $\mu_{q,m}^\infty$-almost everywhere*

$$-\frac{1}{m+1} \leqslant \liminf_{n\to\infty} \frac{L_n^{(m)}(\mathbf{S}) - \frac{mn}{m+1}}{\log_q n} \leqslant \limsup_{n\to\infty} \frac{L_n^{(m)}(\mathbf{S}) - \frac{mn}{m+1}}{\log_q n} \leqslant 1.$$

The following theorem describes the asymptotic behavior of the successive minima profile of random multisequences over $\mathbb{F}_q$.

Theorem 3.1. *For any integer $m \geqslant 1$ and any $i = 1, \ldots, m+1$, we have $\mu_{q,m}^\infty$-almost everywhere*

$$\lim_{n\to\infty} \frac{M_{i,n}^{(m)}(\mathbf{S})}{n} = -\frac{1}{m+1}. \tag{3}$$

Proof. It follows from Proposition 3.2 that $\mu_{q,m}^\infty$-a.e. we have

$$\frac{mn}{m+1} - \frac{2}{m+1}\log_q n \leqslant L_n^{(m)}(\mathbf{S}) \leqslant \frac{mn}{m+1} + \frac{m+2}{m+1}\log_q n \tag{4}$$

for all sufficiently large n. Then for sufficiently large n, with $e_n = \lfloor 6 \log_q n \rfloor$ we have $L_{n+e_n}^{(m)}(\mathbf{S}) > L_n^{(m)}(\mathbf{S})$ since

$$\frac{mn}{m+1} + \frac{m+2}{m+1} \log_q n < \frac{m(n+e_n)}{m+1} - \frac{2}{m+1} \log_q(n+e_n). \tag{5}$$

Thus, if r_n is the least positive integer with $L_{n+r_n}^{(m)}(\mathbf{S}) > L_n^{(m)}(\mathbf{S})$, then $r_n \leqslant 6 \log_q n$. Now $L_{n+r_n}^{(m)}(\mathbf{S}) > L_n^{(m)}(\mathbf{S})$ with $r_n \geqslant 1$ minimal means by Proposition 2.2 that $L_{n+r_n}^{(m)}(\mathbf{S}) = n + r_n + M_{i,n}^{(m)}(\mathbf{S})$ for some i, i.e., $-M_{i,n}^{(m)}(\mathbf{S}) = n + r_n - L_{n+r_n}^{(m)}(\mathbf{S})$. Hence we get

$$\begin{aligned} \left| \frac{-M_{i,n}^{(m)}(\mathbf{S})}{n} - \frac{1}{m+1} \right| &= \left| \frac{n + r_n - L_{n+r_n}^{(m)}(\mathbf{S})}{n} - \frac{1}{m+1} \right| \\ &= \left| \frac{m}{m+1} + \frac{r_n}{n} - \frac{L_{n+r_n}^{(m)}(\mathbf{S})}{n} \right| \\ &\leq \frac{n+r_n}{n} \left| \frac{L_{n+r_n}^{(m)}(\mathbf{S})}{n+r_n} - \frac{m}{m+1} \right| + \frac{2m+1}{m+1} \cdot \frac{r_n}{n}. \end{aligned}$$

By symmetry, this holds $\mu_{q,m}^{\infty}$-a.e. for all $i = 1, \ldots, m+1$, and the result follows from $1 \leqslant r_n \leqslant 6 \log_q n$ and Proposition 3.1. □

For given $m \geq 1$, $n \geq 1$, and any $i = 1, \ldots, m+1$, let $E(M_{i,n}^{(m)}(\mathbf{S}))$ be the expected value of the ith component of the successive minima profile of random m-fold multisequences over $\mathbb{F}_q$ at n.

Theorem 3.2. *For any integer $m \geqslant 1$ and any $i = 1, \ldots, m+1$, we have*

$$\lim_{n \to \infty} \frac{E(M_{i,n}^{(m)}(\mathbf{S}))}{n} = -\frac{1}{m+1}. \tag{6}$$

Proof. By the dominated convergence theorem [6, p. 125] and Theorem 3.1, we obtain

$$\begin{aligned} \lim_{n \to \infty} \frac{-E(M_{i,n}^{(m)}(\mathbf{S}))}{n} &= \lim_{n \to \infty} \int_{(\mathbb{F}_q^m)^\infty} \frac{-M_{i,n}^{(m)}(\mathbf{S})}{n} \, d\mu_{q,m}^{\infty}(\mathbf{S}) \\ &= \int_{(\mathbb{F}_q^m)^\infty} \lim_{n \to \infty} \frac{-M_{i,n}^{(m)}(\mathbf{S})}{n} \, d\mu_{q,m}^{\infty}(\mathbf{S}) \\ &= \int_{(\mathbb{F}_q^m)^\infty} \frac{1}{m+1} \, d\mu_{q,m}^{\infty}(\mathbf{S}) \\ &= \frac{1}{m+1} \end{aligned}$$

for any integer $m \geqslant 1$ and any $i = 1, \ldots, m+1$. □

4. Almost perfect successive minima profile

Sequences with almost perfect linear complexity profile were defined by Niederreiter [8]. Xing [17] extended this concept from the case of single sequences to the case of multisequences and further proposed the following concept of d-perfect multisequences.

Definition 4.1. For a positive integer d, an m-fold multisequence $\mathbf{S}$ is called *d-perfect* if

$$L_n^{(m)}(\mathbf{S}) \geq \frac{m(n+1)-d}{m+1} \qquad \text{for all } n \geqslant 1. \tag{7}$$

In particular, $\mathbf{S}$ is called *perfect* if $\mathbf{S}$ is an m-perfect multisequence.

Feng *et al.* [3] introduced the concept of strongly d-perfect multisequences.

Definition 4.2. An m-fold multisequence $\mathbf{S}$ is called *strongly d-perfect* if

$$\frac{m(n+1)-d}{m+1} \leqslant L_n^{(m)}(\mathbf{S}) \leqslant \frac{mn+d}{m+1} \qquad \text{for all } n \geqslant 1. \tag{8}$$

These authors also showed that d-perfect multisequences are not always strongly d-perfect [3]. It is obvious that strongly d-perfect multisequences are d-perfect.

The description of the successive minima profile of random multisequences in Theorem 3.2 suggests that the components of the successive minima at n are close together. This leads to the following new concept.

Definition 4.3. For a positive integer d, an m-fold multisequence $\mathbf{S}$ has a *d-perfect successive minima profile* if

$$|M_{i,n}^{(m)}(\mathbf{S}) - M_{j,n}^{(m)}(\mathbf{S})| \leq d \qquad \text{for all } n \geq 0 \text{ and all } 1 \leqslant i,j \leqslant m+1. \tag{9}$$

The following result can be obtained immediately.

Proposition 4.1. *For a multisequence $\mathbf{S}$ with a d-perfect successive minima profile, the joint linear complexity increases by at most d for successive values of n.*

Proof. For any $n \geq 0$, by Proposition 2.1 we have

$$L_n^{(m)}(\mathbf{S}) = n + 1 + M_{k_n,n}^{(m)}(\mathbf{S}) \tag{10}$$

with some k_n, $1 \leqslant k_n \leqslant m+1$. By Proposition 2.2, we get

$$L_{n+1}^{(m)}(\mathbf{S}) = n + 1 + M_{h_n,n}^{(m)}(\mathbf{S}) \tag{11}$$

for some h_n, $1 \leqslant h_n \leqslant m+1$. Thus, by (10) and (11) we have

$$L_{n+1}^{(m)}(\mathbf{S}) = L_n^{(m)}(\mathbf{S}) - M_{k_n,n}^{(m)}(\mathbf{S}) + M_{h_n,n}^{(m)}(\mathbf{S}) \leq L_n^{(m)}(\mathbf{S}) + d,$$

which is the desired result. □

Proposition 4.2. *For an m-fold multisequence* **S** *with a d-perfect successive minima profile, we have*

$$-(n+1-L_n^{(m)}(\mathbf{S})) - M_{i,n}^{(m)}(\mathbf{S}) \neq d \tag{12}$$

for all $n \geq 1$ *and all* $1 \leqslant i \leqslant m+1$.

Proof. Suppose that there exist a positive integer n and some integer t, $1 \leqslant t \leqslant m+1$, such that $-(n+1-L_n^{(m)}(\mathbf{S})) - M_{t,n}^{(m)}(\mathbf{S}) = d$. Consider the successive minima at $n-1$. By Proposition 2.2, we have $M_{h_{n-1},n-1}^{(m)} = -(n - L_n^{(m)}(\mathbf{S}))$ for some $1 \leqslant h_{n-1} \leqslant m+1$. Since $M_{t,n}^{(m)}(\mathbf{S}) \neq -(n+1-L_n^{(m)}(\mathbf{S}))$, it belongs to the successive minima at $n-1$. Thus, the successive minima at step $n-1$ must include $-(n - L_n^{(m)}(\mathbf{S}))$ and $M_{t,n}^{(m)}(\mathbf{S})$, and we also have $-(n - L_n^{(m)}(\mathbf{S})) - M_{t,n}^{(m)}(\mathbf{S}) = d+1$, which contradicts the fact that **S** has a d-perfect successive minima profile. □

Proposition 4.3. *The* 1*-perfect successive minima profile for m-fold multisequences* **S** *has the following unique form in multiset notation: if* $n = (m+1)k + r$ *with integers* $k \geqslant 0$ *and* $1 \leq r \leq m$, *then*

$$\mathrm{SM}_n^{(m)}(\mathbf{S}) = \{\underbrace{-k-1,\ldots,-k-1}_{r+1},\underbrace{-k,\ldots,-k}_{m-r}\}, \quad L_n^{(m)}(\mathbf{S}) = n-k; \tag{13}$$

if $n = (m+1)(k+1)$ *with an integer* $k \geqslant 0$, *then*

$$\mathrm{SM}_n^{(m)}(\mathbf{S}) = \{-k-2,\underbrace{-k-1,\ldots,-k-1}_{m}\}, \quad L_n^{(m)}(\mathbf{S}) = n-k-1. \tag{14}$$

Proof. If the successive minima profile of an m-fold multisequence **S** is 1-perfect, then the values of the elements $M_{i,n}^{(m)}(\mathbf{S})$ with $1 \leqslant i \leqslant m+1$ must be $-k-1$ or $-k$ for $n = (m+1)k+r$ with $1 \leqslant r \leqslant m$. By Proposition 4.2, the value of $-(n+1-L_n^{(m)}(\mathbf{S}))$ must be $-k-1$. By Proposition 2.1, the number of occurrences of $-k-1$ in the multiset $\mathrm{SM}_n^{(m)}(\mathbf{S})$ is equal to $r+1$, and so the desired result is obtained. Similarly, we get the result for $n = (m+1)(k+1)$. □

In [17] Xing gave the following necessary and sufficient condition for perfect multisequences.

Proposition 4.4. *An m-fold multisequence* $\mathbf{S}$ *is perfect if and only if*

$$L_n^{(m)}(\mathbf{S}) = \left\lceil \frac{mn}{m+1} \right\rceil \qquad \textit{for all } n \geqslant 1, \tag{15}$$

where $\lceil u \rceil$ denotes the least integer greater than or equal to the the real number u.

Theorem 4.1. *The successive minima profile of a multisequence* $\mathbf{S}$ *is* 1-*perfect if and only if* $\mathbf{S}$ *is perfect.*

Proof. Suppose that the successive minima profile of the m-fold multisequence $\mathbf{S}$ is 1-perfect. Then by Proposition 4.3, we have

$$\left[L_n^{(m)}(\mathbf{S}) = \begin{cases} mk + r = \lceil \frac{mn}{m+1} \rceil & \text{for } n = (m+1)k + r \text{ with } 1 \leqslant r \leqslant m, \\ m(k+1) = \lceil \frac{mn}{m+1} \rceil & \text{for } n = (m+1)(k+1). \end{cases} \right]$$

By Proposition 4.4, $\mathbf{S}$ is perfect.

Conversely, suppose that $\mathbf{S}$ is perfect. By Proposition 4.4, we have $L_n^{(m)}(\mathbf{S}) = \lceil \frac{mn}{m+1} \rceil$ for all $n \geqslant 1$. Thus $L_n^{(m)}(\mathbf{S}) = mk$ and $-(n+1-L_n^{(m)}(\mathbf{S})) = -k-1$ if $n = (m+1)k$ for some integer $k \geq 0$. Since $L_{n+1}^{(m)}(\mathbf{S}) = mk+1$ is greater than $L_n^{(m)}(\mathbf{S})$, there are at least two values $-k-1$ in $\mathrm{SM}_{n+1}^{(m)}(\mathbf{S})$. Continue such an argument until step $n+m$, then we get

$$\mathrm{SM}_{n+m}^{(m)}(\mathbf{S}) = \{\underbrace{-k-1, \ldots, -k-1}_{m+1}\}.$$

In this way, it is easy to determine $\mathrm{SM}_{n+r}^{(m)}(\mathbf{S})$ for $n = (m+1)k + r$ with $1 \leqslant r \leqslant m$, and so the desired result follows for all n. □

In the following we discuss the relationship between strongly almost perfect multisequences and multisequences with almost perfect successive minima profile.

Theorem 4.2. *An m-fold multisequence* $\mathbf{S}$ *with a d-perfect successive minima profile is strongly md-perfect.*

Proof. By Definition 4.3 and Proposition 4.2, we have

$$-d \leq -(n+1-L_n^{(m)}(\mathbf{S})) - M_{i,n}^{(m)}(\mathbf{S}) \leq d-1 \qquad \text{for } n \geq 0 \text{ and } 1 \leqslant i \leqslant m+1. \tag{16}$$

Suppose, by way of contradiction, that there is a positive integer n such that

$$-(n+1-L_n^{(m)}(\mathbf{S})) < -\frac{n+1+md}{m+1}.$$

Then by Proposition 2.1 and (16) we have

$$\begin{aligned}-n-1 = \sum_{i=1}^{m+1} M_{i,n}^{(m)}(\mathbf{S}) &\le m(-(n+1-L_n^{(m)}(\mathbf{S}))+d) - (n+1-L_n^{(m)}(\mathbf{S})) \\ &< -n-1,\end{aligned}$$

which is impossible.

Suppose, again by way of contradiction, that there is a positive integer n such that

$$n+1-L_n^{(m)}(\mathbf{S}) < \frac{n+m+1-md}{m+1}.$$

Then by Proposition 2.1 and (16), we have

$$\begin{aligned}-n-1 = \sum_{i=1}^{m+1} M_{i,n}^{(m)}(\mathbf{S}) &\ge -(n+1-L_n^{(m)}(\mathbf{S})) - m(n+1-L_n^{(m)}(\mathbf{S})+d-1) \\ &> -n-1,\end{aligned}$$

which is also impossible. Therefore we obtain

$$\frac{n+m+1-md}{m+1} \le n+1-L_n^{(m)}(\mathbf{S}) \leqslant \frac{n+1+md}{m+1} \quad \text{for all } n \geqslant 1, \quad (17)$$

and so the result holds by Definition 4.2. □

Theorem 4.2 shows also that the multisequences with almost perfect successive minima profile form a subset of the strongly almost perfect multisequences. Furthermore, we point out that for two integers $m \geqslant 2$ and $d \geqslant 1$, even strongly md-perfect multisequences do not always have a d-perfect successive minima profile by giving a counterexample.

First we construct a successive minima profile by (17) and Proposition 2.2. Let $m = 2$, $d = 4$, and let the successive minima at $n = 0, \ldots, 14$ be $\{-1,0,0\}$, $\{-1,-1,0\}$, $\{-1,-1,-1\}$, $\{-2,-1,-1\}$, $\{-2,-2,-1\}$, $\{-3,-2,-1\}$, $\{-3,-3,-1\}$, $\{-4,-3,-1\}$, $\{-5,-3,-1\}$, $\{-6,-3,-1\}$, $\{-6,-3,-2\}$, $\{-6,-3,-3\}$, $\{-6,-4,-3\}$, $\{-6,-4,-4\}$, $\{-6,-5,-4\}$, respectively. If $n = 3(k+1)$ with $k \ge 4$, the form of $\mathrm{SM}_n^{(m)}(\mathbf{S})$ is the same as (14). If $n = 3k+r$ with $1 \le r \leqslant 2$ and $k \ge 4$, then the form of $\mathrm{SM}_n^{(m)}(\mathbf{S})$ is the same as (13).

According to this successive minima profile, we can construct a 2-fold multisequence $\mathbf{S}$ over $\mathbb{F}_2$ with such a profile, where

$$\mathbf{S} = \begin{pmatrix} 1\,0\,1\,0\,0\,1\,0\,0\,1\,1\,0\,0\,0\,0\,1\cdots \\ 0\,1\,0\,0\,1\,0\,0\,1\,0\,0\,0\,0\,1\,0\,0\cdots \end{pmatrix}.$$

By construction, $\mathbf{S}$ is strongly 8-perfect, but its successive minima profile is not 4-perfect because the successive minima at $n = 9$ is $\{-6, -3, -1\}$.

Acknowledgments. The research of the first author is supported by the National Natural Science Foundation of China (no. 60773141). The research of the second author is partially supported by the MOE-ARF grant R-146-000-066-112.

References

1. E. Dawson and L. Simpson, Analysis and design issues for synchronous stream ciphers, in: H. Niederreiter, ed., *Coding Theory and Cryptology*, 49–90, Singapore: World Scientific, 2002.
2. ECRYPT stream cipher project; available online at `http://www.ecrypt.eu.org/stream`.
3. X. Feng, Q. Wang, and Z. Dai, Multi-sequences with d-perfect property, *J. Complexity* 21 (2005), 230–242.
4. P. Hawkes and G.G. Rose, Exploiting multiples of the connection polynomial in word-oriented stream ciphers, in: T. Okamoto, ed., *Advances in Cryptology – ASIACRYPT 2000*, Lecture Notes in Computer Science, vol. 1976, 303–316, Berlin: Springer, 2000.
5. A.K. Lenstra, Factoring multivariate polynomials over finite fields, *J. Computer and System Sciences* 30 (1985), 235–248.
6. M. Loève, *Probability Theory*, 3rd ed., New York: Van Nostrand, 1963.
7. K. Mahler, An analogue to Minkowski's geometry of numbers in a field of series, *Ann. of Math.* 42 (1941), 488–522.
8. H. Niederreiter, Sequences with almost perfect linear complexity profile, in: D. Chaum and W.L. Price, eds., *Advances in Cryptology – EUROCRYPT '87*, Lecture Notes in Computer Science, vol. 304, 37–51, Berlin: Springer, 1988.
9. H. Niederreiter, Linear complexity and related complexity measures for sequences, in: T. Johansson and S. Maitra, eds., *Progress in Cryptology – INDOCRYPT 2003*, Lecture Notes in Computer Science, vol. 2904, 1–17, Berlin: Springer, 2003.
10. H. Niederreiter, The probabilistic theory of the joint linear complexity of multisequences, in: G. Gong *et al.*, eds., *Sequences and Their Applications – SETA 2006*, Lecture Notes in Computer Science, vol. 4086, 5–16, Berlin: Springer, 2006.
11. H. Niederreiter and L.-P. Wang, Proof of a conjecture on the joint linear complexity profile of multisequences, in: S. Maitra *et al.*, eds., *Progress in*

Cryptology – INDOCRYPT 2005, Lecture Notes in Computer Science, vol. 3797, 13–22, Berlin: Springer, 2005.

12. H. Niederreiter and L.-P. Wang, The asymptotic behavior of the joint linear complexity profile of multisequences, *Monatsh. Math.* 150 (2007), 141–155.
13. W.M. Schmidt, Construction and estimation of bases in function fields, *J. Number Theory* 39 (1991), 181–224.
14. L.-P. Wang and H. Niederreiter, Successive minima profile, lattice profile, and joint linear complexity profile of pseudorandom multisequences, *J. Complexity*, to appear.
15. L.-P. Wang and Y.-F. Zhu, $F[x]$-lattice basis reduction algorithm and multisequence synthesis, *Science in China* (*Series F*) 44 (2001), 321–328.
16. L.-P. Wang, Y.-F. Zhu, and D.-Y. Pei, On the lattice basis reduction multisequence synthesis algorithm, *IEEE Trans. Inform. Theory* 50 (2004), 2905–2910.
17. C.P. Xing, Multi-sequences with almost perfect linear complexity profile and function fields over finite fields, *J. Complexity* 16 (2000), 661–675.

A Construction of Optimal Sets of FH Sequences *

Jianxing Yin
Department of Mathematics
Suzhou University
Suzhou, 215006, China
Email: jxyin@suda.edu.cn

Frequency hopping spread spectrum and direct sequence spread spectrum are two main spread coding technologies. Frequency hopping sequences are needed in FH-CDMA systems. In this paper, a construction of optimal sets of frequency hopping sequences is presented through perfect nonlinear functions. The construction is based on the set-theoretic characterization of an optimal set of FH sequences. In the procedure, the notion of mixed difference functions is proposed and used.

Keywords: Sets of frequency hopping sequence; perfect nonlinear function; construction; optimality.

1. Introduction

Frequency hopping spread spectrum and direct sequence spread spectrum are two main spread coding technologies. Frequency hopping sequences are an integral part of spread-spectrum communication systems such as FH-CDMA systems (for a description of such systems, see [12]). In modern radar and communication systems, frequency-hopping (FH) spread-spectrum techniques have become popular (see [6] for example).

Assume that $F = \{f_0, f_1, \cdots, f_{m-1}\}$ is a set of available frequencies, called an *alphabet*. Let $\mathcal{X}(v; F)$ be the set of all sequences of length v over F. Any element of $\mathcal{X}(v; F)$ is called a *frequency hopping sequence* (FHS) of length v over F. Given two FH sequences, $X = (x_0, x_1, \ldots, x_{v-1})$ and $Y = (y_0, y_1, \cdots, y_{v-1})$, define their Hamming correlation $H_{X,Y}(t)$ to be

$$H_{X,Y}(t) = \sum\nolimits_{0\leqslant i\leqslant v-1} h[x_i, y_{i+t}],$$

*Research is supported by the National Natural Science Foundation of China, Project No. 10671140.

where $0 \leqslant t < v$ if $X \neq Y$ and $0 < t < v$ if $X = Y$, and where

$$h[x, y] = \begin{cases} 1, \text{ if } x = y \\ 0, \text{ otherwise} \end{cases}$$

and all operations among position indices are performed modulo v. If $X = Y$, then $H_{X,Y}(t)$ is the Hamming auto-correlation. If $X \neq Y$, $H_{X,Y}(t)$ is the Hamming cross-correlation.

Example 1.1. Take $v = 10$ and $m = 4$. Label the frequency alphabet $F = \{f_0, f_1, f_2, f_3\}$ by the integers of Z_4. Then $\mathfrak{X}(v; F)$ is the set of all sequences of length 10 whose entries are taken from Z_4. Take

$$X = (3\,0\,0\,0\,1\,2\,\mathbf{0}\,\mathbf{1}\,3\,2) \in \mathfrak{X}(v; F)$$

Let $t = 3 \in Z_{10}$. Cyclic shifting of X to the right for 3 times gives

$$X_3 = (1\,3\,2\,3\,0\,0\,\mathbf{0}\,\mathbf{1}\,2\,0) \in \mathfrak{X}(v; F)$$

The Hamming auto-correlation

$$H_{X,X}(3) = \sum\nolimits_{0 \leqslant i \leqslant 9} h[x_i, x_{i+t}] = 2.$$

In multiple-access spread-spectrum communication systems, mutual interference occurs when two or more transmitters transmit using the same frequency at the same time. FHSs are used to specify which frequency will be used for transmission at any given time. Normally it is desirable to keep the mutual interferences, or the Hamming correlation function as low as possible. For any distinct $X, Y \in \mathfrak{X}(v; F)$, we write

$$\begin{aligned} H(X) &= \max_{1 \leq t < v} \{H_{X,X}(t)\}, \\ H(X,Y) &= \max_{0 \leq t < v} \{H_{X,Y}(t)\}, \\ M(X,Y) &= \max\{H(X), H(Y), H(X,Y)\}. \end{aligned}$$

Following [10], we say that an FHS $X \in \mathfrak{X}(v; F)$ is *optimal* if $H(X) \leq H(X')$ for all $X' \in \mathfrak{X}(v; F)$. A pair of two distinct FHSs $X, Y \in \mathfrak{X}(v; F)$ is termed an *optimal pair* if $M(X,Y) \leq M(X', Y')$ for all $X', Y' \in \mathfrak{X}(v; F)$ with $X' \neq Y'$. A subset $\mathcal{F} \subset \mathfrak{X}(v; F)$ is an *optimal set* if every pair of distinct members of $\mathcal{F}$ is an optimal pair.

Lempel and Greenberger [10] developed the following lower bound for $H(X)$.

Lemma 1.1. *[10] For every FH sequence X of length v over a frequency alphabet F of size m,*

$$H(X) \geqslant \frac{(v - \varepsilon)(v + \varepsilon - m)}{m(v-1)},$$

where ε is the least non-negative residue of v modulo m.

For any given subset $\mathcal{F}$ of $\mathcal{X}(v;F)$ containing N FH sequences, we write

$$M(\mathcal{F}) = \max\left\{\max_{X\in\mathcal{F}} H(X), \max_{X,Y\in\mathcal{F},X\neq Y} H(X,Y)\right\}.$$

Peng and Fan[11] developed the following bounds on $M(\mathcal{F})$, which take into consideration of the number of FH sequences in the set.

Lemma 1.2. *[11] Let $\mathcal{F}\subseteq\mathcal{X}(v;F)$ be a set of N sequences of length v over an alphabet of size m. Define $I=\lfloor vN/m\rfloor$. Then*

$$M(\mathcal{F}) \geq \left\lceil \frac{(vN-m)v}{(vN-1)m} \right\rceil \tag{1}$$

and

$$M(\mathcal{F}) \geq \left\lceil \frac{2IvN-(I+1)mI}{(vN-1)N} \right\rceil. \tag{2}$$

A number of authors have made contributions to the construction of optimal FH sequences. Both algebraic and combinatorial constructions of optimal FH sequences have been given (see, for example, [2,3,7–10,13]). Most of them are concentrated on single optimal FH sequences. The purpose of this paper is to present a construction of optimal sets of FH sequences through perfect nonlinear functions. The construction is based on the set-theoretic characterization of an optimal set of FH sequences. In the procedure, the notion of mixed difference functions is proposed and used. Throughout what follows, we use (v,m,λ)-FHS to denote an FH sequence X of length v over an alphabet of size m whose Hamming auto-correlation $H(X)=\lambda$. We also call a set $\mathcal{F}$ of N FH sequences in $\mathcal{X}(v;F)$ a $(v,N,\lambda;m)$ set of FH sequences, where $\lambda = M(\mathcal{F})$.

2. Mixed Difference Functions

Fuji-Hara, Miao and Mishima [7] characterized a (v,m,λ)-FHS in terms of partition-type cyclic difference packings.

Given a partition $\mathcal{D}=\{D_0,D_1,\cdots,D_{m-1}\}$ of Z_v into m subsets (*called base blocks*), we can define a difference function on $Z_v^*=Z_v\setminus\{0\}$ given by

$$\Phi_{\mathcal{D}}(t) = \sum_{i=0}^{m-1} |D_i\cap(D_i+t)|.$$

Let $\max\{\Phi_{\mathcal{D}}(t)|\ t\in Z_v\setminus\{0\}\}=\lambda$. Then $\mathcal{D}$ is called a $(v,K,\lambda)_m$-PCDP (*partition-type cyclic difference packing*). Here, m is used in the notation to indicate the number of base blocks and $K = \{|D_i| : 0 \leqslant i \leqslant m-1\}$ is the list of the sizes of base blocks. This is to say that a $(v, K, \lambda)_m$-PCDP is a partition of Z_v into m base blocks which satisfies the following property. For any fixed nonzero residue $t \in Z_v$, the equation

$$x - y = t$$

has at most λ solutions (x, y) in the multiset union $\bigcup_{D \in \mathcal{D}}(D \times D)$.

If we label the positions of a (v, m, λ)-FHS X by the elements of Z_v, then, by the above definition, the sets of position indices of m frequencies in X form a $(v, K, \lambda)_m$-PCDP $\mathcal{D}$ with $\Phi_{\mathcal{D}}(t) = H_{X,X}(t)$ for any nonzero $t \in Z_v$. Conversely, if we label the m base blocks of a $(v, K, \lambda)_m$-PCDP by the elements of Z_m and identify the frequency alphabet F with Z_m, then the PCDP gives a (v, m, λ)-FHS in $\mathcal{X}(v; F)$.

This fact reveals that a single FHS can be constructed by a PCDPs. Apparently, the smaller the index λ of a PCDP, the lower the Hamming auto-correlation $H(X)$ of its corresponding FH sequence. For an optimal FH sequence, we need to construct a $(v, K, \lambda)_m$-PCDP so that its index λ is as small as possible for any given value of v and m. Based on Lempel-Greenberger bound on $H(X)$ in Lemma 1.1, Fuji-Hara, Miao and Mishima [7] proved the following result.

Lemma 2.1. *[7] There exists a (v, m, λ)-FHS over the alphabet $F = Z_m$ if and only if there exists a $(v, K, \lambda)_m$-PCDP in $\mathbb{Z}_v$. Furthermore, this FH sequence is optimal if $\lambda = \lfloor v/m \rfloor$ for $v > m$ and if $\lambda = 0$ for $v = m$.*

The following example gives an illustration of Lemma 2.1.

Example 2.1. Take $v = 10$ and $m = 4$. Consider the FH sequence over the alphabet Z_4 given in Example 1.1:

$$X = (3\,0\,0\,0\,1\,2\,0\,1\,3\,2) \in \mathcal{X}(v; F).$$

The sets of position indices of 4 frequencies are as follows:

$$D_0 = \{1, 2, 3, 6\}, D_1 = \{4, 7\}, D_2 = \{5, 9\}, D_3 = \{0, 8\}.$$

It is easy to see that $\mathcal{D} = \{D_0, D_1, D_2, D_3\}$ forms a partition of Z_{10}. Further we can see that it is a $(10, \{4, 2, 2, 2\}, 2)_4$-PCDP in $\mathbb{Z}_{10}$. For any $t \in Z_{10}$, when we cyclically shift X to the right for t times, we obtain

$$H_{X,X}(t)) = \sum\nolimits_{0 \leqslant i \leqslant 9} h[x_i, x_{i+t}] = 2$$

Hence $H(X) = 2$.

The correspondence between an individual FH sequence and a PCDP can be naturally extended to give a set-theoretic interpretation of a set of FH sequences. To do this, we propose the notion of mixed difference functions.

Let N be a positive integer. Let

$$\mathcal{C} = \{\mathcal{D}^{(0)}, \mathcal{D}^{(1)}, \cdots, \mathcal{D}^{(N-1)}\}$$

be a collection of partitions of Z_v into m subsets (*called base blocks*). Write

$$\mathcal{D}^{(r)} = \{D_0^{(r)}, D_1^{(r)}, \cdots, D_{m-1}^{(r)}\}, \quad 0 \leqslant r \leqslant N-1.$$

For any ordered pair (i,j) with $0 \leqslant i < j \leqslant N-1$, defined a difference function on Z_v given by

$$\Phi_{\mathcal{C}}^{(i,j)}(t) = \sum_{k=0}^{m-1} \left| D_k^{(i)} \bigcap (D_k^{(j)} + t) \right| .$$

For any integer r with $0 \leqslant r \leqslant N-1$, defined a difference function on $Z_v \setminus \{0\}$, as before, given by

$$\Phi_{\mathcal{C}}^{(r,r)}(t) = \Phi_{\mathcal{D}^{(r)}}(t) = \sum_{k=0}^{m-1} \left| D_k^{(r)} \cap (D_k^{(r)} + t) \right| .$$

We refer to these $N(N+1)/2$ difference functions defined above as *mixed difference functions* with respect to the given collection $\mathcal{C}$. Since each partition in $\mathcal{C}$ determines uniquely an FH sequence, the collection $\mathcal{C}$ gives a set of N FH sequences in $\mathcal{X}(v;F)$, and vice versa, where the alphabet is regarded as Z_m. For the optimality of the derived set of FH sequences from $\mathcal{C}$, we define

$$\begin{aligned}
\lambda(r) &= \max_{1 \leq t < v} \{\Phi_{\mathcal{C}}^{(r,r)}(t)\}, \quad 0 \leqslant r \leqslant N-1, \\
\lambda(i,j) &= \max_{0 \leq t < v} \{\Phi_{\mathcal{C}}^{(i,j)}(t)\}, \quad 0 \leqslant i < j \leqslant N-1, \\
\mu(i,j) &= \max\{\lambda(i), \lambda(j), \lambda(i,j)\}, \quad 0 \leqslant i < j \leqslant N-1, \\
\lambda &= \max\left\{ \max_{0 \leqslant r \leqslant N-1} \lambda(r), \max_{0 \leqslant i < j \leqslant N-1} \lambda(i,j) \right\}.
\end{aligned}$$

Then $\mathcal{D}^{(r)}$ is a $(v, K_r, \lambda_r)_m$-PCDP, according to the above definition. We say that $\mathcal{C}$ is a $(v, \{K_0, K_1, \cdots, K_{N-1}\}, \{\lambda_0, \lambda_1, \cdots, \lambda_{N-1}\}; \lambda)_m$ collection of N PCDPs. It now turns out that there exists a $(v, N, \lambda; m)$ set of FH sequences in $\mathcal{X}(v;F)$ if and only if there exists a $(v, \{K_0, K_1, \cdots, K_{N-1}\}, \{\lambda_0, \lambda_1, \cdots, \lambda_{N-1}\}; \lambda)_m$ collection of N PCDPs in Z_v under our notations. This gives us an interpretation for a set of FH sequences from set-theoretic perspective. As with individual optimal FH sequence, for an optimal $(v, N, \lambda; m)$ set of FH sequences, we are required to construct a $(v, \{K_0, K_1, \cdots, K_{N-1}\}, \{\lambda_0, \lambda_1, \cdots, \lambda_{N-1}\}; \lambda)_m$ collection $\mathcal{C}$ of N PCDPs so that its index λ is as small as possible. Since the index λ of $\mathcal{C}$ is the same as the Hamming correlation $M(\mathcal{F})$, the Peng-Fan bounds in Lemma 1.2 can be employed as our benchmarks. As noted in [4], a set of FH sequences meeting one of the Peng-Fan bounds must be optimal. We record the above discussion in the following theorem.

Theorem 2.1. *Let $N \geqslant 2$ be an integer. Then there exists a $(v, N, \lambda; m)$ set of FH sequences in $\mathcal{X}(v; F)$ if and only if there exists a $(v, \{K_0, K_1, \cdots, K_{N-1}\}, \{\lambda_0, \lambda_1, \cdots, \lambda_{N-1}\}; \lambda)_m$ collection of N PCDPs in Z_v. Furthermore, this set is optimal if λ meets one of the Peng-Fan lower bounds given in Lemma 1.2.*

3. The Construction

In this section, we present a construction of optimal sets of FH sequences through perfect nonlinear functions. Based on perfect nonlinear functions, Ding et al.[3] presented a construction for an optimal FH sequence. Our construction can be viewed as an extension of theirs.

Let A and B stand for two additive Abelian groups of orders v and m, respectively. Let f be a function from A to B. f is linear if and only if $f(x+y) = f(x) + f(y)$ for all $x, y \in A$. A function $g : A \longrightarrow B$ is called affine if and only if $g = f + b$, where $f : A \longrightarrow B$ is linear and b is a constant. A robust measure of the nonlinearity of a function $f : A \longrightarrow B$ using the derivatives $D_a f(x) = f(x+a) - f(x)$ is given by

$$P_f = \max_{0 \neq a \in A} \max_{b \in B} \frac{|\{x \in A : D_a f(x) = b\}|}{m}.$$

The smaller the value of P_f, the higher the corresponding nonlinearity of f. f is perfect nonlinear if and only if for each nonzero $a \in A$, $f(x+a) - f(x)$ takes on each element of b the same numbers $\frac{v}{m}$ of times when x ranges over all elements of A (see Carlet and Ding [1]).

For an elegant construction and the recent advance on perfect nonlinear functions, the reader is referred to Ding and Yuan.[5] Below we give an example.

Example 3.1. Let p be an odd prime. Define $f : Z_{p^2} \longrightarrow Z_p$ by $f(h+jp) = hj \bmod p$ for $0 \leq h, j \leq p-1$. Then f has perfect nonlinearity with respect to $A = (Z_{p^2}, +)$ and $B = (Z_p, +)$.

Now take A and B to be the additive groups of Z_v and Z_m, respectively. Consider a surjection $f : A \longrightarrow B$ which is perfect nonlinear. Define

$$D_j = f^{-1}(j) = \{x \in Z_v : \ f(x) = j\} \ (j \in Z_m);$$
$$\mathcal{D} = \{D_j : \ j = 0, 1, \cdots, m-1\}.$$

Then $\mathcal{D}$ forms a partition of Z_v and we have

$$\begin{aligned}
&\sum_{j=0}^{m-1} |D_j \cap (D_{j+b} + t)| \\
&= \left| \bigcup_{j=0}^{m-1} (D_j \cap (D_{j+b} - a)) \right| \\
&= \left| \bigcup_{j=0}^{m-1} \{x \in Z_v : f(x) = j \text{ and } f(x+a) = j+b\} \right| \\
&= |\{x \in Z_v : D_a f(x) = b\}| \\
&= v/m,
\end{aligned}$$

for any $b \in Z_m$ and $t \in Z_v \setminus \{0\}$. We further define

$$\begin{aligned}
D_j^{(r)} &= D_{j+r}, \quad j = 0, 1, \cdots m-1, \\
\mathcal{D}^{(r)} &= \{D_0^{(r)}, D_1^{(r)}, \cdots, D_{m-1}^{(r)}\}, \quad 0 \leqslant r \leqslant m-1, \\
\mathcal{C} &= \{\mathcal{D}^{(0)}, \mathcal{D}^{(1)}, \cdots, \mathcal{D}^{(m-1)}\},
\end{aligned}$$

where all operations among the subscripts are performed modulo m. Taking $b = 0$ in the above equalities, we obtain that

$$\begin{aligned}
\Phi_{\mathcal{C}}^{(r,r)}(t) &= \sum_{k=0}^{m-1} \left| D_k^{(r)} \cap (D_k^{(r)} + t) \right| \\
&= \quad v/m \text{ (a constant)},
\end{aligned}$$

for any $r \in Z_m$ and any $t \in Z_v \setminus \{0\}$. Hence, $\mathcal{D}^{(r)}$ is a $(v, K_r, v/m)_m$-PCDP in Z_v. We have also

$$\Phi_{\mathcal{C}}^{(i,j)}(t) = \sum_{k=0}^{m-1} \left| D_k^{(i)} \bigcap (D_k^{(j)} + t) \right| = v/m \text{ (also a constant)},$$

for any ordered pair (i, j) with $0 \leqslant i < j \leqslant m-1$ and any $t \in Z_v \setminus \{0\}$. By the definition of $\mathcal{C}$,

$$\Phi_{\mathcal{C}}^{(i,j)}(0) = \sum_{k=0}^{m-1} \left| D_k^{(i)} \bigcap D_k^{(j)} \right| = 0,$$

for any ordered pair (i, j) with $0 \leqslant i < j \leqslant m-1$. It follows that $\mathcal{C}$ is a $(v, \{K_0, K_1, \cdots, K_{m-1}\}, \{\lambda_0, \lambda_1, \cdots, \lambda_{m-1}\}; v/m)_m$ collection of m PCDPs in Z_v, where $\lambda_0 = \lambda_1 = \cdots = \lambda_{m-1} = v/m$. By Theorem 2.1, it derives a $(v, m, v/m; m)$ set of FH sequences in $\mathcal{X}(v; F)$. For this set, we have $N = m$ and both lower bounds in Lemma 1.2 are equal to v/m. Hence it is optimal.

Example 3.2. Take $p = 3$ in Example 3.1. Then $v = 9$ and $m = 3$. It is readily calculated that

$$D_0 = \{0, 1, 2, 3, 6\},$$
$$D_1 = \{4, 8\},$$
$$D_2 = \{5, 7\}.$$

We then have

$$\begin{aligned}
\mathcal{D}^{(0)} &= \{D_0^{(0)} = \{0,1,2,3,6\}, D_1^{(0)} = \{4,8\}, \quad D_2^{(0)} = \{5,7\}\},\\
\mathcal{D}^{(1)} &= \{D_0^{(1)} = \{5,7\}, \quad D_1^{(1)} = \{0,1,2,3,6\}, D_2^{(1)} = \{4,8\}\},\\
\mathcal{D}^{(2)} &= \{D_0^{(2)} = \{4,8\}, \quad D_1^{(2)} = \{5,7\}, \quad D_2^{(2)} = \{0,1,2,3,6\}\}.
\end{aligned}$$

The collection $\mathcal{C} = \{\mathcal{D}^{(0)}, \mathcal{D}^{(1)}, \mathcal{D}^{(2)}\}$ is a $(9, \{K_0, K_1, K_2\}, \{\lambda_0, \lambda_1, \lambda_2\}; 3)_3$ collection of 3 PCDPs in Z_9, where $K_0 = \{5, 2, 2\}$, $K_1 = \{2, 5, 2\}$, $K_3 = \{2, 2, 5\}$ and $\lambda_0 = \lambda_1 = \lambda_2 = 3$. By Theorem 2.1, it derives a $(9, 3, 3; 3)$ set $\mathcal{F}$ consisting of the following 3 FH sequences over Z_3:

$$X = \{0, 0, 0, 0, 1, 2, 0, 2, 1\},$$
$$Y = \{1, 1, 1, 1, 2, 0, 1, 0, 2\},$$
$$Z = \{2, 2, 2, 2, 0, 1, 2, 1, 0\}.$$

It can be checked that $M(X, Y) = M(Y, Z) = M(Z, X) = 3$. Hence $M(\mathcal{F}) = 3$ which is equal to the lower bounds in Lemma 1.2.

The foregoing can be summarized in the following theorem.

Theorem 3.1. *Suppose that there exists a surjection f from Z_v to Z_m which is perfect nonlinear. Then so does a $(v, \{K_0, K_1, \cdots, K_{m-1}\}, \{\lambda_0, \lambda_1, \cdots, \lambda_{m-1}\}; v/m)_m$ collection of m PCDPs in Z_v, or equivalently, an optimal $(v, m, v/m; m)$ set of FH sequences over alphabet Z_m of length v. Here, $\lambda_0 = \lambda_1 = \cdots = \lambda_{m-1} = v/m$.*

From Theorem 3.1, all existing and potential perfect nonlinear functions over cyclic groups can be utilized to produce directly optimal sets of FH sequences. We remark that the mixed difference functions with respect to the collection of PCDPs from a perfect nonlinear function take a constant value v/m. Our construction implies that for an optimal set of FH sequences, the mixed difference functions are not necessarily to be constant. This suggests that one may utilize other tools, instead of perfect nonlinear functions, to search the desired collections of PCDPs in order to construct new optimal sets of FH sequences.

References

1. C. Carlet and C. Ding, *Highly nonlinear mappings*, *J. Complexity* **20** (2004), 205–244.
2. W. Chu and C. J. Colbourn, *Optimal frequency-hopping sequences via cyclotomy*, *IEEE Trans. Inform. Theory* **51** (2005), 1139–1141.
3. C. Ding, M. Moisio and J. Yuan, *Algebraic Constructions of Optimal Frequency Hopping Sequences*, *IEEE Trans. Inform. Theory* **53** (2007), 2606–2610.
4. C. Ding and J. Yin, *Optimal Sets of Frequency Hopping Sequences*, submitted.
5. C. Ding and J. Yuan, *A family of skew Hadamard difference sets*, *J. Combinatorial Theory Ser. A* **113** (2006), 1526–1535.
6. P. Fan and M. Darnell, *Sequence Design for Communications Applications*, Research Studies Press LTD, Taunton, England, 1996.
7. R. Fuji-Hara, Y. Miao and M. Mishima, *Optimal frequency hopping sequences: a combinatorial approach*, *IEEE Trans. Inform. Theory* **50** (2004), 2408-2420.
8. G. Ge, R. Fuji-Hara and Y. Miao, *Further combinatorial constructions for optimal frequency hopping sequences*, *J. Combinatorial Theory Ser. A* **113** (2006), 1699-1718.
9. P. V. Kumar, *Frequency-hopping code sequence designs having large linear span*, *IEEE Trans. Inform. Theory* **34** (1988), 146-151.
10. A. Lempel and H. Greenberger, *Families of sequences with optimal Hamming correlation properties*, *IEEE Trans. Inform. Theory* **20** (1974), 90-94.
11. D. Peng and P. Fan, *Lower bounds on the Hamming auto- and cross correlations of frequency-hopping sequences*, *IEEE Trans. Inform. Theory* **50** (2004), 2149-2154.
12. R. A. Scholtz, *The spread spectrum concept*, *IEEE Trans. Commun.* **25** (1977), 748-755.
13. P. Udaya and M. N. Siddiqi, *Optimal large linear complexity frequency hopping patterns derived from polynomial residue class rings*, *IEEE Trans. Inform. Theory* **44** (1998), 1492-1503.

AUTHOR INDEX